Matrix Metalloproteinases

Biology of Extracellular Matrix Series

Editor
ROBERT P. MECHAM

Robert P. Mecham: REGULATION OF MATRIX ACCUMULATION

Thomas N. Wight and Robert P. Mecham: BIOLOGY OF PROTEOGLYCANS

Richard Mayne and Robert E. Burgeson: STRUCTURE AND FUNCTION OF COLLAGEN TYPES

Deane R. Mosher: FIBRONECTIN

W. Steven Adair and Robert P. Mecham: ORGANIZATION AND ASSEMBLY OF PLANT AND ANIMAL EXTRACELLULAR MATRIX

Linda J. Sandell and Charles D. Boyd: EXTRACELLULAR MATRIX GENES

John A. McDonald and Robert P. Mecham: RECEPTORS FOR EXTRACELLULAR MATRIX

David D. Roberts and Robert P. Mecham: CELL SURFACE AND EXTRACELLULAR GLYCOCONJUGATES

Peter D. Yurchenco, David E. Birk, and Robert P. Mecham: EXTRACELLULAR MATRIX ASSEMBLY AND STRUCTURE

David A. Cheresh and Robert P. Mecham: INTEGRINS: MOLECULAR AND BIOLOGICAL RESPONSES TO THE EXTRACELLULAR MATRIX

MATRIX METALLOPROTEINASES

Edited by **WILLIAM C. PARKS**

Departments of Medicine and Cell Biology and Physiology
Washington University School of Medicine
St. Louis, Missouri

ROBERT P. MECHAM

Department of Cell Biology and Physiology
Washington University School of Medicine
St. Louis, Missouri

ACADEMIC PRESS
San Diego London Boston
New York Sydney Tokyo Toronto

Front cover photographs: (Top) *In situ* hybridization showing expression of collagenase-1 (MMP-1) by adventitial fibroblasts in a developing pulmonary artery from a 210-day fetal calf. Photomicrograph provided by William C. Parks. (Center) As detected by *in situ* hybridization, matrilysin (MMP-7) is prominently expressed by Paneth cells in the crypts of mouse ileum. Photomicrograph from Wilson *et al.*, 1995, used with permission. (Bottom) At the point of epidermal separation in blister formation, gelatinase-B (MMP-9) is produced and released by eosinophils. Photomicrograph provided by Mona Ståhle-Bäckdahl.

Wilson, C. L., Heppner, K. J., Rudolph, L. A., and Matrisian, L. M. (1995). The metalloproteinase matrilysin is preferentially expressed by epithelial cells in a tissue-restricted pattern in the mouse. *Mol. Biol. Cell* **6,** 851– 869.

Academic Press
a division of Harcourt Brace & Company
525 B Street, Suite 1900, San Diego, California 92101-4495, USA
http://www.apnet.com

Academic Press Limited
24-28 Oval Road, London NW1 7DX, UK
http://www.hbuk.co.uk/ap/

International Standard Book Number: 0-12-545090-7

Printed and bound in the United Kingdom
Transferred to Digital Printing, 2011

Contents

72-kDa Gelatinase (Gelatinase A): Structure, Activation, Regulation, and Substrate Specificity
ANITA E. YU, ANNE N. MURPHY, and WILLIAM G. STETLER-STEVENSON

Gelatinase B: Structure, Regulation, and Function
THIENNU H. VU and ZENA WERB

Matrilysin
CAROLE L. WILSON and LYNN M. MATRISIAN

Macrophage Elastase (MMP-12)
STEVEN D. SHAPIRO and ROBERT M. SENIOR

Membrane-Type Matrix Metalloproteinases and Cell Surface-Associated Activation Cascades for Matrix Metalloproteinases

VERA KNÄUPER and GILLIAN MURPHY

Substrate Specificity and Mechanisms of Substrate Recognition of the Matrix Metalloproteinases

VERA IMPER and HAROLD E. VAN WART

Synthetic Inhibitors of Matrix Metalloproteinases

PETER D. BROWN

Matrix Metalloproteinases in Tissue Repair
WILLIAM C. PARKS, BARRY D. SUDBECK, GLENN R. DOYLE, and
ULPU K. SAARIAHLO-KERE

Regulation of Matrix Metalloproteinase Gene Expression
M. ELIZABETH FINI, JEFFERY R. COOK, ROYCE MOHAN, and
CONSTANCE E. BRINCKERHOFF

Contributors

Numbers in parentheses indicate the pages on which the authors' contributions begin.

CONSTANCE E. BRINCKERHOFF (299), Departments of Medicine and Biochemistry, Dartmouth Medical School, Hanover, New Hampshire 03755

PETER D. BROWN (243), Department of Clinical Research, British Biotech Pharmaceutical Ltd., Oxford OX4 5LY, United Kingdom

JEFFERY R. COOK (299), Vision Research Laboratories of the New England Medical Center and Departments of Ophthalmology and Anatomy and Cellular Biology, Tufts University School of Medicine, Boston, Massachusetts 02111

GLENN R. DOYLE (263), Departments of Medicine and Cell Biology and Physiology, Washington University School of Medicine, St. Louis, Missouri 63110

M. ELIZABETH FINI (299), Vision Research Laboratories of the New England Medical Center and Departments of Ophthalmology and Anatomy and Cellular Biology, Tufts University School of Medicine, Boston, Massachusetts 02111

VERA IMPER (219), Faculty of Pharmacy and Biochemistry, University of Zagreb, 10000 Zagreb, Croatia

JOHN J. JEFFERY (15), Department of Biochemistry and Molecular Biology, Albany Medical College, Albany, New York 12208

VERA KNÄUPER (199), Strangeways Research Laboratory, Cambridge, and School of Biological Sciences, University of East Anglia, Norwich NR4 7TJ, United Kingdom

LYNN M. MATRISIAN (149), Department of Cell Biology, Vanderbilt University School of Medicine, Nashville, Tennessee 37232

ROYCE MOHAN (299), Vision Research Laboratories of the New England Medical Center and Departments of Ophthalmology and Anatomy and Cellular Biology, Tufts University School of Medicine, Boston, Massachusetts 02111

ANNE N. MURPHY (85), Department of Biochemistry, George Washington University Medical School, Washington, DC 20037

GILLIAN MURPHY (199), Strangeways Research Laboratory, Cambridge, and School of Biological Sciences, University of East Anglia, Norwich NR4 7TJ, United Kingdom

HIDEAKI NAGASE (43), Department of Biochemistry and Molecular Biology, University of Kansas Medical Center, Kansas City, Kansas 66160

WILLIAM C. PARKS (263), Departments of Medicine and Cell Biology and Physiology, Washington University School of Medicine, St. Louis, Missouri 63110

ULPU K. SAARIAHLO-KERE (263), Department of Dermatology, University of Helsinki, Helsinki, Finland

ROBERT M. SENIOR (185), Department of Medicine, Washington University School of Medicine at Barnes-Jewish Hospital, St. Louis, Missouri 63110

STEVEN D. SHAPIRO (185), Departments of Medicine and Cell Biology, Washington University School of Medicine, St. Louis, Missouri 63110

WILLIAM G. STETLER-STEVENSON (85), Extracellular Matrix Pathology Section, Laboratory of Pathology, National Cancer Institute, National Institutes of Health, Bethesda, Maryland 20892

BARRY D. SUDBECK (263), Departments of Medicine and Cell Biology and Physiology, Washington University School of Medicine, St. Louis, Missouri 63110

HAROLD E. VAN WART (219), Inflammatory Diseases Unit, S3-1, Roche Bioscience, Palo Alto, California 94304

THIENNU H. VU (115), Department of Medicine, Division of Pulmonary and Critical Care Medicine, University of California, San Francisco, California 94143

ZENA WERB (115), Department of Anatomy, University of California, San Francisco, California 94143

CAROLE L. WILSON (149), Dermatology Division, Washington University School of Medicine, St. Louis, Missouri 63110

J. FREDERICK WOESSNER, JR. (1), Department of Biochemistry and Molecular Biology, University of Miami School of Medicine, Miami, Florida 33101

ANITA E. YU (85), Extracellular Matrix Pathology Section, Laboratory of Pathology, National Cancer Institute, National Institutes of Health, Bethesda, Maryland 20892

Preface

The spatially and temporally precise removal and remodeling of connective tissue are critical to several developmental, homeostatic, and reparative processes. Matrix turnover requires the activity of many different endopeptidases acting on a variety of compositionally distinct proteins, and consequently, it is not surprising that over the past decades numerous proteinases of distinct gene families have been characterized and implicated as serving a catalytic function during tissue remodeling. For example, the large and physiologically important serine proteinase family, which includes leukocyte elastase, plasminogen, and its activators, among many other enzymes, mediates a variety of activities, from clot dissolution to tissue destruction. The matrix metalloproteinases (MMPs), which compose the matrixin subfamily of the large metalloproteinase family (see Chapter 1) and which are the focus of this book, have a specialized function in turnover of the extracellular matrix. As suggested throughout this volume, however, the activity of these enzymes may also be involved in other functions, such as protein processing and activation.

In recent years, MMPs have gained considerable attention in many studies on normal tissue events, inflammation, and disease processes, and this enhanced interest is probably due to several factors. For the most part, MMPs are produced or activated when needed, and expression of these enzymes provides a reliable indicator of ongoing tissue remodeling. Thus, for investigators, these enzymes are models of gene products that are accurately regulated and precisely targeted to specific extracellular substrates by a wide variety of cells during numerous normal tissue processes, such as wound healing, bone resorption, and morphogenesis. In contrast, exuberant production of MMPs is a hallmark of many destructive diseases, such as arthritis and chronic ulcerations, and of many disease-related processes, such as inflammation, metastasis, and angiogenesis, and aberrant regulation of MMP production is thought to be a primary mechanism contributing to disease progression and injury. As such, understanding how MMP expression and activation are controlled and identifying which proteinases are produced by which cells under defined conditions will have a great impact on our understanding of normal biology and disease pathogenesis.

The metalloproteinase field has been expanded by a number of important and interesting discoveries. For example, activation of latent tumor necrosis factor-α (TNF-α), selectin shedding, cleavage of collagen N-propeptide, and processing of laminin-5 chains are all conferred by metalloproteinases but by enzymes that are members of subfamilies distinct from MMPs. In this volume, we focus on the matrix metalloproteinases, which are believed to act in the extracellular space and to have a specialized role in the degradation of connective tissue proteins. Although the ADAMs, such as TNF convertase and others, constitute an emerging and very interesting field in themselves, we have not included these in this volume, as they have not been implicated as having a direct role in matrix turnover. Stromelysin-3 (MMP-11) is also not included, because it apparently has no significant activity for matrix components.

This book begins with a chapter by J. Frederick Woessner, Jr., who outlines the distinct and specific features of the matrixin subfamily of metalloproteinases. The book then provides up-to-date reviews on the structure, biochemistry, function, and molecular biology of each of the principal MMPs written by investigators who are actively studying these enzymes (Chapters 2–8). Knowledge of the crystal structure and distinct substrate specificity of different MMPs, most of which has been gained over the past few years, has provided needed information for cogent design of synthetic inhibitors of metalloproteinase activity, and recent work in this field is discussed in chapters by Harold Van Wart and Peter Brown (Chapters 9 and 10). This book also includes two chapters on regulation of MMP expression at the level of the gene and by influences arising in the tissue environment (Chapters 11 and 12).

We thank the many authors for their thoughtful and timely contributions, and we hope this book proves helpful for the many investigators and students who have an interest in metalloproteinase biology.

William C. Parks
Robert P. Mecham

The Matrix Metalloproteinase Family

J. Frederick Woessner, Jr.

Department of Biochemistry and Molecular Biology, University of Miami School of Medicine, Miami, Florida 33101

I. Introduction

Seldom has a field had such a clearly defined beginning as did the field of matrix metalloproteinases (MMPs). The first report of an enzyme from vertebrate (as opposed to bacterial) sources that was capable of attacking the triple helix of native type I collagen was published in 1962 by Jerome Gross and Charles Lapiere. In this first report they demonstrated that the enzyme activity, secreted by cultured tissue fragments of tail fin skin from resorbing tadpole tails in metamorphosis, was a true collagenase acting on collagen at 27°C at neutral pH. They also found activity produced by cultured chick embryo skin, postpartum rat uterus, and mouse and rat bone. Activity was not found in tissue extracts. In retrospect this is now known to be due both to the ability of MMPs to bind to the extracellular matrix and to the stimulation of high enzyme secretion that occurs when tissues are excised and cultured. Also in retrospect we see that at least collagenase 1 (MMP-1) and 3 (MMP-13) were present and possibly collagenase 4 (MMP-18) in these first experiments. Within a short period of time these studies had been broadened by the work of many groups to include various

1

human sources including neutrophils (MMP-8) and to show that collagenase cut the triple helix at a point one-quarter of the distance in from the C-terminal end and that the activity was metal dependent.

Since Gross and Lapiere's 1962 report, there has been a tremendous efflorescence of this field; in preparing for a more detailed overview I have compiled a bibliography of 8500 items, with the current rate of publication approaching 1000 items per year. The reason for this interest is pointed out most clearly by Table I, which shows some of the points of involvement of the MMPs in normal biologic and pathologic processes. Cancer and its metastatic spread have been of the greatest current widespread interest. Not only are a great many different MMPs involved in the processes of carcinogenesis and its accompanying remodeling or destruction of the extracellular matrix, but there is also

TABLE I

NORMAL AND PATHOLOGICAL PROCESSES IN WHICH MMPS
ARE IMPLICATED

Normal	Pathological
Development	**Tissue Destruction**
Blastocyst implantation	Rheumatoid arthritis
Embryonic development	Osteoarthritis
Nerve growth	Cancer invasion
Growth plate cartilage removal	Cancer metastasis
Skeletal, bone growth	Decubitus ulcer
Nerve outgrowth	Gastric ulcer
Enamel maturation	Corneal ulceration
Primary tooth resorption	Periodontal disease
Reproduction	**Fibrotic Diseases**
Endometrial cycling	Liver cirrhosis
Graafian follicle rupture	Fibrotic lung disease
Luteolysis	Otosclerosis
Cervical dilatation	Atherosclerosis
Postpartum uterine involution	Multiple sclerosis
Mammary gland morphogenesis	
Mammary gland involution	
Rupture of fetal membranes	
Maintenance	**Weakening of Matrix**
Remodeling of bone	Dilated cardiomyopathy
Hair follicle cycle	Epidermolysis bullosa
Wound healing	Aortic aneurysm
Angiogenesis	
Apoptosis	
Nerve regeneration	
Macrophage function	
Neutrophil function	

hope of countering this destruction by various approaches such as systemic administration of inhibitors of MMPs, gene therapy to knock out enzymes, or overexpression of natural inhibitors such as the TIMPs (tissue inhibitors of metalloproteinases).

II. THE POSITION OF THE MMPs WITHIN THE CLASS OF METALLOPROTEASES

The proteases comprise both exopeptidases and endopeptidases (proteinases). These hydrolytic enzymes can be divided into four classes based on the catalytic group at their active center: serine/threonine, cysteine, aspartic, and metallo. It is the class of metalloproteases that concerns us here; most of its members depend on zinc for their catalytic action. A new compendium of proteases (Barrett et al., 1998) describes 200 metalloproteases, of which only 14 are MMPs. Therefore, it is important to understand the relationship of the MMPs to this larger class. A useful, but still provisional, guide has been provided by Rawlings and Barrett (1995), who divide the class into clans (based on similarity of protein fold) and families (based on evolutionary relationships). Currently the metallo class comprises eight clans and some 40 families (Barrett et al., 1998). For example, enzymes in clan MA have the zinc-binding motif HEXXH and the third zinc ligand is Glu. Thermolysin provides the X-ray structure for this clan and 5 families are typified by thermolysin, mycolysin, neprilysin, membrane alanyl aminopeptidase, and peptidyl-dipeptidase A (Rawlings and Barrett, 1995); there are currently 32 species of enzymes in this clan (Barrett et al., 1998). This HEXXH motif is widespread and occurs in several additional clans.

The clan containing the MMP family is clan MB in which the third zinc ligand is not Glu but rather a third His residue in the consensus sequence HEXXHXXGXXH. The two major families in this clan are M12, with its subfamilies astacin and reprolysin, and M10, with its subfamilies serralysin and matrixin (MMP). These two families contain more than 60 members and there are two further families of one member each (autolysin and snapalysin; Barrett et al., 1998). The X-ray structure of this clan is now known for a number of species in the four subfamilies. Bode et al. (1994) have given the name metzincins to this group because all contain a conserved Met residue to the carboxy side of the zinc site, which produces a turn in the protein chain that provides the base of the active center binding pocket. However, the sequence similarities are not otherwise very close between families M10 and M12. The similarities of the binding pocket have the important consequence that hydroxamate inhibitors designed to block the action of MMPs are frequently found to effectively block members of the M10 family. A classic example is the inhibition of tumor necrosis factor α (TNF-α) convertase, an enzyme now known to belong to the ADAM family within the reprolysin subfamily (ADAM-17; Black *et al.*, 1997). I suspect that the "aggrecanase" activity important in cartilage degradation may also prove to be such an example.

It can be seen that less than half of the metallo class is accounted for by the first two clans. Further clans include enzymes with the HEXXH motif and a different third ligand for zinc, the motif HXXEH (pitrilysin), and a variety of additional three-ligand arrangements involving various combinations of three residues selected from His, Asp, and Glu, as well as some cases in which there are two zinc atoms and some in which the ligands remain unknown. Full details may be found in Barrett *et al.* (1998). The scheme is provisional in that we do not yet know the three-dimensional structure of many of the enzymes, and the ligands that bind zinc have not been confirmed in many cases. Note that there is no uniform practice in this field for the naming of families, so the widely used terms *MMP family* or *matrixin family* actually describe a subfamily.

An earlier set of criteria (Woessner, 1994) for assignment of a new enzyme to the matrixin family included the display of proteolytic activity, function outside the cell, possession of conserved sequence around cysteine in the propeptide (PRCGxPD) and a zinc-binding consensus sequence of HEXGHXXGXXHS/T. However, members of family 10 also meet all of these criteria except the last. Today, a better criterion would be a cDNA sequence that is sufficiently close to that of collagenase to permit assignment to the matrixin subfamily. Many synthetic inhibitors of MMPs, particularly in the hydroxamate series, inhibit members of family 10 equally well as noted earlier. However, TIMP-1 does not appear to block any members of that family, but does appear to inhibit all MMPs. This provides a second criterion.

III. THE MATRIXIN SUBFAMILY

A. *Members of the Subfamily*

At the time of this writing (late 1997), 17 enzymes have received MMP numbers. The numbers are not being assigned by an official governing body, so some confusion arises as individual authors assign numbers they believe come next (e.g., Cossins *et al.*, 1996). It was suggested at the 1997 Gordon conference on MMPs that I make the assignments in the future. This plan is feasible only if the new enzymes are discovered by those already working in the field. The MMP numbers are often convenient to use as a shorthand when speaking or writing, but their utility is somewhat diminished by the large number of enzyme species that are coming to light. Table II indicates the most commonly used names, including those recommended by the International Union of Biochemistry and Molecular Biology (through MMP-12). Three enzymes reported earlier (MMP-4, -5, -6) were later found to correspond to known enzymes, so these three numbers have been discontinued and remain vacant. The enzymes are placed into arbitrary groups that originally arose from considerations of the substrates cleaved. This is not a very sound basis, because we have very little information about the natural substrates of any of these enzymes (see later discussion). Stromelysin 3, while having some substrates in common with the other two stromelysins, is very distantly related and is activated through a furin site and only indirectly through disruption of the cysteine switch. The membrane-type MMPs have been grouped on the basis of possession of a transmembrane domain. Table II also notes a few of the many synonyms for these enzymes and indicates additional features of some MMPs.

B. Evolutionary Relationships among the MMPs

A dendrogram showing the relationships among the MMPs known from humans is presented in Fig. 1. The figure is limited to human enzymes because the more than 65 known sequences from all species produce a tree that is difficult to see for the forest of entries. Only the approximately 170 residues of the catalytic domain for each enzyme have been used in deriving this tree. The earliest branches include MMP-19 and the four membrane-type MMPs. These MT-MMPs form a tight cluster except for MT4-MMP, which is somewhat earlier than the others. A metalloproteinase of the human ovary has been reported as a gene sequence (Acc. No. D83647), but nothing is known of its properties. As alluded to earlier, stromelysin 3 arose quite early relative to the other two stromelysins. Stromelysin 3 and the four MT-MMPs have in common the RXKR/RRKR furin cleavage site, suggesting that they may be activated while still in the cell, whereas the remaining enzymes require proteolytic cleavage of their propeptide after the zymogen leaves the cell. It is interesting to speculate that simpler organisms regulated their matrix largely by contact, through cell surface MMPs; then as more extensive and complex extracellular matrices evolved, it became advantageous to secrete the MMPs for action at a distance from the cell.

The last 10 enzymes form a cluster of the more modern and more commonly known MMPs. The two gelatinases are first in this group; they contain additional fibronectin-like domains. Matrilysin, the small-

TABLE II

MEMBERS OF THE MATRIXIN FAMILY

Group name	MMP number	EC number	M_r latent/active	Notes
Collagenase				
Collagenase 1	MMP-1	EC 3.4.24.7	52,000 42,000	**Interstitial collagenase**
Collagenase 2	MMP-8	EC 3.4.24.34	85,000 64,000	**Neutrophil collagenase**
Collagenase 3	MMP-13		52,000 42,000	Rodent interstitial collagenase
Collagenase 4	MMP-18		53,000 42,000	*Xenopus*
Gelatinase				
Gelatinase A	MMP-2	EC 3.4.24.24	72,000 66,000	Type IV collagenase
Gelatinase B	MMP-9	EC 3.4.24.25	92,000 84,000	Type V collagenase
Stromelysin				
Stromelysin 1	MMP-3	EC 3.4.24.17	57,000 45,000	Transin
Stromelysin 2	MMP-10	EC 3.4.24.22	54,000 44,000	Transin-2
Stromelysin 3[1]	MMP-11	EC 3.4.24.	64,000 46,000	RXKR furin cleavage
Membrane-type				
MT1-MMP	MMP-14		66,000 54,000	Transmembrane domain and RRKR furin cleavage site
MT2-MMP	MMP-15		72,000 60,000	
MT3-MMP	MMP-16		64,000 53,000	
MT4-MMP	MMP-17		57,000 53,000	
Others				
Matrilysin	MMP-7	EC 3.4.25.33	28,000 19,000	Lacks hemopexin
Metalloelastase	MMP-12	EC 3.4.24.65	54,000 22,000	**Macrophage elastase**
(No trivial name)[2]	MMP-19		54,000 45,000	
Enamelysin[3]	MMP-20		54,000 22,000	
Nonmammalian				
Xenopus XMMP[4]			70,000 53,000	Cys in catalytic domain
Envelysin[5]			63,000 48,000	Sea urchin
Soybean MMP[6]			? 19,000	Protein sequencing

Note: The values of M_r, except for MMP-8, are based on cDNA sequence; glycosylation may increase these values. Values for the active forms of MT-MMPs assume cleavage at the furin site. Names in bold are those recommended by the IUBMB. Certain of these enzymes do not receive further attention in the individual chapters; reference to these is as follows: [1] Basset *et al.*, 1990; [2] Cossins *et al.*, 1996; Pendas *et al.*, 1997; [3] Bartlett *et al.*, 1996; [4] Yang *et al.*, 1997; [5] Lepage and Gache, 1990; [6] McGeehan *et al.*, 1992.

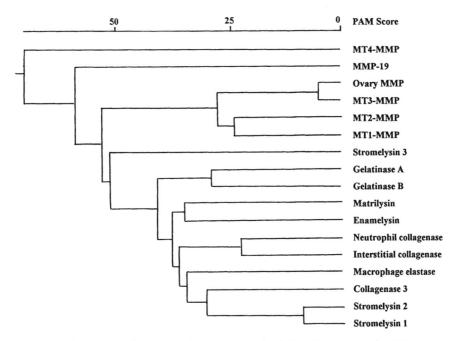

FIG. 1. Dendrogram illustrating the evolutionary relationships among the 17 known matrix metalloproteinases found to date in humans. The sequences were aligned by using the PILEUP program for the catalytic domain only (exclusive of fibronectin repeats). The tree was generated by the KITSCH algorithm. (Figure kindly provided by Dr. Neil Rawlings.)

est MMP, appears next; the absence of a hemopexin-like domain in this enzyme is probably due to a deletion, rather than to an evolutionary origin prior to the addition of hemopexin. Collagenases 1 and 2 are fairly close, followed shortly by macrophage elastase; stromelysins 1 and 2 come last. The last 7 enzymes are closely related in their domain structure, although macrophage elastase loses its hemopexin domain upon activation.

IV. THE DOMAIN STRUCTURE OF THE MATRIXINS

The matrixins form an interesting group of enzymes in that there is a central catalytic domain to which have been added a variety of additional domains or short inserts. Matrilysin represents the "minimal" enzyme—it consists of a signal peptide, a propeptide, and the catalytic domain. No one enzyme has all of the possible building blocks. If one examines the MMPs starting from the N terminus, the following features are seen:

Signal Peptide   This is typically a stretch of 17–20 residues, rich in hydrophobic amino acids, that serves as a signal for secretion into the endoplasmic reticulum for eventual export from the cell. All of the MMPs except for MMP-17 (Puente *et al.*, 1996) possess a signal peptide.

Propeptide   This region contains about 80 amino acids, typically with an N-terminal hydrophobic residue. There is a highly conserved PRCXXPD sequence near the C-terminal end of this segment; this provides the cysteine residue that makes contact with the catalytic zinc atom and maintains the enzyme in its zymogen form. This cysteine is found in all MMPs, including those that have the furin-cleavage site.

Furin-Cleavage Site Insert   This stretch of about nine residues includes the consensus sequence of RXKR/RRKR that leads to intracellular cleavage by furin. MMP-11, -14, -15, -16, and -17 possess this sequence. In the remaining enzymes, a cleavage by external proteases occurs in the middle of the propeptide, partially exposing the zinc and leading to autolytic cleavage of the remainder of the propeptide. The exact site of final cleavage may vary within a given enzyme leading to different degrees of activation, particularly in the case of MMP-1. XMMP contains a much longer insert of 37 residues (similar in sequence to vitronectin) that ends in the RRKR motif (Yang *et al.*, 1997).

Catalytic Domain   This domain typically contains about 160–170 residues, including sites for the binding of calcium ions and the structural zinc atom. The 50–54 residues at the C-terminal end of the catalytic domain include the site of binding of the catalytic zinc. This involves the highly conserved HEXGHXXGXXHS/T sequence mentioned earlier. MMP-17, however, has Val in place of Ser (Puente *et al.*, 1996). The zinc-binding region is somewhat independent of the remainder of the catalytic domain because various insertions can occur between these two portions.

Fibronectin-Like Repeats   There are three repeats of the fibronectin type II domain in MMP-2 and MMP-9, inserted in the catalytic domain just ahead of the 50-residue zinc-binding region. These specialized structures aid the binding of enzyme to gelatin substrates.

Hinge Region   The catalytic domain is connected to the following hemopexin domain by a linker region usually referred to as the hinge region. It ranges in length from 0 to 75 residues. The longest hinge is found in MMP-9 and shows considerable homology to type V collagen in that it is rich in proline. MMP-7, having no hemopexin domain, has no need for a hinge and XMMP also lacks this insert. A typical hinge contains about 16 residues including a number of proline residues. MMP-19 has a highly acidic region DEEEEETE within its linker (Cossins *et al.*, 1996; Pendas *et al.*, 1997).

Hemopexin Domain   This domain of about 200 residues contains four repeats that resemble hemopexin and vitronectin. There is a Cys residue at either end; these join and the resultant domain folds into a four-bladed propeller structure. All MMPs except MMP-7 contain this structure, but it does not appear to be essential for catalytic activity. Many truncated forms of MMPs have been produced that lack this domain and all retain activity. However, substrate specificity for macromolecules may be greatly affected (see review by Murphy and Knäuper, 1997). The binding of TIMP is also assisted by this domain and, in the case of gelatinases, there is binding of TIMP to this domain even when the enzymes are in their zymogen form.

Membrane Insertion Extension   Although most MMPs have their C terminus at the end of the hemopexin domain, the four MT-MMPs have a further extension that governs insertion of these proteases into the cell membrane. Its length ranges from about 80 to 110 residues. The membrane-spanning region of about 20 residues is about 20 residues in from the C terminus, leaving this short segment within the cytoplasm. MMP-19 has a 36-residue extension beyond the hemopexin domain but this is not membrane inserted.

V.   THE BIOLOGICAL ROLE OF THE MATRIXINS

A.   *Do the MMPs Play a Major Role in Degrading the Extracellular Matrix?*

Table I provides a good overview of the vast number of biologic and pathologic processes in which it is believed the MMPs play an important, or even indispensable, role. However, it must be admitted that firm proof for such involvement is very sparse. The general sorts of proofs that are offered include the use of *in situ* hybridization to show that mRNA for a given MMP is present at the site of tissue remodeling, use of immunohistochemistry to demonstrate that the enzyme protein is present, and localization of specific degradation products at the site (e.g., the cleaved fragments of collagen). These criteria, however, merely indicate guilt by association; they do not generally prove that a specific MMP is responsible for a specific effect. Now that we know of 17 MMPs in humans, it must be admitted that no one has examined a particular case of matrix remodeling for each activity nor established the contribution of each enzyme to the process. Although one can demonstrate the cleavage of collagen type II in arthritic cartilage (Dodge *et al.*, 1991), for example, we now know that the identical specific cleavage can be produced by MMP-1, MMP-2, MMP-8, MMP-13, and MMP-14 and there is some evidence that each of these enzymes can be produced by chondrocytes.

Krane (1994) has suggested criteria for proving the role of an MMP in a remodeling process:

- Remodeling can be blocked with a drug or antibody specific to the MMP.
- Remodeling can be reproduced by overexpression of the MMP gene in transgenic animals.
- Remodeling can be abolished by deleting the MMP gene.
- Spontaneous mutations can be identified and the phenotypes characterized.
- Mutations can be induced in the gene that reproduces the remodeling process.

Progress in carrying out this program of establishing the roles of the various MMPs has been slow and somewhat disappointing. Shapiro (1997) has recently reviewed the results with transgenic mice—there are knock-outs for MMP-3, MMP-7, MMP-9 and MMP-12 and overexpressers of MMP-1 and MMP3. In most cases these have not led to sharply defined phenotypes and they have not led to death of the organism. A problem appears to be the high redundancy of function of the MMPs, so that when one enzyme is knocked out another becomes more highly expressed to compensate for the loss. However, Shapiro has established quite clearly that the mouse macrophage, with a somewhat limited repertoire of MMPs, requires MMP-12 (macrophage elastase) in order to produce emphysema in mice induced to inhale cigarette smoke (Hautamaki et al., 1997). A similar problem arises in the use of specific inhibitors to block the action of an MMP. In general, the active centers and bond specificity of the various MMPs are fairly similar. For example, the collagenase substrate DNP–Pro–Leu–Gly*Ile–Ala–Gly–Pro–D–Arg is cleaved by all the MMPs tested to date. Furthermore, MMP inhibitors currently in use such as Batimastat are hydroxamate compounds; these are found to inhibit members of the M10 family, such as TNF-α convertase, as well as MMPs (Black et al., 1997). However, progress is being made in making such inhibitors ever more specific as can be seen in Chapters 10 and 11.

B. What Are the True Substrates for Each MMP?

It is very common to find reviews containing a table similar to Table II in which there is a list of substrates for each MMP. A still more extensive compendium of substrates is provided by Chandler et al. (1997). However, I have avoided this practice because it is my opinion that almost nothing is known about the natural substrates of the MMPs and that such a table is very misleading. Investigators tend to test only those matrix components ready to hand. Because many are cell biologists, their shelves contain collagens I and IV, fibronectin, laminin, and nidogen. However, I estimate that there are more than 100 known macromolecular components of the extracellular matrix—about 30–40 each of collagens, proteoglycans, and glycoproteins. Examination of the digestion of this vast array of potential substrates by the 15 MMPs is barely under way.

To establish a natural enzyme–substrate relationship, it is necessary to show that the substrate is cleaved *in vivo* at a certain point that matches the specificity of the enzyme and is unlikely to be due to other proteases. The case on which the greatest attention has been lavished is that of the cleavage of the large aggregating proteoglycan aggrecan, found in articular cartilage. A good deal of circumstantial evidence was built up by showing that stromelysin 1 could readily digest the protein core of aggrecan and that stromelysin was elevated in osteoarthritic cartilage (Dean *et al.*, 1989). However, more detailed study of the cleavage of aggrecan by various MMPs showed that every one of those tested, including collagenase 1, 2, and 3, could produce the same specific cleavage at the bond DIPEN*FFGVG (Fosang *et al.*, 1996). Moreover, when aggrecan degradation products were isolated from cartilage and synovial fluid, the products had arisen from the cleavage of an ITEGE*ARGSV bond (Sandy *et al.*, 1992). It therefore appears that a novel protease, aggrecanase, is required for this cleavage and that this is a metalloprotease but probably not an MMP.

This case is sufficient to illustrate the problem. The redundancy of specificity of the MMPs makes it difficult to pin down which one did what, and very few cases have had the cleavage sites of the substrate examined *in vivo*. Even in a clear case such as the interstitial collagens, which show the specific cleavage by collagenases of their Gly–Ile bond, a cleavage unlikely to be produced by other proteases in the tissue, one is left with five contenders for the role of cleaving enzyme. However, in the transgenic mouse in which this collagen cleavage site is mutated, collagen is still degraded through cleavage of the telopeptides. In this case, MMP-13 appears to be the enzyme that is capable of this cleavage (Krane *et al.*, 1996).

C.   Why So Many MMPs?

In view of the redundancy of function of many of the MMPs, one may ask why so many enzymes have evolved. In spite of the difficulty of proving what each one does, I believe that the multiplicity of forms underlines the extreme importance of the MMPs for the normal morphogenesis, maintenance, and repair of the matrix. The individual cell is very highly dependent on its battery of MMPs to control its environment, to move through it, and to maintain its protective cocoon. The knock-outs seem to tell us that the cell can make do without one of these enzymes by mobilizing one or more of the remaining MMPs to take over the function of the deficient enzyme. Furthermore, we should not think that each cell possesses the entire battery of enzymes. So far as is known, enamelysin (MMP-20; Bartlett *et al.*, 1996) is found only in developing tooth enamel. MMP-8 and MMP-12 are largely confined to the neutrophil and macrophage, respectively. MMP-7 appears to be restricted to epithelial cells (Wilson and Matrisian, 1996). The MMPs must be considered one more example of the profligacy of nature.

VI.   WHAT THE FUTURE MIGHT HOLD

The imminent arrival of the millenium prompts one to prognosticate about the prospects for the MMP field. Activity in this area of research has now reached a fever pitch and, with the clear involvement of the MMPs in such major disease problems as cancer, arthritis, and atherosclerosis, it seems unlikely that a sudden decline in interest in the MMPs will occur. Diseases will remain a major driving force because economics and politics will favor the distribution of resources in that area. Current work on the development of specific inhibitors will intensify. Both the working out of the detailed specificity requirements of each MMP and the development of highly specific inhibitors will be facilitated by the increasing use of combinatorial chemistry. However, there appear to be limits to the specificity that might be achieved, with consequent unintended side effects, so increasing attention will be focused on regulating MMP activity through genetic techniques such as antisense methods and through the use of specific drugs/factors that can regulate the expression of each MMP. This will require more detailed knowledge of promoters and cell factors governing expression. Intervention may also be attempted at the level of proteolytic activation of the MMP zymogens.

With respect to fundamental problems that remain to be answered, I am interested in the binding of MMPs to the cell and substratum. It is very difficult to extract most of the MMPs from tissues due to various types of anchoring. It is probably crucial for the cell to keep the MMPs in its vicinity to govern their activity, to keep track of how much enzyme is out there, and to prevent the enzymes from washing away with the blood until needed. A number of the MMPs appear to be attached to the cell surface bound to receptors, inserted into the membrane, or localized to invadopodia. This permits the cell to effect proteolysis in a specific direction following a regulated process of surface activation.

Much more needs to be learned about how the cell senses the proteolytic activity in the outside world—through receptors, integrins, or similar signaling mechanisms and through feedback from substrate fragments interacting with the cell. We need to learn more about how the cell regulates the activation of MMPs already released from the cell. Finally, with the detailed structure of MMP–TIMP complexes now in hand, we can begin to explore in more detail how the TIMPs interact with the enzymes. The full story here can only emerge when we see the full structure of the gelatinases with TIMPs in place on both the proenzyme and the active enzyme.

In summary, the past 36 years have seen tremendous advances in our understanding of the structure and activity of the MMPs. However, the areas of ignorance appear to be almost infinite in extent, promising many exciting years ahead in the MMP field.

ACKNOWLEDGMENTS

The author is supported by grant AR-16940 from the National Institutes of Health. Dr. Neil D. Rawlings, The Babraham Institute, Cambridge, UK, generously provided Fig. 1.

REFERENCES

Bartlett, J. D., Simmer, J. P., Xue, J., Margolis, H. C., and Moreno, E. C. (1996). Molecular cloning and mRNA tissue distribution of a novel matrix metalloproteinase isolated from porcine enamel organ. *Gene* **183,** 123–128.

Barrett, A. J., Rawlings, N. D., and Woessner, J. F. (1998). "Handbook of Proteolytic Enzymes," Academic Press, London.

Basset, P., Bellocq, J. P., Wolf, C., Stoll, I., Hutin, P., Limacher, J. M., Podhajcer, O. L., Chenard, M. P., Rio, M. C., and Chambon, P. (1990). A novel metalloproteinase gene specifically expressed in stromal cells of breast carcinomas. *Nature* **348,** 699–704.

Black, R. A., Rauch, C. T., Kozlosky, C. J., Peschon, J. J., Slack, J. L., Wolfson, M. F., Castner, B. J., Stocking, K. L., Reddy, P., Srinivasan, S., Nelson, N., Boiani, N., Schooley, K. A., Gerhart, M., Davis, R., Fitzner, J. N., Johnson, R. S., Paxton, R. J., March, C. J., and Cerretti, D. P. (1997). A metalloproteinase disintegrin that releases tumour-necrosis factor-alpha from cells. *Nature* **385,** 729–733.

Bode, W., Reinemer, P., Huber, R., Kleine, T., Schnierer, S., and Tschesche, H. (1994). The X-ray crystal structure of the catalytic domain of human neutrophil collagenase inhibited by a substrate analogue reveals the essentials for catalysis and specificity. *EMBO J.* **13,** 1263–1269.

Chandler, S., Miller, K. M., Clements, J. M., Lury, J., Corkill, D., Anthony, D. C., Adams, S. E., and Gearing, A. J. (1997). Matrix metalloproteinases, tumor necrosis factor and multiple sclerosis: An overview. *J. Neuroimmunol.* **72,** 155–161.

Cossins, J., Dudgeon, T. J., Catlin, G., Gearing, A. J., and Clements, J. M. (1996). Identification of MMP-18, a putative novel human matrix metalloproteinase. *Biochem. Biophys. Res. Commun.* **228,** 494–498.

Dean, D. D., Martel-Pelletier, J., Pelletier, J.-P., Howell, D. S., and Woessner, J. F., Jr. (1989). Evidence for metalloproteinase and metalloproteinase inhibitor imbalance in human osteoarthritic cartilage. *J. Clin. Invest.* **84,** 678–685.

Dodge, G. R., Pidoux, I., and Poole, A. R. (1991). The degradation of type II collagen in rheumatoid arthritis: an immunoelectron microscopic study. *Matrix* **11,** 330–338.

Fosang, A. J., Last, K., Knäuper, V., Murphy, G., and Neame, P. J. (1996). Degradation of cartilage aggrecan by collagenase-3 (MMP-13). *FEBS Lett.* **380,** 17–20.

Gross, J., and Lapiere, C. M. (1962). Collagenolytic activity in amphibian tissues: A tissue culture assay. *Proc. Natl. Acad. Sci. USA* **48,** 1014–1022.

Hautamaki, R. D., Kobayashi, D. K., Senior, R. M., and Shapiro, S. D. (1997). Requirement for macrophage elastase for cigarette smoke-induced emphysema in mice. *Science* **277,** 2002–2004.

Krane, S. M. (1994). Clinical importance of metalloproteinases and their inhibitors. *Ann. N.Y. Acad. Sci.* **732,** 1–10.

Krane, S. M., Byrne, M. H., Lemaître, V., Henriet, P., Jeffrey, J. J., Witter, J. P., Liu, X., Wu, H., Jaenisch, R., and Eeckhout, Y. (1996). Different collagenase gene products have different roles in degradation of type I collagen. *J. Biol. Chem.* **271,** 28509–28515.

Lepage, T., and Gache, C. (1990). Early expression of a collagenase-like hatching enzyme gene in the sea urchin embryo. *EMBO J.* **9,** 3003–3012.

McGeehan, G., Burkhart, W., Anderegg, R., Becherer, J. D., Gillikin, J. W., and Graham, J. S. (1992). Sequencing and characterization of the soybean leaf metalloproteinase. Structural and functional similarity to the matrix metalloproteinase family. *Plant Physiol.* **99,** 1179–1183.

Murphy, G., and Knäuper, V. (1997). Relating matrix metalloproteinase structure to function: Why the "hemopexin" domain? *Matrix Biol.* **15,** 511–518.

Pendas, A. M., Knäuper, V., Puente, X. S., Llano, E., Mattei, M. G., Apte, S., Murphy, G., and López-Otín, C. (1997). Identification and characterization of a novel human matrix metalloproteinase with unique structural characteristics, chromosomal location, and tissue distribution. *J. Biol. Chem.* **272,** 4281–4286.

Puente, X. S., Pendas, A. M., Llano, E., Velasco, G., and López-Otín, C. (1996). Molecular cloning of a novel membrane-type matrix metalloproteinase from a human breast carcinoma. *Cancer Res.* **56,** 944–949.

Rawlings, N. D., and Barrett, A. J. (1995). Evolutionary families of metallopeptidases. *Methods Enzymol.* **248,** 183–228.

Sandy, J. D., Flannery, C. R., Neame, P. J., and Lohmander, L. S. (1992). The structure of aggrecan fragments in human synovial fluid. Evidence for the involvement in osteoarthritis of a novel proteinase which cleaves the Glu 373–Ala 374 bond of the interglobular domain. *J. Clin. Invest.* **89,** 1512–1516.

Shapiro, S. D. (1997). Mighty mice: Transgenic technology "knocks out" questions of matrix metalloproteinase function. *Matrix Biol.* **15,** 527–533.

Wilson, C. L., and Matrisian, L. M. (1996). Matrilysin: an epithelial matrix metalloproteinase with potentially novel functions. *Int. J. Biochem. Cell Biol.* **28,** 123–136.

Woessner, J.F., Jr. (1994). The family of matrix metalloproteinases [review]. *Ann. N.Y. Acad. Sci.* **732,** 11–21.

Yang, M. Z., Murray, M. T., and Kurkinen, M. (1997). A novel matrix metalloproteinase gene (XMMP) encoding vitronectin-like motifs is transiently expressed in *Xenopus laevis* early embryo development. *J. Biol. Chem.* **272,** 13527–13533.

Interstitial Collagenases

John J. Jeffrey

*Department of Biochemistry and Molecular Biology, Albany Medical College, Albany,
New York 12208*

I. Introduction

Since interstitial collagenases were last reviewed by this author in this series (Jeffrey, 1986), the area of interstitial collagenase chemistry and biology—in parallel with that of the matrix metalloproteinases in general—has undergone explosive growth. The emergence of our knowledge of the collagenases as functional proteins, as well as the information provided by molecular biological studies of these molecules, has provided quantum leaps in our knowledge of the structure, chemistry, enzymology, and biology of these enzymes. So much information has emerged in the last 8 to 10 years that it is impossible in a chapter such as this to thoroughly explore every aspect of the subject in the detail that one would want. Thus, in this chapter, massive selectivity was exercised in presenting information in any kind of useful fashion. Accordingly, studies that speak definitively of the overall chemistry or biology of the interstitial collagenases have been given preference over more circumscribed investigations of phenomena in individual tissue physiology or disease state pathologies. As a result, a great deal of excellent and important work has been neglected, for which the author appologizes. Constraints, unpleasant as they are, are nevertheless necessary to attempt to present a coherent overview of the state of the field.

In the introduction to the previous review of interstitial collagenases, the point was made that the specification of the precise three-dimensional structure of the connective tissues places unusual, perhaps

Matrix Metalloproteinases

unique, requirements on biology. The distances over which this precision is required to extend in tissues is the analogue of light years in physics, and the necessity of the maintenance of directionality ("up," "down," "left," "left," "right," "how far?") add a further demand on biological information transfer. The intervening years have yielded little in the elucidation of these fundamental biological issues, but with the enormous increase in knowledge of the number and nature of the molecules involved in these processes, some level of information regarding these questions can be expected by the time this subject is revisited. This chapter intends to provide the general reader with a broad overview of our knowledge of the nature of interstitial collagen degradation and the enzymes that catalyze it.

II. MATRIX METALLOPROTEINASE-1: THE ORIGINAL INTERSTITIAL COLLAGENASE

The first vertebrate collagenase both purified to homogeneity as a protein and cloned as a cDNA was that from human fibroblasts (Stricklin et al., 1977; Goldberg et al., 1986). This enzyme, bearing the designation matrix metalloproteinase-1 (MMP-1), has served as the prototype for all other interstitial collagenases. In view of that historical fact, the characteristics of MMP-1 are reviewed in some detail and are then used as a template for comparison with other interstitial collagenases.

A. Chemistry of MMP-1

The enzyme is secreted from the cell as a pair of zymogens of M_r approximately 52,500 (Stricklin et al., 1977). The two secreted proteins apparently differ from each other only in that one is glycosylated and the other either less so or not at all. The precise chemistry of the glycosyl moiety in the glycosylated form of the protein has not been defined, nor has its biological function. All that is known at this time is that there is no significant effect of glycosylation on enzymatic activity (Stricklin et al., 1978). In all other respects examined to date, such as chromatographic behavior, sedimentation equilibrium, and substrate specificity, the two forms of the proenzyme behave identically. Given the fact, however, that the presence of differentially glycosylated forms of MMP-1 has been observed in a variety of species (e.g., Roswit et al., 1983), it seems fair to speculate that some function—unknown as yet—is served by the presence of the glycosyl residues.

The interstitial collagenases so far identified belong to a class of proteinases known as the matrix metalloproteinases (MMPs) (Woessner, 1991). A number of excellent reviews of this ever-expanding gene family have been published (Sang and Douglas, 1996; Birkedal-Hansen *et al.*, 1993; Dioszegi *et al.*, 1995; Nagase, 1994), to which the reader is referred for further specific information. Briefly, MMP-1, and all members of the MMP family, including interstitial collagenases, share a very similar domain structure, as illustrated in Fig. 1 (Birkedal-Hansen *et al.*, 1993), composed of a single peptide, a proenzyme domain, a catalytic domain, a short (usually 16–17 amino acids in length) "hinge" region, and a C-terminal domain resembling hemopexin (Li *et al.*, 1995; Faber *et al.*, 1995), originally identified as a heme-binding protein (Jenne and Stanley, 1987). All the members of the MMP gene family except one, matrilysin, contain a similar domain and, interestingly, a similar hemopexin domain has also been identified in the cell-matrix adhesion protein vitronectin (Hunt *et al.*, 1987; Jenne, 1991). All members of the MMP family contain Zn^{2+} at the catalytic site (Vallee and Auld, 1992; Birkedal-Hansen *et al.*, 1993) and, in addition, require Ca^{2+} for stability and activity (Seltzer *et al.*, 1976; McKerrow, 1987; Zhang *et al.*, 1997). The intrinsic zinc at the catalytic site is chelated by the imidazole nitrogens of three histidine residues together with the sulfur of a free cysteine, all located in the proenzyme domain (Sanchez-Lopez *et al.*, 1993). The amino acid motif containing the critical histidines, HexGHxxGxxH, is not restricted to the interstitial collagenases but is present in highly conserved form in all MMPs (Sang and Douglas, 1996). In addition, the amino acid sequence surrounding a cysteine residue also believed to be required for latency (PRCGVPD) is highly

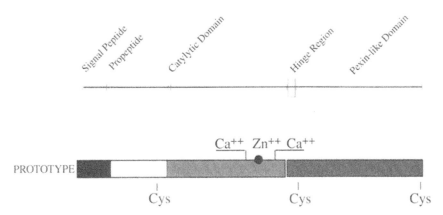

FIG. 1. Representation of a prototype matrix metalloproteinase. All the interstitial collagenases contain variations of the domains indicated in this representation. (Adapted from Birkedal *et al.*, 1993.)

conserved in all members of the MMP family (Sang and Douglas, 1996). A second molecule of zinc has also been shown to be present in the enzyme (Bode, 1994; Lovejoy et al., 1994); this second zinc, however, is believed not to be involved in the catalytic activity of the enzyme but rather to play a structural role, serving to stabilize the enzyme.

Similarly, Ca^{2+} has long been implicated in the activity of MMP-1, and recent studies indicate that, like the second zinc atom, calcium is involved in stabilizing the structure of the enzyme, in both the catalytic- and the C-terminal hemopexin domains (Li et al., 1995; Gomis-Ruth et al., 1996).

A number of X-ray crystallographic studies have now been performed on various forms of MMP-1 from a variety of species. The first complete crystallographic analysis of full-length MMP-1 was that of the porcine enzyme (Li et al., 1995), crystallized as a complex with a low-molecular-weight inhibitor. For the purposes of this discussion, the structure presented consists of three important domains: (1) the amino-terminal catalytic domain, which, not surprisingly, bears strong resemblances to the catalytic domains of other MMPs crystallized to date; (2) the linker region, rich in proline, at residues 261–277, highly exposed to the environment; and (3) the hemopexin domain, revealed for the first time as four antiparallel beta-sheet subdomains, arranged in a four-bladed propeller structure stabilized by one of the disulfide bridges in MMP-1, that between Cys_{278}, which is at the very beginning of the domain, and Cys_{466}, at the C-terminal end of the protein. A good feel for the geometry of the four-bladed structure can be obtained from the stereo $C\alpha$ trace diagram taken from Li et al. (1995). A number of studies have shown that the hemopexin domain is crucial to the nature of the catalytic activity of MMP-1. Truncated MMP-1, which lacks the hemopexin domain, displays no activity against native collagen, although it retains enzymatic activity against low-molecular-weight peptide substrates (Murphy et al., 1992; Knauper et al., 1993). Similarly, substituting the hemopexin region of MMP-3 (stromelysin) for the native domain of MMP-1 results in essentially total loss of collagenolytic activity (Sanchez-Lopez et al., 1993). Thus, the hemopexin domain is a critical specifier of the enzymatic specificity of MMP-1, but the reason for this remarkably effective specification is not clear. Preliminary modeling studies by Li et al. (1995) and Bode (1994) to assess the possibility that the hemopexin domain directs the favorable binding of triple helical collagen to the enzyme and of the scissile bond to the active site suggest that no particularly favorable binding motif exists. To compound this conundrum further, other workers have pointed out, on the basis of crystallographic studies, that the active site is situated in the enzyme such that it would be predicted to be relatively inaccessible to a peptide bond in triple helical collagen. Thus, the active site of interstitial collagenase, as currently viewed, appears to be an inhospitable one for accommodating the triple helical collagen molecule for the purpose of catalysis.

Nevertheless, cleavage not only occurs, but appears to occur in an apparently *en bloc* fashion. That is, all three chains appear to be cleaved in a given substrate molecule before the products of any cleaved molecule are released (Welgus *et al.*, 1981a). In addition, no evidence exists for a preferential cleavage of one of the α chains of a heterotrimeric collagen substrate vis à vis another. In unpublished experiments, this author has incubated native type I collagen $[(\alpha_1)_2(\alpha_2)]$ in solution with a molar excess of MMP-1 at low temperature to allow binding and then examined the ratio of cleavage products of the chains in the substrate at short times of incubation. In such a situation, it was hoped that if a preference for one chain versus another existed, the ratio of the three-quarter-length products derived from the α chains of the parent molecule would differ from 2:1, depending on what preference exists. No deviation from a 2:1 ratio was ever observed. Again, this result is consistent with the presence of a functional *en bloc* cleavage of collagen by MMP-1. It also appears, from the use of type I collagen mutated at the collagenase cleavage site in only the α_1 chain, that the cleavage of the normal α_2 chain in the same molecule cannot be catalyzed (Krane *et al.*, 1996). Furthermore, a preliminary report (Byrne *et al.*, 1992) indicates that the cleavable bonds in the α chains of a triple-helical collagen molecule must be in their native register for any cleavage to occur. That is, if the cleavable bond of one chain is moved with respect to the analogous bonds in the other two chains, no collagenase cleavage is observed. In other words, all three chains must be cleavable, or no chain is cleaved.

It is, therefore, not easy to readily visualize a mechanism to accommodate all of these data, simply from the structural knowledge of MMP-1 available at this time. An intriguing hypothesis has been presented by DeSouza *et al.* (1996), who have identified a sequence motif in the hinge region of MMP-1 (GRSQNPVQPIGPQTP that theoretical considerations suggest would, by virtue of its repetitive PXX sequences, adopt a helical configuration similar to that of the collagen helix itself: a polyproline II-like helix. Based on this conformational analysis of the hinge region sequence, these authors propose that this region interacts with the collagen molecule at or near the cleavage site. This interaction is hypothesized to destabilize further the cleavage site area, long proposed to possess a "looser" helical structure as a reason for its collagenase susceptibilility (Miller *et al.*, 1976). In the model of DeSouza *et al.,* a possible induced relaxation of the substrate helix occurs, allowing only one chain of the triple helix to become available to the active site of the molecule, rather than the difficult achievable fit of all three chains. In addition, this hypothesis contains a tacit imbedded principle: that of multidomain-induced conformational changes resulting in a transition to an ultimately favorable environment for collagenolysis. Although this proposal remains theoretical at present, it should provide the basis for further experiments to elucidate the processes whereby interstitial collagenases such as MMP-1 are able to cleave their substrate at such a specific locus and in such a unique way.

B. Catalytic Activity of MMP-1

This general topic has been the subject of numerous studies and reviews, and it is not the purpose of this chapter to exhaustively revisit the details of these many investigations. Rather, it is intended to present some basic facts and features of MMP-1 activity and to attempt to put them in a rational biological context.

MMP-1 degrades types I, II, III, and X collagens (Welgus *et al.,* 1981a), type I gelatin, the antiprotease α_1-antitrypsin as well as its own zymogen in at least two places (see Birkedal, 1993, for further detail), and it is likely that other substrates exist for this enzyme as well. Its activity on the interstitial collagens has been most extensively studied; the enzyme cleaves the triple helix of types I, II, and III collagen exclusively at the Gly_{775}–$Leu/Ileu_{776}$ peptide bond, very close to three-quarters of the distance from the amino terminus of the substrate molecule (Miller *et al.,* 1976).

Aside from the considerations discussed in the previous section with respect to structural features of the enzyme molecule with potential relevance to catalytic activity, the nature of the cleavage site in native collagen that renders it susceptible to catalysis is essentially unknown. It has been proposed that the helix is weaker in this region by virtue of a low average hydroxyproline content compared to elsewhere in the collagen molecule, but no experimental evidence exists to confirm this hypothesis. If indeed the helix is less rigid in this area, then this region should be more susceptible to proteolysis. However, a wide array of enzymes of varying classes fails to cleave any bond in this area of the molecule, and MMP-1 cleaves only one of these bonds, with no evidence of "nibbling" in the hypothetically weak helix area (Welgus *et al.,* 1980; Welgus *et al.,* 1981a).

One situation, that of the cleavage of type III collagen in some species (the human is one), is particularly instructive. Trypsin makes a single cleavage in these collagens—although it should be emphasized that this cleavage is never observed in type I collagen—at an Arg–Gly bond, eight residues C-terminal to the collagenase cleavage site (Miller *et al.*, 1976). Although this experimental finding is consistent with the weak helix hypothesis, type III collagens from some species are not similarly susceptible to trypsin (Welgus *et al.*, 1985)—chick type III, for example. The same bond cleaved by trypsin in type III collagen exists in these type III collagens as well as in the type I collagens of the same species, yet is not cleaved by trypsin in these molecules (Welgus, Burgeson, *et al.*, 1985). Thus, one is left with a murky picture of the requirement for substrate determinants of cleavage. One finding does stand out in this unclear area, however. Krane and colleagues (1996) have shown, as mentioned earlier, that all three chains of a collagen molecule must be cleavable to allow cleavage of any chain and that the three scissile bonds must be in "native" register for cleavage to take place (Byrne *et al.*, 1992). If the scissile bond of one chain is moved in relation to the analogous bonds in the other chains in a collagen molecule, cleavage is again inhibited. Thus, although the mechanistic implications of these data are not known, it is clear that stringent spatial and structural requirements must be met for collagenolysis to proceed.

The dynamics of the action of MMP-1 on various collagenous substrates was dealt with at length in the previous review of this subject by this reviewer (Jeffrey, 1986). To summarize, however, the following points can be useful: (1) MMP-1 cleaves all three classical interstitial collagens in solution, displaying classical Michaelis–Menten kinetics (Welgus *et al.*, 1981a; Fields *et al.*, 1987). Values for K_m ranged from 1 to 2 μM, indicating a rather high affinity of enzyme for substrate. The trade-off for this high affinity appears to be a very slow catalytic rate: the V_{max} for the activity of MMP-1 on type I collagen in solution at 25°C is approximately 25 mol^{-1} hr^{-1} (Welgus *et al.*, 1981a). This value translates to the cleavage of a single collagen α chain per minute per enzyme molecule, a very low rate indeed in enzymology, especially considering the high affinity of the enzyme for its substrate. Of further interest was the large disparity in the rates at which MMP-1 degraded the three principal interstitial collagens. Type III was degraded some 10 times faster than type I collagen, and more than 100 times more rapidly than type II collagen (Welgus *et al.*, 1981a). Indeed, type II was degraded so slowly that it gave rise to speculation that MMP-1 might not be involved in the degradation of cartilage collagen at all. This latter speculation turned out to be a fortunate guess, with the discovery of MMP-13 and its presence in cartilage together with MMP-8 (*vide infra*).

The interaction of MMP-1 with its various collagenous substrates has been the subject of extensive investigation, and the picture that emerges implicates the availability of water to the site of peptide bond hydrolysis as a critical determinant in the catalytic efficiency of the enzyme (Welgus *et al.,* 1981b; Jeffrey *et al.,* 1983). As previously described, the exclusion of water from collagenous substrates as they gain higher levels of organization massively affects the rate of collagen degradation by MMP-1. Thus, when this enzyme is examined for its ability to degrade gelatin chains, even though its catalytic activity (i.e., V_{max}) is rather low, it behaves *energetically* as a "normal" enzyme (Welgus *et al.,* 1981b). That is, it displays a temperature dependence characteristic of most enzymes in biology, with a so-called Q_{10} of approximately 2. In other words, for a change in reaction temperature of 10°C, the rate of catalysis changes approximately 2-fold, indicating an activation energy (E_A) of approximately 10 kcal/mol. When, however, the cleavage of the same chains as part of a native, triple helical collagen molecule is assessed, the energetics of collagenolysis change drastically. In this setting, the rate of collagenolysis varies much more drastically with temperature, and E_A, as referred to earlier, rises to values in the vicinity of 40 kcal/mol. Such values specify a change in reaction rate of approximately 10-fold (as opposed to 2-fold for a normal enzyme) for every 10° of temperature change. When the triple helical molecules are aggregated into native fibrils, the energetics change dramatically once again. The activation energy of MMP-1 on fibrillar type I collagen is of the order of 100 kcal/mol, which specifies a change of some 200-fold for every 10° of temperature change, an unprecedented situation in enzymology. This value indicates that the rate of collagenolysis of fibrillar type I collagen triples for every 2° of rise in temperature! Another remarkable phenomenon relating the availability of water and the collagenolytic activity of MMP-1 on fibrillar collagen is observed when D_2O is substituted for H_2O in the incubation mixture. Instead of the usual reduction in reaction rate of 2-fold or less, the rate decreases by 10-fold or more. Because this value cannot be explained in terms of classical isotope effects, it was hypothesized that this remarkable D_2O effect was another example of the difficulty of getting water to the site of peptide bond hydrolysis (Jeffrey *et al.,* 1983).

Thus, the overall conclusion from a number of studies bearing on this situation was that the availability of the water required for the hydrolysis of the scissile peptide bond in collagen becomes more and more the rate-limiting step as the molecule becomes more organized. The aggregation of monomers to form the mature collagen fibril, with the exclusion of water that accompanies this process, presents the most difficult substrate of all for MMP-1. A diagrammatic representation of this process, as derived from the studies described, is illustrated in Fig. 2.

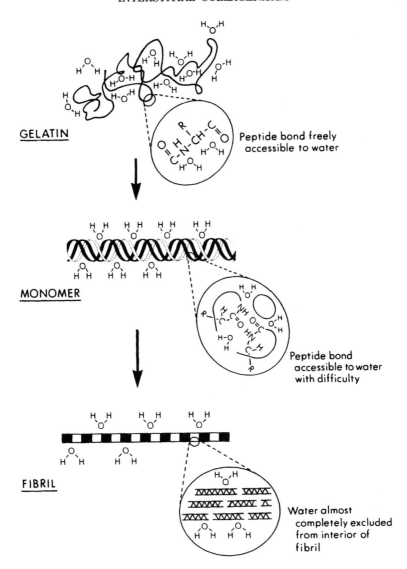

FIG. 2. Representation of the relationship of the availability of water to the ease of peptide bond hydrolysis by interstitial collagenase in collagenous substrates at progressive levels of organization. As the level of organization increases, the difficulty of bringing water to the site of hydrolysis increases drastically. [From Jeffrey, J.J. (1986). The biological regulation of collagenase activity. *In* "Regulation of Matrix Accumulation" (R.P. Mecham, ed.), pp. 53–92, Academic Press, New York, with permission.]

When the data describing the energetics of MMP-1 activity are taken together with the considerations presented by the structural analyses of MMP-1, it is tempting to paint a picture of a combination of scissile bonds in a hydrophobic environment, requiring energy to transport the water of hydrolysis to the correct site. Together with major induced-fit changes in the enzyme molecule to fully enable an active site that appears to have difficulty accessing its substrate the result is a low rate of proteolysis. Given the fact that processes of development, metamorphosis, morphogenesis, repair, and involution take place over long periods of biological time—all requiring spatial and temporal precision to ensure correct outcomes—it may be that nature has accepted these barriers to collagenolysis as an appropriate price to pay for spatial precision.

C. Activation of proMMP-1

Studies of MMP-1 as a protein and as an enzyme revealed that proMMP-1 can be activated by a wide array of enzymes, reagents, and manipulations. Thus, for example, enzymes as different as trypsin, kallikrein, cathepsin G, and plasminogen were shown by a number of laboratories to activate the zymogen to an active form of the enzyme, with an accompanying loss of about 10 kDa in mass (see Birkedal-Hansen et al., 1993, for a more comprehensive discussion). At the same time, a variety of apparently unrelated small molecules were found, as a group, to accomplish proMMP-1 activation as well. These included KSCN and KI at high concentrations ($\sim 3M$), sodium dodecyl sulfate (SDS) and, most interestingly, a number of congeners of a series of organomercurial compounds, including phenyl mercuric chloride, p-hydroxy mercuribenzoate, and aminophenyl mercuric acetate. The earliest study to address the nature of the activation of the proenzyme by organomercurials strongly suggested that a conformational activation was occurring in the presence of these compounds (Stricklin et al., 1983). Thus, it was proposed that the initial active form of the enzyme was that of an "active zymogen," and that this conformationally active form subsequently underwent autolytic cleavage to the truncated active form. Because the autolytic process was independent of the concentration of enzyme protein, it was further proposed that this cleavage was an intramolecular, rather than an intermolecular process. In the intervening years, this hypothesis has been largely confirmed, in more specific terms, as the sequence of the molecule has become available. Cleavage sites of the various proteases that activate the zymogens have been identified. They are clustered at the center of the pro-piece (residues 35–40). In a sense, this region acts in a manner analogous to the "bait" region of α_2-macroglobulin and allows a number of unrelated proteinases to catalyze this initial cleavage (see Birkedal-Hansen et *al.*, 1993, for more detail). Following the action of a proteinase, the resultant truncated zymogen is able to cleave the remainder of its own pro-piece, resulting in the production of the mature form of the enzyme.

The ability of the truncated zymogen to catalyze an intramolecular cleavage has been attributed to the interruption of the $Cys–Zn^{2+}$ bond, giving rise to the oft-invoked "cysteine switch" mechanism of activation, wherein the switch is closed when the bond is intact and open when the bond is disrupted, allowing proteolysis to occur (Van Wart and Birkedal-Hansen, 1990). Attractive as this hypothesis is, it has not been unequivocally demonstrated to be operative, and some existing data cannot be fully explained by the universal participation of the cysteine switch mechanism in proMMP-1 activation. The most difficult data to reconcile with this mechanism come from reports from Chen *et al.* (1993) and Birkedal-Hansen's laboratory (Galazka *et al.*, 1996), in which the cysteine in proMMP-3 (i.e., stromeylsin) putatively involved in the cysteine switch mechanism was mutated to a variety of other amino acids. Under these conditions, the proenzyme—absent the cysteine—remained latent and, more importantly, could still be activated by organomercurials. These data are not consistent with the existence of cysteine chelated to the active site zinc as a primary regulator of proMMP-3 latency, and these investigators speculated that the crucial effect of the chemical activators of MMPs is the disruption of a salt bridge between other amino acids of the pro-piece and the active portion of the molecule (Galazka *et al.*, 1996). Unfortunately, an identical study has not been done using proMMP-1, but it is of interest that, in the initial study of organomercurial activation of proMMP-1, Stricklin *et al.* (1983) were unable to detect alkylation of the sulfhydryl of the cysteine by *p*-chloromercuribenzoate (pCMB), a compound that was effective in generating the "active zymogen" form of the enzyme. This compound has an extremely high affinity for free sulfhydryl groups (and for this reason was used for many years to determine the number of free cysteines in proteins); nevertheless, neither a well-established spectrophotometric method nor radiolabeling with $[^{14}C]$-pCMB was successful in demonstrating alkylation of the –SH group of what we now know is the only unpaired cysteine in proMMP-1.

All of these data, coupled with the ability of a number of chemically unrelated compounds to produce conformational activation, suggest the possibility that this process is not as simple as once thought. What these compounds do have in common, as a group, is that they all are chaotropes. That is, they all display the ability to disrupt the structure of water and, as such, might be expected to exert powerful effects on the conformation of the structure of proteins. Thus, it is possible that conformational distortion alone, without the need for a *direct* rupture of a co-ordinate covalent bond such as $Cys-Zn^{2+}$, is sufficient for the conformational activation of proMMP-1 and possibly other MMPs as well.

This general question of proenzyme activation is raised in view of the fact that the precise mechanism of *in vivo* activation of proMMP-1 has never been established in any biological setting. A number of suggestions have been made, for example, that the plasminogen cascade serves as a major physiological activator of interstitial collagenase (He *et al.*, 1989), but it is important to bear in mind that these proposals are based on test tube observations, and because the actual *in vivo* form of activated MMP-1 is unknown, the possibility of multiple pathways of activation must be entertained. This might even include conformational activation in biology: A report in the literature (Tyree *et al.*, 1981) describes the apparent existence of proteins in conditioned culture medium of cells or tissue fragments which display the characteristics of conformational activators of human proMMP-1. Thus, in the presence of these "factors," the proenzyme displays full activity at the time of addition of the activating factor, and examination of the resultant fully active enzyme shows that it retains the molecular weight of the zymogen. Unfortunately, purification of these putative proteins has not been further documented, so it is unclear whether they do indeed serve the same function in biology. Nevertheless, the possibility that nature makes use of the propensity of the MMPs to undergo conformational activation cannot be ruled out.

To further illustrate the potential complexity of the processes of MMP-1 activation, the example of the potential participation of stromelysin (MMP-3) in the activation of proMMP-1 is instructive. Stromelysin is capable of catalyzing a further cleavage in the initial autocatalytically truncated form of the proMMP-1 as initiated, for example, by trypsin, resulting in a "mature" enzyme slightly different from that produced entirely by autolysis (Suzuki *et al.*, 1990). The resulting product (FVL-collagenase) is nearly 10 times more active than the autolytically derived product: VL-collagenase (Suzuki *et al.*, 1990). This remarkable change in the activity of the enzyme, as a result of the action of stromelysin, suggests the possibility that, in nature, a given amount of collagenase can be manipulated to exhibit a wide range of activity. As an example, in preliminary cell culture experiments (Jeffrey, unpublished observations) the production of stromelysin by human fibroblasts clearly modifies the specific activity of MMP-1, again over a range of at least 10-fold. Of potentially more biological importance, however, is our observation in this system (Jeffrey, Ehlich, and Roswit, unpublished) that the inhibition of the production of stromelysin is some 10-fold more sensitive to glucocorticoids than is the analogous inhibition of MMP-1 itself by these steroids. In such a setting, the *activity* of a given *amount* of MMP-1 can be imagined to be regulated over a very wide range by such regulatory molecules as steroids, depending on the relative amounts of stromelysin and collagenase in a given biological setting. Thus, it is attractive to hypothesize that the collagenolytic potential in a given tissue can be regulated significantly by mechanisms that are designed to modify the specific activity of a constant amount of collagenase.

The main point of the foregoing discussion is that the general question of how MMPs are activated in specific settings *in vivo* remains perhaps the major unknown area of matrix metalloproteinase biology today. The answer to this question presents a formidable challenge to the field, given the low levels of expression of an enzyme such as MMP-1 in general, and the likelihood of the focal nature of that expression, the determination of the characteristics of the true native active form of the enzyme will be difficult indeed. For example, the amount of collagen degraded in the uterus of a rat during postpartum involution is approximately 0.1 μmol. Clearly the amount of collagenase, acting catalytically as it does, must be considerably less, presenting a massive challenge to our present methodology of protein characterization.

D. Collagenase-3: MMP-13 and Rodent Interstitial Collagenase

As indicated earlier in the section devoted to MMP-1 specificity, the latter enzyme was found to have such a low activity on type II collagen that its usefulness as an enzyme that could manage the collagen phenotype of cartilage was called into question (Welgus *et al.*, 1981a). Indeed, it was hypothesized that a cartilage-specific collagenase might exist for this particular biological purpose. At the same time, an interstitial collagenase was purified from rat myometrial smooth muscle cells (Roswit *et al.*, 1983). This enzyme catalyzed cleavage at the 3/4:1/4 site of the major interstitial collagens, but with significant quantitative differences from the cleavages catalyzed by human MMP-1 (Welgus *et al.*, 1983). Although the rat enzyme displayed similar affinities (K_m) for collagen substrates, the massive differences in V_{max} displayed by human MMP-1 were not evidenced in the activity of the rat enzyme. Thus, the values for V_{max} dispalyed by rat interstitial collagenase were essentially identical for the three major interstitial collagens. Of special interest was the fact that type II collagen was degraded as efficiently as were types I and III. Recent observations (Jeffrey, Wilcox, Hambor and Mitchell, unpublished) suggest that this enzyme is apparently the only interstitial collagenase in the rodent genome. If this is indeed the case, the equal efficiency of cleavage by the enzyme of all the interstitial collagens would allow this MMP to effectively manage collagenolysis in a wide variety of tissues.

Interestingly, the rat enzyme displayed high levels of gelatinolytic activity, in marked contrast to human MMP-1. Although the two enzymes displayed similarities in the bonds cleaved in denatured collagen substrates (Gly-Leu, Gly-Ileu, Gly-Phe, Gly-Ala), the rat enzyme displayed a distinctly higher specific activity against gelatin than against native collagen, while the reverse was true for human MMP-1 (Welgus et al., 1985).

In 1990, the rat enzyme was cloned (Quinn et al., 1990), and it became clear that this collagenase and human MMP-1 were very different in homology (less than 50% similarity exists), although the major modules characteristic of MMPs (Quinn et al., 1990; Sang and Douglas, 1996) were maintained. Furthermore, this enzyme appeared to be produced by a variety of mesenchymal cells in the rat—osteoblasts, fibroblasts, and smooth muscle cells—indicating that the enzyme was not specialized to a single cell type (Quinn et al., 1990). This difference between rodent and human collagenases remained a puzzle, especially when mouse interstitial collagenase was cloned and found to be essentially identical to the rat enzyme (Henriet et al., 1992). By contrast, the interstitial collagenases cloned from rabbit and pig sources displayed much higher homology to human MMP-1 than to the rodent collagenases (see Sang and Douglas, 1996, for detailed homology comparisons).

Finally, Freije et al. (1994) cloned a human homologue of rodent collagenase from a breast carcinoma cDNA library. The homology between this second human enzyme of resident cell origin and the rodent mesenchymal enzyme is nearly 90%, although some significant sequence differences nevertheless exist, particularly in the proenzyme domain. This enzyme has been designated as MMP-13 or, alternatively, human collagenase-3. Tissue distribution of MMP-13 was originally viewed as quite limited; indeed, initially it appeared to be specific to breast carcinoma cells in situ, but more recently it has become apparent that chondrocytes are a major source of this collagenase. Two groups have now identified MMP-13 in both normal and osteoarthritic articular chondrocytes (Mitchell et al., 1996; Reboul et al., 1996). Furthermore, the enzyme has been examined for collagen substrate specificity (Mitchell et al., 1996; Billinghurst et al., 1997; Knauper et al., 1996) and found to cleave type II collagen—the major interstitial collagen of cartilage—at a much higher rate than types I and III. Thus, this enzyme appears to fulfill the role, only a matter of speculation 15 years ago, of a type II selective collagenase. This finding has understandably engendered considerable excitement among investigators attempting to understand the pathophysiology of rheumatoid arthritis and osteoarthritis, pathologic conditions in which the degradation of type II collagen has long been postulated to play a major role.

Human MMP-13 is very similar, but by no means identical, to its rodent homologues (Sang and Douglas, 1996). By way of similarity, it appears that human MMP-13 displays the ability to cleave type II collagen at least as efficiently as types I and III. In the case of the rat homologue, the catalytic efficiency against type II collagen is approximately equal to that exhibited against types I and III. In the case of human MMP-13, the catalytic efficiency of the enzyme against type II collagen is some 10-fold higher than the efficiency of the enzyme on types I and II. Suprisingly, given the apparent universality of bond cleavage shared by interstitial collagenases, MMP-13 catalyzes an additional cleavage in the triple helix of type II collagen, one triplet aminoterminal to the primary cleavage site (Billinghurst *et al.*, 1997). The biological significance of this unanticipated extra cleavage in cartilage collagen is unknown. In addition, it has been found that human MMP-13 degrades denatured collagen extremely efficiently, as had previously been shown for rat collagenase. Indeed, the *gelatinolytic* activity of both rat collagenase and human MMP-13 is substantially greater than the *collagenolytic* activity of either enzyme (Welgus *et al.*, 1985; Mitchell *et al.*, 1996). The full biological implication of this capability of MMP-13 homologues is unclear at this time, but a report from Fosang *et al.* (1996) indicates that MMP-13 has the ability to cleave aggrecan, a possible indicator of the existence of yet additional substrates for this subfamily of enzymes. In further support of this notion, Krane and coworkers (1996) have shown that rodent collagenases, but not human MMP-1, have the ability to cleave the telopeptide region of type I collagen in addition to the classical 775–776 cleavage site. This activity appears to be determined by the amino-terminal domain of the rodent collagenases. Chimeric collagenases with MMP-1 N-terminal domains fail to catalyze this cleavage, whereas chimeras containing *rodent* (i.e., MMP-13) N-terminal domains do catalyze this extra cleavage in the collagen molecule. These authors suggest that this additional proteolytic capability of MMP-13 might allow some level of collagen degradation to proceed during development in the transgenic mouse. Thus, it is possible that enzymes of this subgroup of interstitial collagenases are designed to degrade a wider variety of substrates than MMP-1 homologues. In general then, the catalytic activities of both rodent collagenase and MMP-13 display the similarities one might predict from the extensive amino acid homologies they share.

X-ray crystallographic studies of the C-terminal (hemopexin) domain of MMP-13 complexed with an active site inhibitor have been performed (Gomis-Ruth *et al.*, 1996). The structure of this domain in MMP-13 is very similar to that present in porcine MMP-1, taking the form of a four-bladed propeller consisting of four antiparallel beta-sheet domains. No difference could be identified as providing a likely basis for the difference in collagen substrate specificities of the two MMPs, even though this domain is critical for the activity of both enzymes. Thus, the structural determinants of the specificity of the interstitial collagenases remain elusive. The activation of MMP-13 homologues display marked similarities to the analogous processes involved in MMP-1 activation. Although a thorough examination of the ability of chaotropes to activate MMP-13 has not been performed, it is clear that aminophenyl mercuric acetate (APMA) is an effective activator of the human enzyme, whereas trypsin appears to degrade the enzyme (Knauper et al., 1996). In the rodent, on the other hand, our laboratory has consistently observed that trypsin is an extremely reliable activator of rat interstitial collagenase, while APMA appears to promote loss of activity by degradation of the enzyme protein (Jeffrey and Roswit, unpublished observations). A thorough study, under carefully controlled conditions, is clearly required to obtain an appropriate comparison of the two molecules. For now it is sufficient to conclude that they share far more similarities than differences. A noteworthy finding regarding human MMP-13 is that it is apparently activated by the membrane-bound MT1-MMP, a property not shared by human MMP-1 (Knauper et al., 1996). This finding suggests the possibility of localized, pericellular activity of MMP-13 by virtue of its ability to be activated by a protease associated with the cell membrane. This could be important in the setting of breast carcinoma, in which human MMP-13 was first observed. Somewhat strangely, new knowledge of the role of this protease in the tumorigenic process—whether in breast or other carcinomas—has not been forthcoming.

The rodent MMP-13 homologues have been utilized to develop a fascinating and informative animal molecule, that bids fair to illustrate the *in vivo* role of interstitial collagenase in development as well as in normal adult life. Krane and colleagues have used a targeted mutagenic strategy to develop a mouse whose homozygotic offspring contain only type I collagen, which is mutated at the 775–776 cleavage site (Liu *et al.*, 1995) of the α chains. Careful studies have revealed that his mutation completely prevents degradation of type I collagen by interstitial collagenases, of both human and rodent sources. These homozygotes readily conceive and deliver normal litters of apparently normal offspring. This phenomenon alone was surprising to many in the community, given the tacit assumption over the years that collagenase would be required for many of the events that constitute normal connective tissue development; such is apparently not the case.

On the other hand, phenotypic aberrations do occur in the homozygous mice: principal among them is the inability to properly involute the uterus *postpartum*. This massive, and relatively rapid, physiological removal of collagen has stood for years as a principal paradigm for the action of collagenolytic enzymes (Jeffrey, 1991). In the mutant mice, however, massive nodules of unresorbed collagen remain in the myometrium for long periods after parturition, adding further evidence for the requirement for interstitial collagenase in post-partum uterine involution.

In addition to this impressive consequence of the existence of nondegradable collagen, the homozygous mice slowly develop other connective tissue aberrations as they age. Abnormal thickening of the papillary dermis occurs, as does kyphosis of the spinal column. Last, contractures develop in limbs, which appear to be the result of the failure of tendons to be able, by virtue of normal remodeling processes, to accommodate the growth of bones during postnatal development (Krane, personal communication). These slowly developing connective tissue defects are hypothesized to be the result of very slow and/or very focal processes of interstitial collagenase activation and function. Such processes would be difficult to detect with current techniques designed to assess major processes of tissue collagen degradation, but are likely to represent the biologically significant role of interstitial collagenases in mammalian growth and development. It will be exciting to follow further phenotypic consequences of this crucial targeted mutation in mammalian type I collagen as they are explored in a variety of other tissues.

The discovery of a human homologue of previously defined rodent collagenases has helped to both clarify and confuse our understanding of the strategy adopted by biology for the degradation of interstitial collagens. On the one hand, the finding of this enzyme in cartilage—both normal and pathologic—provides the possibility of an answer to the long-awaited question of how the degradation and remodeling of type II collagen are managed in mammalian systems. On the other hand, the question of why species such as human—and many others—have two mesenchymal interstitial collagenases, whereas rodents appear to manage essentially the same biological processes with only one, remains a paradox. Clearly, this area of interstitial collagenase biology will be a major focus of research in the future. Of particular interest will be further information on the involvement of MMP-13 in tumors, given that the original localization of this enzyme indicated that breast carcinoma cells were a major, and originally apparently the only, source of the enzyme in a first approximation of biological localization.

E. Human Neutrophil Collagenase: MMP-8

This enzyme represents the third major mammalian interstitial collagenase—although historically the second to have been described—and remains one of the most difficult for which to assign specific physiologic or pathologic roles. Neutrophil collagenase has the distinction of being the only interstitial collagenase to be stored in cells—in this case in the specific granules of the neutrophil—rather than being synthesized and released on biological demand. This enzyme was first described in the late 1960s (Lazarus *et al.*, 1968), purified in limited amounts in the 1980s (Hasty *et al.*, 1987), and cloned and subsequently expressed as a pure protein in the early 1990s (Hasty *et al.*, 1990). The results of all these studies clearly put neutrophil collagenase in the MMP family and establish it as a member of the interstitial collagenase family, with the designation MMP-8.

Curiously, another circulating cell, the monocyte/macrophage, fails to produce MMP-8; rather it produces MMP-1 in an "on-demand" fashion and, unlike the neutrophil, does not store the enzyme (Wahl 1977; Welgus, Campbell, *et al.*, 1985). As a protein, the neutrophil interstitial collagenase is almost identical in size to other interstitial enzymes as a protein, but is considerably more highly glycosylated. Thus, fully glycosylated, the proenzyme form of MMP-8 is approximately 60 kDa; activation results in the loss of approximately 20,000 Da of mass, and the active form of the enzyme is approximately 40 kDa. Again, in the case of the neutrophil collagenase, the function of the glycosylation is unknown. However, again in common with MMP-1, it appears not to affect catalytic activity (see Birkedal-Hansen, 1983, for further details).

Neutrophil collagenase displays a substrate specificity profile different from that of MMP-1 (Hasty *et al.*, 1987). The preference for type III relative to type I collagen, which exists for MMP-1, is reversed in the case of MMP-8, which digests soluble type I collagen at a significantly higher rate than type III. Of particular interest in this area is the ability of MMP-8 to cleave type II collagen—the predominant collagen of cartilage—at a markedly higher rate than does MMP-1. The molecular determinants of these different specificities for two very similar interstitial collagenases remain to be elucidated.

Of potential major significance, however, relating to the ability of MMP-8 to degrade type II collagen at appreciable rates is the recent finding that MMP-8 is produced by normal human articular chondrocytes in the presence of low concentrations of interleukin-1 (Cole *et al.*, 1996). It is clear from this study that it was the chondrocyte, and not contaminating leukocytes, that produced the MMP-8 and that the chondrocyte produces the enzyme in an on-demand fashion rather than storing it in granules as is the case in the neutrophil. Thus, clearly, the neutrophil is not the sole source for this interstitial collagenase, and it will be exciting to follow the emerging biology of this MMP in the cartilage. Together with the presence of MMP-13 in cartilage (see earlier discussion), the potential for type II collagen degradation is considerable, particularly in pathologic states. In addition, although some controversy exists as to exclusivity, MMP-8 has been shown to possess the ability to degrade aggrecan, a major structural glycosaminoglycan in cartilage (Fosang *et al.*, 1996; Arner *et al.*, 1997). The cleavage specificity of MMP-8 on the interglobular domain of aggrecan appears to differ from the specificity of the major, although as yet unidentified, cartilage aggrecanase (Arner *et al.*, 1997); nevertheless, the presence of this enzyme in osteoarthritis or rheumatoid arthritis may signify an even more important role for MMP-8 in the pathophysiology of these diseases.

Table 1 is a selected compilation of kinetic data, comparing the ability of the three interstitial collagenases to degrade a variety of collagen types. Although not complete, the values in the table illustrate the notions discussed previously. By way of summary, the catalytic efficiencies (k_{cat}/K_m) for human MMP-13 and human MMP-8 on type II collagen are massively higher than that of human MMP-1 on this cartilage collagen. This, together with the recent localization of both enzymes in chondrocytes d further suggests that they have been adapted for handling the normal development and repair of cartilage

TABLE I

SELECTED KINETIC PARAMETERS OF INTERSTITIAL COLLAGENASES ON COLLAGEN SUBSTRATES

Enzyme	Collagen substrate	K_m (μM)	K_{cat} (h^{-1})	k_{cat}/K_m ($\mu M^{-1}h^{-1}$)
Human MMP-1	Human I	0.8	53	67
	Human II	2.1	1.0	0.48
	Human III	1.7	350	206
Human MMP-13	Human I	n.d.	n.d.	50*
	Human II	2	23	11.5
	Human III	n.d.	n.d.	n.d.
Rat MMP-13	Human I	0.9	10.7	12.3
	Human II	0.9	14.2	15.7
	Human III	1.7	20.2	11.9
Human MMP-8	Human I	1	450	450
	Human II	2	150	75
	Human III	2.2	200	80

All values obtained from studies on collagens in solution at 25°C, using the method of Welgus *et al.*, 1981.

*Calculated from the specific activity presented in Knauper *et al.*, 1997, assuming a K_m of ~1 μM.

collagen; their presence also allows for significant untoward collagenolysis in pathological states as discussed previously.

MMP-8 contains all the modules that have become hallmarks of
matrix metalloproteinases: a proenzyme domain, followed by a catalytic
domain, the hinge region, and then by the characteristic C-terminal
hemopexin-like domain. The catalytic site and cysteine switch regions
display high homology to the analogous regions of all the MMPs, and
structure–activity relationships are similar to those displayed by the
other interstitial collagenases. Thus, for example, C-terminal truncation results in the loss of triple helical degrading activity, but retention
of activity on other substrates (again, see Sang and Douglas, 1996, and
Birkedal-Hansen, 1993, for further detail).

Of particular interest with respect to domain requirements for substrate specificity in the MMPs as a class, the neutrophil collagenase
has been used rather extensively by two groups to examine this issue,
particularly with respect to the involvement of the hinge region in
determining collagenolytic activity. By using a variety of chimeric molecules containing deletions and/or swapped domains from stromelysin,
it has been shown (Hirose et al., 1993) that a 62-amino-acid sequence
in the hinge region of MMP-8 was required for collagenolytic activity.
Substituting the analogous, although longer, sequence from stromelysin failed to restore activity. Knauper et al. (1997) have reexamined this
same region. Using alanine-scanning mutagenesis techniques, these
workers found that mutating four proline residues in the hinge region
resulted in profound reductions in collagenase activity. The results of
this study are in substantial concurrence with the earlier proposal by
DeSouza et al. (1996), which is discussed previously in some detail.
Briefly, this proposal hypothesizes that the hinge region of interstitial
collagenase adopts a polyproline II-like structure which interacts in a
critically important way with the constituent chains of native collagen
substrates. It is possible that a destabilization of the substrate occurs,
allowing collagenolysis to proceed in the protected environment produced by the trapping of the substrate between the hemopexin domain
and the active site region of the enzyme. Undoubtedly, this exciting
possibility will receive considerably more attention in future studies
of structural requirements for collagenolysis.

The activation of neutrophil collagenase, released as it is as a proenzyme upon neutrophil degranulation, has been the subject of considerable study over the years. Activation can be achieved in vitro by a
number of proteases and organomercurials and by autolysis, at sites
identical to or very similar to those observed for the same processes in
MMP-1. In addition, a number of studies have strongly implicated the
role of products of the myeloperoxidase pathway in activated neutro-

phils, such as hypochlorous acid and monochloroamines in the *in vivo* activation of the enzyme (Test and Weiss 1986; Saari *et al.*, 1992). A number of *in vitro* studies support the ability of HOCl and chloramines such as taurine to activate neutrophil collagenase. In addition, cathepsin G, released upon neutrophil activation, has been shown to activate the proenzyme (Capodici *et al.*, 1989), and considerable discussion has ensued as to which pathway is significant in biology. Claesson *et al.* (1996) have provided evidence that both pathways may play a role. When neutrophil activation is performed under aerobic conditions, significant HOCl-mediated activation occurs quite rapidly. Subsequent to this initial activation a further activation occurs, most likely mediated by cathepsin G. Prevention of cathepsin G activity under these conditions has little effect on the initial enzyme activation. Under *anaerobic* conditions, however, although the magnitude of activation is reduced from aerobic levels, all the activation can be prevented by serine proteinase inhibitors. Thus, these authors suggest that the two mechanisms—HOCl/chloramines and proteolytic/cathepsin G—exist to provide for effective activation in both highly oxygenated and relatively anaerobic tissue milieu. The potential availability of multiple pathways of activation may be mirrored by the analogous setting observed with proMMP-1, in which so many pathways have been shown to activate the proenzyme effectively. It may be that here, as well as in the neutrophil, biology provides a form of parallel processing to ensure that sufficient enzyme will be made available under potentially variable tissue conditions.

III. "Nontraditional" Interstitial Collagenases

The original central dogma of this area of extracellular matrix chemistry was that only interstitial collagenase could cleave the native collagen triple helix. A small, but significant tear in that seamless theory was provided by Miller and colleagues (1976), who found that trypsin and elastase could catalyze a single cleavage in type III collagen. High concentrations of these enzymes were required, and the finding was viewed more as a tool to characterize the region of cleavage in the substrate than as a harbinger of new enzymes that might cleave native collagen. Recently, however, two reports have appeared that convincingly indicate that two enzymes, MMP-2 and MT1-MMP, can catalyze the characteristic G_{775}–I/L_{776} cleavage in type I collagen. Aimes and Quigley (1995) have reported that human MMP-2, or 72-kDa gelatinase, free of the TIMP-2 that is normally associated with this enzyme during its biosynthesis, catalyzes this characteristic cleavage in soluble calf and chick type I collagens and appears to degrade native fibrillar type I collagen as well. Interestingly, the homologous 92-kDa gelatinase (MMP-9), similarly TIMP-free, fails to catalyze these cleavages. The enzyme displays values of k_{cat} that are quite close to those exhibited by MMP-1, the original interstitial collagenase. Values for K_m for MMP-2, with soluble collagen as substrate, are some five- to eight-fold higher; that is, the affinity of this enzyme is considerably lower than that displayed by other, so-called "classical" interstitial collagenases ($\sim 1 \ \mu M$). The significance of this disparity in biology is not clear. In earlier studies by Welgus et al. (1980) it was calculated that the effective concentration of collagen in fibrils, that is, that available to an enzyme in solution, is approximately 1 μM. This derives from the observation that the enzyme has access only to the molecules on the surface of the fibril. Assuming a K_d of the enzyme for its substrate of approximately micromolar, only half of what one imagines to be quite a high concentration of enzyme in biological terms ($\sim 50 \ \mu g/mL$) would be bound to collagen fibrils under equilibrium conditions. Thus, an enzyme with a K_d eight-fold higher would be bound to a correspondingly lesser degree. The challenge for the future will be to determine whether there are biological settings in which the concentration of MMP-2 is indeed high enough to allow for significant binding to fibrillar substrates. In addition, the study of Aimes and Quigley (1995) indicates that only MMP-2 free of its normally associated TIMP-2 is able to degrade native collagen. This finding is intriguing in that it suggests the possibility that directed biological mechanisms exist whereby the extent of the MMP-2/TIMP complex could be modulated in vivo. Again, future studies will be required to explore this possibility.

Finally, it has been recently shown that two members of the membrane-bound MMP family—MT1-MMP and a truncated version of the enzyme lacking the transmembrane domain, produced in recombinant form—both cleave native collagen, again into the classical 3/4:1/4 fragments (Ohuchi et al., 1997). Heretofore, these molecules have been thought to interact only with proMMP-2 and proMMP-13 for the purpose of activating these enzymes. The findings of Ohuchi et al. now provide the possibility that pericellular collagenolysis by these molecules could play a role in biology. Note that, in the case of full-length MT1-MMP, the values of K_m and V_{max} for this enzyme on type I collagen are very similar to those displayed by MMP-1, indicating a considerable potential for collagenolysis at or near the cell surface. One experiment that was not performed in these studies was to examine the potential of native MT1-MMP, bound to membranes, to exhibit the same collagenolytic capability; only purified recombinant forms of the enzymes were studied. If the native membrane-bound form does, in fact, exhibit the ability to degrade native collagen, it would provide an even more exciting setting for this family of proteases.

IV. SUMMARY AND FUTURE PERSPECTIVES

The growth in our knowledge of the chemistry and biology of the matrix metalloproteinases in general has been explosive in the last several years, and increases in our knowledge of the interstitial collagenases have paralleled that growth. The identification, in just the last 3 or 4 years, of new interstitial collagenases and old enzymes with newly identified interstitial collagenolytic activities expands the biological repertoire of interstitial collagenolysis, requiring considerable reevaluation and experimental approaches to assessing the roles of these proteases in appropriate systems. Molecular biological technology has allowed for the cloning and expression of all the known interstitial enzymes—both "classical" and "nonclassical"—and has provided pure proteins as well as specific antibodies to the enzymes. This technology has allowed the identification of domains responsible for specific biochemical characteristics of the enzymes and of domains that are crucial for one or another aspect of their function. Thus, the era of the "impure reagent" is past; with the tools now at hand investigators will be able to examine complex biological and pathological settings at a new, more precise level of approximation.

X-ray crystallography has, in the past 5 years, become an essential tool in our understanding of not only the overall structure of the interstitial collagenases, but perhaps more importantly, the relationships between domains in the enzymes. Extremely valuable insights have been gained into the role of the C-terminal hemopexin domain, for example. It is fair to say that extension of these studies will shed yet more light on the specific mechanism by which the interstitial collagenases manage the cleavage of the chains of native triple helical collagen. Crystallographic studies have also been invaluable in facilitating the design of inhibitors that are selective for one MMP over another. The ultimate goal is to achieve specificity of inhibition of individual matrix metalloproteinases, both to elucidate better the role of each in a variety of physiologic processes and in disease states as well. Even at the state of the art today, there is promise for therapeutic value from some of the inhibitors available now; this situation is bound to improve in the next few years, an exciting prospect indeed.

One crucial area of MMP biology yet to be elucidated—and the interstitial collagenases are no exception—is the definition of the mechanism or mechanisms by which the pro-form of the enzyme is activated. As noted earlier, this will be a formidable task, but one that nevertheless would open new horizons, both in our understanding of the basic biology of the enzymes and in our ability to modulate their activity in rational ways.

Finally, transgenic and targeted mutational technology, as typified by the studies of Liu *et al.* (1995), should provide major insights into the roles of these enzymes in developmental and pathologic processes. This especially promising area bids fair to provide massive increases in our knowledge of the sequelae of slow, low-level, long-term processes of collagen degradation *in vivo,* where much of the turnover of collagen appears to take place.

In closing, the following observation was made concerning the attainment of the ultimate goals in our understanding of collagen degradation:

> as unattainable as they may appear at this time, one must have a certain faith that, as the molecules involved in these processes continue to be identified, purified and described in precise chemical terms, we are ipso facto approaching these ultimate answers. Twenty-five years ago workers in this field despaired of even finding a true collagenolytic enzyme; the breadth and depth of developments since that time should encourage us to think that today's insoluble problems will bring a set of solutions as exciting as those that have appeared since that time.

At the end of this chapter, as the developments in this field that have occurred since the mid-1980s are discussed, it is appropriate to echo and reemphasize those same thoughts and expectations.

REFERENCES

Aimes, R.T., Quigley, J.P. (1995). Matrix metalloproteinase-2 is an interstitial collagenase—Inhibitor-free enzyme catalyzes the cleavage of collagen fibrils and soluble native type-1 collagen generating the specific 3/4 length and 1/4 length fragments. *J. Biol. Chem.* **270,** 5872–5876.

Arner, E.C., Decicco, C.P., Cherney, R., *et al.* (1997). Cleavage of native cartilage aggrecan by neutrophil collagenase (MMP-8) is distinct from endogenous cleavage by aggrecanase. *J. Biol. Chem.* **272,** 9294–9299.

Billinghurst, R.C., Dahlberg, L., Ionescu, M., *et al.* (1997). Enhanced cleavage of type II collagen by collagenases in osteoarthritic articular cartilage. *J. Clin. Invest.* **99,** 1534–1545.

Birkedal-Hansen, H., Moore, W.G.I., Bodeen, M.K., *et al.* (1993). Matrix metalloproteinases: A review. *Crit. Rev. Oral Biol. Med.* **4,** 197–250.

Bode, W. (1994). The X-ray crystal structure of the catalytic domain of human neutrophil collagenase inhibited by a substrate analogue reveals the essentials for catalysis and specificity. *EMBO J.* **13,** 1263–1269.

Byrne, M.H., Wu, H., Birkhead, J.R., *et al.* (1992). Sliding the collagenase cleavage site in the alpha-1(I) chain out of phase with the alpha-2(I) chain confers collagenase resistance. *J. Bone Min. Res.* **7,** s131.

Capodici, C., Muthukumaran, G., Amoruso, M.A. (1989). Activation of neutrophil collagenase by cathepsin G. *Inflammation* **13**, 245–258.

Chen, L.-C., Noelken, M.E., Nagase, H. (1993). Disruption of the cysteine-75 and zinc ion coordination is not sufficient to activate the precursor of matrix metalloproteinase-3 (stromelysin-1). *Biochemistry* **32**, 10289–10295.

Claesson, R., Karlsson, M., Zhang, Y.Y., *et al.* (1996). Relative role of chloramines, hypochlorous acid, and proteases in the activation of human polymorphonuclear leukocyte collagenase. *J. Leuk. Biol.* **60**, 598–602.

Cole, A.A., Chubinskaya, S., Schumacher, B., *et al.* (1996). Chondrocyte matrix metalloproteinase-8: Human articular chondrocytes express neutrophil collagenase. *J. Biol. Chem.* **271**, 11023–11026.

DeSouza, S.J., Pereira, H.M., Jacchier, S., *et al.* (1996). Collagen/collagenase interaction—Does the enzyme mimic the conformation of its own substrate? *FASEB J.* **10**, 927–930.

Dioszegi, M., Cannon, P., and Van Wart, H.E. (1995). Vertebrate collagenases. *Methods Enzymol.* **248**, 413–431.

Faber, H., Groom, C.R., Baker, H.M., *et al.* (1995). 1.8 A crystal structure of the C-terminal domain of rabbit hemopexin. *Structure* **3**, 551–559.

Fields, G.B., Birkedal-Hansen, H., and Van Wart, H.E. (1987). Sequence specificity of human skin fibroblast collagenase. *J. Biol. Chem.* **262**, 6221–6226.

Fosang, A.J., Last, K., Knauper, V., *et al.* (1996). Degradation of cartilage aggrecan by collagenase-3 (MMP-13)L. *FEBS Lett.* **380**, 17–20.

Freije, J.M., Diez-Itza, I., Balbin, M., *et al.* (1994). Molecular cloning and expression of collagenase-3, a novel human matrix metalloproteinase produced by breast carcinomas. *J. Biol. Chem.* **269**, 16766–16773.

Galazka, G., Windsor, L.J., Birkedal-Hansen, H., *et al.* (1996). APMA (aminophenylmercuric acetate) activation of stromelysin-1 involves protein interactions in addition to those with cysteine-75 in the propeptide. *Biochemistry* **35**, 11221–11227.

Goldberg, G.I., Wilhelm, S.M., Kronberger, A., *et al.* (1986). Human fibroblast collagenase: Complete primary structure and homology to an oncogene transformation-induced rat protein. *J. Biol. Chem.* **261**, 6600–6605.

Gomis-Ruth, F.X., Gohlke, U., Betz, M., *et al.* (1996). The helping hand of collagenase-3 (MMP-13): 2.7 A crystal structure of its C-terminal haemopexin-like domain. *J. Mol. Biol.* **264**, 556–566.

Hasty, K.A., Jeffrey, J.J., Hibbs, M.S., *et al.* (1987). The collagen substrate specificity of neutrophil collagenase. *J. Biol. Chem.* **262**, 48–52.

Hasty, K.A., Pourmotabedd, T.F., Goldberg, G.I., *et al.* (1990). Human neutrophil collagenase—A distinct gene product with homology to other matrix metalloproteinases. *J. Biol. Chem.* **265**, 1421–1424.

He, C.S., Wilhelm, S.M., Pentland, A.P., *et al.* (1989). Tissue cooperation in a proteolytic cascade activating human interstitial collagenase. *Proc. Nat'l. Acad. Sci. USA* **86**, 2632–2636.

Henriet, P., Rousseau, G.G., Eeckhout, Y. (1992). Cloning and sequencing of mouse collagenase cDNA—Divergence of mouse and rat collagenases from the other mammalian collagenases. *FEBS Lett.* **310**, 175–178.

Hirose, T., Patterson, C., Pourmotabbed, T., *et al.* (1993). Structure–function relationship of human neutrophil collagenase—Identification of regions responsible for substrate specificity and general proteinase activity. *Proc. Natl. Acad. Sci. USA* **90**, 2569–2573.

Hunt, L.T., Barker, W.C., and Chen, H.R. (1987). A domain structure common to hemopexin, vitronectin, interstitial collagenase and a collagenase homolog. *Protein Seq. Data Anal.* **1**, 21–26.

Jeffrey, J.J. (1986). The biological regulation of collagenase activity. *In* "Regulation of Matrix Accumulation" (R.P. Mecham, ed.), pp. 53–92, Academic Press, New York.

Jeffrey, J.J. (1991). Collagen and collagenase—Pregnancy and parturition. *Semin. Perinatol.* **15,** 118–126.

Jeffrey, J.J., Welgus, H.G., Burgeson, R.E., *et al.* (1983). Studies on the activation energy and deuterium effect of human skin collagenase on homologous collagen substrates. *J. Biol. Chem.* **258,** 11123–11127.

Jenne, D. (1991). Homology of placental protein 11 and pea seed albumin 2 with vitronectin. *Biochem. Biophys. Res. Commun.* **176,** 1000–1006.

Jenne, D., and Stanley, K.K. (1987). Nucleotide sequence and organization of the human S-protein gene: Repeating peptide motifs in the pexin family and a model for their evolution. *Biochemistry* **26,** 6735–6742.

Knauper, V., Docherty, A.J., Smith, B., *et al.* (1997). Analysis of the contribution of the hinge region of human neutrophil collagenase (HNC, MMP-8) to stability and collagenolytic activity by alanine scanning mutagenesis. *FEBS Lett.* **405,** 60–64.

Knauper, V., Osthues, A., Declerck, Y.A., *et al.* (1993). Fragmentation of human polymorphonuclear leukocyte collagenase. *Biochem. J.* **291,** 847–854.

Knauper, V., Will, H., Lopez-Otin, C., *et al.* (1996). Cellular mechanisms for human procollagenase-3 (MMP-13) activation: Evidence that MT1-MMP (MMP-14) and gelatinase-a (MMP-2) are able to generate active enzyme. *J. Biol. Chem.* **271,** 17124–17131.

Krane, S.M., Byrne, M.H., Lemaitre, V., *et al.* (1996). Different collagenase gene products have different roles in degradation of type-1 collagen. *J. Biol. Chem.* **271,** 8509–8515.

Lazarus, G.S., Brown, R.S., Daniels, J., *et al.* (1968). Human granulocyte collagenase. *Science* **159,** 1483–1485.

Li, J., Brick, P., O'Hare, M.C., *et al.* (1995). Structure of full-length porcine synovial collagenase reveals a C-terminal domain containing a calcium-linked, four-bladed b-propeller. *Structure* **3,** 541–549.

Liu, X., Wu, H., Byrne, M., *et al.* (1995). A targeted mutation at the known collagenase cleavage site in mouse type I collagen impairs tissue remodeling. *J. Cell Biol.* **130,** 227–237.

Lovejoy, B., Cleasby, A., Hassell, A.M., *et al.* (1994). Structure of the catalytic domain of fibroblast collagenase complexed with an inhibitor. *Science* **263,** 375–377.

McKerrow, J.H. (1987). Human fibroblast collagenase contains an amino acid sequence homologous to the zinc-binding site of serratia protease. *J. Biol. Chem.* **262,** 5943.

Miller, E.J., Finch, J.E., Chung, E., *et al.* (1976). Specific cleavage of the native type III collagen molecule with trypsin. Similarity of the cleavage products to collagenase-produced fragments and primary structure at the cleavage site. *Arch. Biochem. Biophys.* **173,** 631–637.

Miller, E.J., Harris, E.D., Chung, E., *et al.* (1976). Cleavage of type I and II collagens with mammalian collagenase: Site of cleavage and primary structure at the NH2 terminal portion of the smaller fragment released from both collagens. *Biochemistry* **15,** 787–792.

Mitchell, P.G., Magna, H.A., Reeves, L.M., *et al.* (1996). Cloning, expression, and type-II collagenolytic activity of matrix metalloproteinase-13 from human osteoarthritic cartilage. *J. Clin. Invest.* **97,** 761–768.

Murphy, G., Allan, J.A., Willenbrock, F., *et al.* (1992). The role of the C-terminal domain in collagenase and stromelysin specificity. *J. Biol. Chem.* **267,** 9612–9618.

Murphy, G., Reynolds, J.J., Bretz, U., *et al.* (1977). Collagenase is a component of the specific granules of human neutrophil leukocytes. *Biochem. J.* **162,** 195–197.

Nagase, H. (1994). Matrix metalloproteinases: A mini-review. *Extracell. Matrix Kidney Contrib. Nephrol.* 85–93.

Ohuchi, E., Imai, K., Fujii, Y., *et al.* (1997). Membrane type-1 matrix metalloproteinase digests interstitial collagens and other extracellular matrix macromolecules. *J. Biol. Chem.* **272**, 2446–2451.

Quinn, C.O., Scott, D.K., Brinckerhoff, C.E., *et al.* (1990). Rat collagenase—Cloning, amino acid sequence comparison and parathyroid hormone regulation in osteoblastic cells. *J. Biol. Chem.* **265**, 2342–2347.

Reboul, P., Pelletier, J.P., Tardif, G., *et al.* (1996). The new collagenase, collagenase-3, is expressed and synthesized by human chondrocytes but not by synoviocytes. A role in osteoarthritis. *J. Clin. Invest.* **97**, 2011–2019.

Roswit, W.T., Halme, J., and Jeffrey, J.J. (1983). Purification and properties of rat uterine procollagenase. *Arch. Biochem. Biophys.* **225**, 285–295.

Saari, H., Suomalinin, K., Lindy, O., *et al.* (1990). Activation of latent human neutrophil collagenase by reactive oxygen species and serine proteases. *Biochem. Biophys. Res. Comm.* **171**, 979–987.

Sanchez-Lopez, R., Alexander, C.M., Behrendt, O., *et al.* (1993). Role of zinc-binding-encoded and hemopexin domain-encoded sequences in the substrate-specificity of collagenase and stromelysin-2 as revealed by chimeric proteins. *J. Biol. Chem.* **268**, 7238–7247.

Sang, Q.A., and Douglas, D.A. (1996). Computational sequence analysis of matrix metalloproteinases. *J. Protein Chem.* **15**, 137–160.

Seltzer, J.L., Welgus, H.G., Jeffrey, J.J., *et al.* (1976). The function of calcium ion in the action of mammalian collagenase. *Arch. Biochem. Biophys.* **173**, 355–361.

Stricklin, G.P., Bauer, E.A., Jeffrey, J.J., *et al.* (1977). Human skin collagenase: Isolation of precursor and active forms from both fibroblast and organ cultures. *Biochemistry* **16**, 1607–1615.

Stricklin, G.P., Eisen, A.Z., Bauer, E.A., *et al.* (1978). Human skin fibroblast collagenase: Chemical properties of precursor and active forms. *Biochemistry* **17**, 2331–2337.

Stricklin, G.P., Jeffrey, J.J., Roswit, W.T., *et al.* (1983). Human skin fibroblast procollagenase: Mechanisms for activation by organomercurials and trypsin. *Biochemistry* **22**, 61–70.

Suzuki, K., Enghild, J.J., Morodomi, T., *et al.* (1990). Mechanisms of activation of tissue procollagenase by matrix metalloproteinase 3 (stromelysin). *Biochemistry* **29**, 10261–10270.

Test, S.T., Weiss, S.J. (1986). The generation and utilization of chlorinated oxidants by human neutrophils. *Adv. Free Rad. Biol. Med.* **2**, 91–116.

Tyree, B., Seltzer, J.L., Halme, J., *et al.* (1981). The stoichiometric activation of human skin fibroblast procollagenase by factors present in human skin and rat uterus. *Arch. Biochem. Biophys.* **208**, 440–443.

Vallee, B.L., and Auld, D.S. (1992). Active zinc binding sites of zinc metalloenzymes. Matrix metalloproteinases and inhibitors. *Matrix* (Suppl.) **1**, 5–19.

Van Wart, H.E., Birkedal-Hansen, H. (1990). The cysteine switch: A principle of regulation of metalloproteinase activity with potential applicability to the entire matrix metalloproteinase gene family. *Proc. Natl. Acad. Sci. USA* **87**, 5578–5582.

Wahl, L.M. (1977). Hormonal regulation of macrophage collagenase activity. *Biochem. Biophys. Res. Commun.* **74**, 838–845.

Welgus, H.G., Burgeson, R.E., Wootton, J.A.M., *et al.* (1985). Degradation of monomeric and fibrillar type III collagens by human skin collagenase—Kinetic constants using different animal substrates. *J. Biol. Chem.* **260**, 1052–1059.

Welgus, H.G., Campbell, E.J., Cury, J.D., Eisen, A.Z., Senior, R.M., Wilhelm, S.M., and Goldberg, G.I. (1990). Neutral metalloproteinases produced by human mononuclear phagocytes—enzyme profile, expression and regulation during development. *J. Clin. Invest.* **86**, 1556–1564.

Welgus, H.G., Grant, G.A., Sacchettini, J.C., *et al.* (1985). The gelatinolytic activity of rat uterus collagenase. *J. Biol. Chem.* **260,** 3601–3606.

Welgus, H.G., Jeffrey, J.J., Eisen, A.Z. (1981b). Human skin fibroblast collagenase: Assessment of activation energy and deuterium isotope effect with collagenous substrates. *J. Biol. Chem.* **256,** 9516–9521.

Welgus, H.G., Jeffrey, J.J., and Eisen, A.Z. (1981a). The collagen substrate specificity of human skin fibroblast collagenase. *J. Biol. Chem.* **256,** 9511–9515.

Welgus, H.G., Jeffrey, J.J., Stricklin, G.P., *et al.* (1980). Characteristics of the action of human skin fibroblast collagenase on fibrillar collagen. *J. Biol. Chem.* **255,** 6806–6813.

Welgus, H.G., Kobayashi, D.K., Jeffrey, J.J. (1983). The collagen substrate specificity of rat uterine collagenase. *J. Biol. Chem.* **258,** 4162–4165.

Wilson, C.L., and Matrisian, L.M. (1996). Matrilysin—An epithelial matrix metalloproteinase with potentially novel functions (review). *Int. J. Biol. Chem.* **28,** 123–136.

Woessner, J.F. (1991). Matrix metalloproteinases and their inhibitors in connective tissue remodeling. *FASEB J.* **5,** 2145–2154.

Zhang, Y.N., Dean, W.L., and Gray, R.D. (1997). Co-operative binding of Ca2+ to human interstitial collagenase, assessed by circular dichroism, fluorescence, and catalytic activity. *J. Biol. Chem.* **272,** 1444–1447.

Stromelysins 1 and 2

Hideaki Nagase

Department of Biochemistry and Molecular Biology, University of Kansas Medical Center, Kansas City, Kansas 66160

I. Introduction

Stromelysins 1 and 2 are closely related metalloproteases that belong to the matrixin family (Woessner, 1991; Nagase, 1996). They degrade various components of the extracellular matrix, but not the triple helical regions of interstitial collagens. These properties distinguish them from collagenases, which share structural homology with stromelysins. A noncollagenolytic metalloprotease activity was first recognized in the extract of human articular cartilage by Sapolsky *et al.* (1974) and in the conditioned medium of rabbit synovial fibroblasts by Werb and Reynolds (1974). The enzyme was subsequently purified by Galloway *et al.* (1983) from the conditioned medium of rabbit bone explants and it was called *proteoglycase* because it digested the core protein of cartilage proteoglycans at neutral pH. Later, the enzyme was renamed *stromely-*

Matrix Metalloproteinases

sin, denoting a stromal cell-derived metalloproteinase that degrades extracellular matrix (Chin *et al.,* 1985).

On the other hand, Vater *et al.* (1983) reported *collagenase activator* of a doublet of 52 and 53 kDa from rabbit synovial fibroblasts and it was shown to be the precursor of a metalloproteinase. A similar activator, called *collagenase activator protein* was isolated in bovine articular cartilage (Treadwell *et al.,* 1986). In 1985, Matrisian *et al.* isolated a cDNA clone whose message was induced in rat fibroblasts by epidermal growth factor or viral transformation. The protein, called *transin,* was expressed and shown to have metalloproteinase activity (Matrisian *et al.,* 1986). In 1986, Okada *et al.* purified two isoforms of a metalloproteinase of 28 and 45 kDa from the medium of human rheumatoid synoviocytes that were stimulated with macrophage-conditioned medium. They refer to this metalloproteinase as *matrix metalloproteinase 3 (MMP-3)* to distinguish it from collagenase (MMP-1) and gelatinase (MMP-2), which are found in the same medium. By cDNA cloning of rabbit collagenase activator (Fini *et al.,* 1987), human stromelysin (Whitham *et al.,* 1986; Wilhelm *et al.,* 1987), and human MMP-3 (Saus *et al.,* 1988) and by immunochemical studies, it was shown that they are the same metalloendopeptidase. In 1988, Muller *et al.* identified a cDNA clone that is closely related (78% identical in amino acid sequence) to stromelysin from the human breast cancer cell cDNA library. They called it *stromelysin 2.* The original stromelysin, now called *stromelysin 1,* and stromelysin 2 are designated MMP-3 and MMP-10, respectively, following the numerical distinction of matrix metalloproteinases in the matrixin family (Nagase *et al.,* 1992). An acid metalloproteinase purified from human cartilage was first designated as MMP-6 (Azzo and Woessner, 1986), but it was proven to be MMP-3 (Wilhelm *et al.,* 1993).

This chapter reviews structure and function, gene regulation, and biological and pathological roles of stromelysins 1 and 2. The purification procedures, enzyme properties, and assays of human stromelysins 1 and 2 have already been reviewed (Nagase, 1995).

II. STRUCTURE AND SUBSTRATE SPECIFICITY

A. *Structure*

Stromelysin 1 (MMP-3) and stromelysin 2 (MMP-10) are synthesized as pre-proenzymes and secreted from cells as proenzyme forms (proMMP-3 and proMMP-10). The primary structure of MMP-3 was deduced from cDNA clones of human (Whitham *et al.,* 1986; Wilhelm *et al.,* 1987; Saus *et al.,* 1988), rabbit (Fini *et al.,* 1987), rat (Matrisian *et al.,* 1985), and mouse (Ostrowski *et al.,* 1988; Hammani *et al.,* 1992), and those of MMP-10 are reported for human (Muller *et al.,* 1988) and rat (Breathnach *et al.,* 1987). Human proMMP-3 and proMMP-10 consist of a propeptide (82 amino acids; 79% identical), a catalytic domain (165 amino acids; 86% identical), a proline-rich "hinge region" (25 amino acids; 52% identical); and a C-terminal domain (188 amino acids; 75% identical) (see Fig. 1). The propeptide sequences contain the "cysteine switch" sequence PRCGVPD conserved in all matrixins, and the catalytic domains have the zinc-binding motif **HEXXHXXGXXH** conserved among metzincin metalloproteinases (Bode *et al.,* 1993; Stöcker *et al.,* 1995). The C-terminal domains have a sequence that is

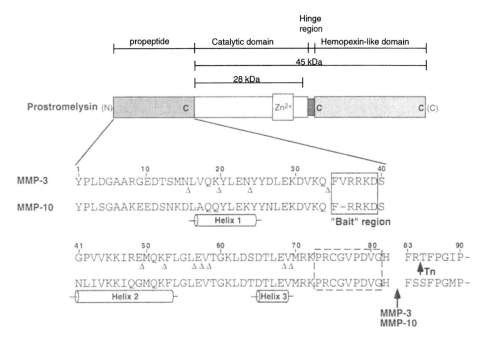

FIG. 1. Domain structures of prostromelysins 1 and 2 and the sites cleaved in the propeptide during activation. The box denotes the proteinase susceptible bait region. The conserved "cysteine switch" sequence PRCGVPD is boxed with a dashed line. The regions corresponding to α-helices are indicated (Becker *et al.,* 1995). Cleavages induced by APMA treatment are shown by Δ (Nagase *et al.,* 1990; Cameron *et al.,* 1995). The final activation is mediated by the action of MMP-3 or MMP-10 intermediates generated by proteinases or APMA, and a 45-kDa stromelysin is generated. The 45-kDa form undergoes autolysis to the 28-kDa enzyme (Nagase *et al.,* 1990; Suzuki *et al.,* 1997). Trypsin (Tn) cleaves the Arg[84]–Thr[85] bond and generates a 45-kDa species that exhibits about 20% of the activity (Benbow *et al.,* 1996).

similar to that of hemopexin and vitronectin. Stromelysin 3 (MMP-11) (Basset *et al.*, 1990) diverges significantly from stromelysins 1 and 2 in amino acid sequence (Murphy, G. J. P., *et al.*, 1991) and in enzymic activity.

The crystal structure of the C-terminal domain-truncated proMMP-3 has revealed that the pro-domain and the catalytic domain have separate folding units (Becker *et al.*, 1995). The pro-domain consists of three α-helices and an extended peptide around the PRCGVPD (73–79) sequence. The overall folding of the catalytic domain of MMP-3 is very similar to interstitial collagenase (MMP-1) and neutrophil collagenase (MMP-8) (see Stöcker *et al.*, 1995; Browner *et al.*, 1995). It consists of five-stranded β-sheet, three α-helices and connecting loops. It contains two molecules of zinc and at least two molecules of calcium (Stöcker *et al.*, 1995). One zinc molecule is located at the active site of the enzyme interacting with side chains of His^{201}, His^{205}, and His^{211}, and the other is a structural zinc interacting with side chains of Asp^{153}, His^{151}, His^{166}, and His^{179} (Becker *et al.*, 1995). Studies by Wetmore and Hardman (1996) propose that the second zinc is critical for maintaining the active form of MMP-3 by pulling the contiguous loops, which form the second zinc and first calcium-binding site (His^{151} through His^{166}), away from the active site and helping to organize the active site pocket, particularly the S_1' site, for substrate binding. About 2 mM Ca^{2+} is required to stabilize proMMP-3 and active MMP-3 (Housley *et al.*, 1993). The solution structure of the catalytic domain of MMP-3 has been also determined by nuclear magnetic resonance (NMR) imaging (Gooley *et al.*, 1994; Van Doren *et al.*, 1995). Biochemical and biophysical studies of proMMP activation suggest that the side chain of Cys in the conserved PRCGVPD motif interacts with the catalytic Zn^{2+} as a fourth ligand to maintain the latency of proMMPs (Springman *et al.*, 1990; Van Wart and Birkedal-Hansen, 1990; Salowe *et al.*, 1992; Holz *et al.*, 1992). The crystal structure of the C-terminal-truncated proMMP-3 (Becker *et al.*, 1995) indicates that the region from Lys^{72} to Val^{77} of KPRCGVPD occupies the active site cleft of the catalytic domain in a manner similar to that of a substrate forming identical β-strand-like hydrogen bonds. However, the direction of this peptide is opposite from that of the substrate. The Cys^{75} of this region interacts with the catalytic zinc as predicted by other studies. A similar three-dimensional structure can be predicted for proMMP-10 because they are closely related in the primary structure. The three-dimensional structures of the C-terminal domains of stromelysins have not been determined, but they are likely to be similar to those of collagenases (Li *et al.*, 1995; Gomis-Rüth *et al.*, 1996) and gelatinase A (MMP-2) (Libson *et al.*, 1995; Gohlke *et al.*, 1996), which consist of four units of four-stranded antiparallel β-sheet stabilized on its fourfold axis by a cation (thought to be a calcium ion), forming a four-bladed β-propeller structure. The presence of the C-terminal hemopexin-like domain is essential for collagenolytic activities of collagenases, but it does not influence the activity of MMP-3 on various substrates (Okada *et al.*, 1986). ProMMP-3 secreted from human fibroblasts is partially glycosylated and a doublet of the 57-kDa (unglycosylated) and 59-kDa (glycosylated) forms is detected (Wilhelm *et al.*, 1987; Okada *et al.*, 1988).

B. Substrate Specificity

MMP-3 digests various components of extracellular matrix. The information for MMP-10 is limited, but the enzyme also degrades a similar repertoire of matrix components although the catalytic efficiency is lower compared with stromelysin 1 (Nicholson *et al.*, 1989; Murphy *et al.*, 1991a; Nagase, 1995). The MMP-3 cleavage sites of various protein substrates are summarized in Table I. Other natural substrates include tenascin (Imai *et al.*, 1994; Siri *et al.*, 1995), vitronectin (Imai *et al.*, 1995), perlecan (Whitelock *et al.*, 1996), versican (Perides *et al.*, 1995), laminin (Okada *et al.*, 1986), elastin (Murphy *et al.*, 1991a), and interleukin 1β (Ito *et al.*, 1996). MMP-10 cleaves cartilage link protein at the His[16]–Ile[17] and Leu[25]–Leu[26] bonds (Nguyen *et al.*, 1993). However, the activity of MMP-10 on fibronectin is negligible (Suzuki *et al.*, 1997). MMP-3 exhibits an acid pH optimum activity around 5.5–6.0 for digestion of aggrecan and synthetic substrates, but it retains about 30–50% of the activity at pH 7.5 (Harrison *et al.*, 1992; Wilhelm *et al.*, 1993). MMP-10 has optimal activity against Azocoll and synthetic substrates at around pH 7.5–8.0 (Suzuki *et al.*, 1997).

In general, MMP-3 preferentially cleaves the peptide bond with a hydrophobic residue at the P_1' position (Table I). However, the catalytic efficiency depends on the length of the peptide substrate. A peptide containing only three residues in the P site (the N-terminal side of the scissile bond) and two residues in the P' site (the C-terminal side of the scissile bond) is not readily cleaved unless the N-terminal and C-terminal ends are blocked (Table II) (Niedzwiecki *et al.*, 1992). This suggests the enzyme has an extended substrate-binding site. MMP-3 accommodates substrates with aliphatic and aromatic residues at the P_1' site, but MMP-1 fails to cleave substrates with aromatic residues at this site. This is due to the fact that MMP-3 has a larger and deeper S_1' pocket than MMP-1. The residue at the P_1 position is not involved significantly in contacting with the enzyme, but the catalytic efficiency increases when the P_1 position has a charged group. The preferred residues at P_3, P_2 and P_2' possessions are Pro, Leu, and aromatic resi-

TABLE I

HUMAN STROMELYSIN 1 CLEAVAGE SITES IN NATURAL SUBSTRATES

Substrate	Source	P_4	P_3	P_2	P_1	+	P_1'	P_2'	P_3'	P_4'	Reference
Aggrecan core protein	Human	I	P	E	N^{341}	—	F^{342}	F	G	V	Flannery et al., 1992
Cartilage link protein	Human	R	A	I	H^{16}	—	I^{17}	Q	A	E	Nguyen et al., 1989
Collagen α_1(II)	Bovine	A	G	G	A^{115}	—	Q^{116}	M	G	V	Wu et al., 1991
		Q	M	G	V^{119}	—	M^{120}	Q	G	P	
Collagen α_1(IV)	Bovine	(G	P	P	G^{1341}[a]	—	L^{1342}	K	G	L	Mott et al., 1997
Collagen α_2(IV)	Bovine	(G	P	I	G^{1430}[a]	—	F^{1431}	E	G	E	Mott et al., 1997
Collagen α_1(IX)	Bovine	(L	A	A	S^{597}[a]	—	L^{598}	K	R	P	Wu et al., 1991
Collagen α_2(IX)	Bovine	(E	V	A	S^{597}[a]	—	A^{598}	K	R	E	Wu et al., 1991
Collagen α_1(XI)	Bovine	Q	A	Q	A^{482}	—	I^{483}	L	Q	Q	Wu et al., 1991
Fibronectin	Human	P	F	S	P^{689}	—	L^{690}	V	A	T	Wilhelm et al., 1993
α_2-Macroglobulin	Human	G	P	E	G^{679}	—	L^{680}	R	V	G	Enghild et al., 1989
		R	V	G	F^{684}	—	Y^{685}	E	S	D	
Ovostatin	Chicken	L	N	A	G^{677}	—	F^{678}	T	A	S	Enghild et al., 1989
α_1-Proteinase inhibitor	Human	E	A	I	P^{357}	—	M^{358}	S	I	P	Mast et al., 1991
α_1-Antichymotrypsin	Human	L	L	S	A^{360}	—	L^{361}	V	E	T	Mast et al., 1991
Antithrombin III	Human	I	A	G	R^{393}	—	S^{394}	L	N	P	Mast et al., 1991
ProMMP-1	Human	D	V	A	Q^{80}	—	F^{81}	V	L	T	Suzuki et al., 1990
ProMMP-3	Human	D	T	L	E^{68}	—	V^{69}	M	R	K	Nagase et al., 1990
		D	V	G	H^{82}	—	F^{83}	R	T	F	
ProMMP-8	Human	D	S	G	G^{78}	—	F^{79}	M	L	T	Knäuper et al., 1993
ProMMP-9	Human	R	V	A	E^{40}	—	M^{41}	R	G	E	Ogata et al., 1992
		D	L	G	R^{87}	—	F^{88}	Q	T	F	

48

Substrate	Species					—					Reference
ProMMP-13	Human	S	F	F	G⁵⁷	—	L⁵⁸	E	V	T	Knäuper et al., 1996a
Substance P		K	P	Q	Q⁶	—	F⁷	F	G	L	Harrison et al., 1989
Insulin B chain		L	V	E	A¹⁴	—	L¹⁵	Y	L	V	Wilhelm et al., 1993
		E	A	L	Y¹⁶	—	L¹⁷	V	C—SO₃	G	
ProMMP-7	Human	D	V	A	E⁷⁷	—	Y⁷⁸	S	L	F	Imai et al., 1995
IGF-BP-3	Human	L	R	A	Y⁹⁹	—	L¹⁰⁰	L	L	P	Fowlkes et al., 1994
Nidogen	Mouse	F	L	A	D⁸²	—	L⁸³	D	T	T	Mayer et al., 1993
		G	L	G	N⁹¹	—	V⁹²	Y	Y	Y	
		V	V	F	S³⁵¹	—	Y³⁵²	N	T	G	
		Y	A	P	P⁸⁹⁶	—	I⁸⁹⁷	N	H	Q	
		G	R	A	S⁹⁷⁶	—	L⁹⁷⁷	N	H	G	
		E	P	T	T⁹⁸⁴	—	I⁹⁸⁵	I	R	Q	
		N	P	R	G¹⁰³⁸	—	I¹⁰³⁹	V	T	D	
		K	T	N	S¹¹⁴²	—	V¹¹⁴³	I	A	N	
SPARC (BM-40, osteonectin)	Human	T	V	A	E²¹	—	V²²	T	E	V	Sasaki et al., 1997
Fibulin-2	Mouse	H	P	V	E¹⁹⁸	—	L¹⁹⁹	L	A	R	Sasaki et al., 1996
		P	P	S	S²⁰⁹	—	L²¹⁰	R	V	T	
		P	P	A	P²³³	—	V²³⁴	Q	Q	K	
		R	V	S	E⁵⁴³	—	M⁵⁴⁴	E	M	A	
Fibrin γ-chain	Human	H	L	G	G⁴⁰⁴	—	A⁴⁰⁵	K	Q	A	Bini et al., 1996

[a] Human sequences.

TABLE II

Hydrolysis of Synthetic Peptides by MMP-3[a]

P6	P5	P4	P3	P2	P1	~	P1'	P2'	P3'	P4'	P5'	k_{cat} (s⁻¹)	K_m (μM)	k_{cat}/K_m (s⁻¹ M⁻¹)	Reference
		Dnp	Pro	Tyr	Ala	~	Tyr	Tyr	Met	Arg		—	—	2,400[b]	Netzel-Arnett et al., 1991
		Dnp	Pro	Leu	Gly	~	Leu	Trp	Ala	D-Arg	NH₂	—	—	2,200	Knight et al., 1992
Mca	Pro	Lys	Pro	Gln	Gln	~	Phe	Phe	Gly	Leu	Lys(Dnp)–Gly	0.53	50	10,900	Nagase et al., 1994
		Mca	Pro	Leu	Gly	~	Leu	Dpa	Ala	Arg	NH₂	—	—	23,000	Knight et al., 1992
Dnp–Arg	Pro	Lys	Pro	Leu	Ala	~	Nva	Trp	NH₂			—	—	45,000[c]	Niedzwiecki et al., 1992
Mca–Arg	Pro	Lys	Pro	Val	Glu	~	Nva	Trp	Arg	Lys(Dnp)	NH₂	1.3[c]	20[c]	65,700[c]	Nagase et al., 1994
Mca–Arg	Pro	Lys	Pro	Val	Glu	~	Nva	Trp	Arg	Lys(Dnp)	NH₂	5.4	25	218,000	Nagase et al., 1994
		Ac	Pro	Phe	Glu	~	Leu	Arg	NH₂			10	80	126,000	Smith et al., 1995
Arg	Pro	Lys	Pro	Gln	Gln	~	Phe	Phe	Gly	Leu	Met–NH₂	—	—	1790[c]	Niedzwiecki et al., 1992
Arg	Pro	Lys	Pro	Gln	Gln	~	Phe	Phe	Gly	Leu	Met–NH₂	—	—	800[c]	Niedzwiecki et al., 1992
		Lys	Pro	Gln	Gln	~	Phe	Phe	Gly	Leu	Met–NH₂	—	—	290[c]	Niedzwiecki et al., 1992
		Lys	Pro	Gln	Gln	~	Phe	Phe	Gly	Leu	Met–NH₂	—	—	<3[c]	Niedzwiecki et al., 1992
		Ac	Pro	Gln	Gln	~	Phe	Phe	Gly	Leu	Met–NH₂	—	—	500[c]	Niedzwiecki et al., 1992
Arg	Pro	Lys	Pro	Gln	Gln	~	Phe	Phe	Gly	Leu		—	—	1300[c]	Niedzwiecki et al., 1992
Arg	Pro	Lys	Pro	Gln	Gln	~	Phe	Phe	Gly	Leu		—	—	790[c]	Niedzwiecki et al., 1992
Arg	Pro	Lys	Pro	Gln	Gln	~	Phe	Phe				—	—	<3[c]	Niedzwiecki et al., 1992
Arg	Pro	Lys	Pro	Gln	Phe	~	NH₂					—	—	1900[c]	Niedzwiecki et al., 1992

[a] Assays were performed at 37°C, pH 7.5, unless otherwise noted. Dnp, 2,4-dinitrophenyl; Mca, 7-methoxycoumarin-4-yl; Dpa, N-3-(2,4-dinitrophenyl)-L-2,3-diaminopropionyl; Nva, norvaline; Ac, acetyl; —, not determined.

[b] Assayed at 23°C.

[c] Assayed at 25°C.

dues, respectively. (Niedzwiecki et al., 1992; Nagase and Fields, 1996). Examples of synthetic substrates for MMP-3 are shown in Table II.

III. ACTIVATION MECHANISMS

A. Activation by Nonproteolytic Agents

Most proMMPs are activated by nonproteolytic agents such as mercurial compounds [e.g., HgCl, 4-aminophenylmercuric acetate (APMA)], thiol-modifying agents (e.g., iodocatamide, N-ethylmaleimide, oxidized glutathione, HOCl), and denaturants (e.g., urea, SDS, NaSCN). ProMMP-3 and proMMP-10 are activated by APMA in a similar kinetics (Okada et al., 1988; Suzuki et al., 1997). Treatment of proMMP-3 with APMA results in propeptide processing in a stepwise manner. Several intermediates are initially generated, which are then processed to a fully active 45-kDa MMP-3 consisting of the catalytic domain and the hemopexin-like C-terminal domain. The formation of an initial intermediate is thought to be the result of an intramolecular reaction of proMMP-3, because this reaction follows first-order kinetics (Okada et al., 1988) and it is not inhibited by an active site-directed synthetic inhibitor (Cameron et al., 1995). The final activation step involves the cleavage of the His[82]–Phe[83] bond by a bimolecular reaction of the MMP-3 intermediates (Okada et al., 1988; Nagase et al., 1990) (Fig. 1). On prolonged incubation, the 45-kDa MMP-3 slowly converts to the 28-kDa form by autolytically losing the C-terminal hemopexin-like domain. Incubation of proMMP-3 at 55°C results in formation of active 45- and 28-kDa forms of MMP-3 (Koklitis et al., 1991). This reaction is thought to be a bimolecular process (Wetmore and Hardman, 1996). To explain the proMMP activation process by various nonproteolytic agents VanWart and colleagues proposed the "cysteine switch" hypothesis (Springman et al., 1990; Van Wart and Birkedal-Hansen, 1990). In this model, the interaction of the unique Cys in the pro-domain with the catalytic Zn^{2+} is thought to transiently dissociate and react with SH reagents, which prevents the reassociation of Cys and Zn^{2+}. Denaturants disturbs the Cys–Zn^{2+} interaction and the removal of these reagents results in autoprocessing of the propeptide.

To investigate the involvement of Cys-Zn^{2+} interaction in the activation process of proMMPs, Chen et al. (1993) specifically modified Cys[75] of proMMP-3 by alkylating agents and examined the activation of proMMP-3. Their studies showed that the disruption of Cys[75]–Zn^{2+} alone was not sufficient to activate proMMP-3, probably because of a stronger interaction between the pro-domain and the catalytic domain. Activation of proMMP-3 with chemically modified Cys[75] required further treatment with APMA or a proteinase. These observations were further substantiated by mutagenesis studies by Galazka *et al.* (1996) of proMMP-3 substituted with Ser or His for Cys[75]. These mutants were secreted from cells as proenzyme and activated by APMA or trypsin. By contrast, mutation of Tyr[20] or Leu[21] to Ala results in spontaneous activation of proMMP-3 (Freimark *et al.*, 1994). From those studies, it may be concluded that the activation of proMMP-3 by a mercurial compound is initiated by perturbation of the zymogen, rather than by its reaction with the SH group of Cys[75]. A similar effect of APMA without the reaction with Cys in the propeptide was observed during the activation of proMMP-2 (progelatinase A) (Itoh *et al.*, 1995). McLaughlin and Weiss (1996) reported that endothelial-cell-stimulating angiogenesis factor (ESAF) of molecular mass of approximately 600 Da activates proMMP-1, proMMP-2, and proMMP-3, but this activation mechanism is not understood.

B. Activation by Proteinases

A number of proteinases can activate proMMP-3, resulting in the active 45-kDa form with Phe[83] at the N-terminus. However, the activator proteinases do not directly cleave the His[82]–Phe[83] bond of proMMP-3. Instead, they attack the proteinase susceptible "bait" region located in the loop between the first and the second α-helices (Fig. 1). The removal of the first α-helix presumably destabilizes the Cys[75]–Zn[2+] interaction, resulting in an activated intermediate. This process also causes conformational changes around the His[82]–Phe[83] bond, which renders a bimolecular reaction of MMP-3 intermediates; the His[82]–Phe[83] bond of the native proMMP-3 is not cleaved by MMP-3 unless a part of the propeptide is removed (Nagase *et al.*, 1990). The activation of many secreted proMMPs follows the "stepwise activation" mechanism (see Nagase, 1997, for a review). Thus, the bait region sequence dictates which proteinases can trigger the activation of a particular proMMP. The sequence FVRRKD (34–39) of proMMP-3 in this region allows a large number of endopeptidases such as plasmin, plasma kallikrein, thermolysin, chymotrypsin, and leukocyte elastase to activate this zymogen as it accommodates their substrate specificity. For example, chymotrypsin and leukocyte elastase cleave the Phe[34]–Val[35] and Val[35]–Arg[36] bonds, respectively. On the other hand, proMMP-10, which lacks Val[35], is not activated by leukocyte elastase, but it is readily activated by plasmin, trypsin, and chymotrypsin (Fig. 1) (Suzuki *et al.*, 1997). Although many proteinases can activate prostromelysins, MMPs fail to activate those zymogens, even though the Gln[33]–Phe[34] bond is a potential candidate for MMP cleavage. Inability of MMPs to activate them is most likely due to structural constraints around the bait region.

The stepwise activation processes of proMMPs suggest that MMP activities are controlled by the endogenous inhibitors such as α_2-macroglobulin or tissue inhibitors of metalloproteinases (TIMPs) before MMPs are fully activated. Indeed, the binding of activation intermediates of MMP to a TIMP has been demonstrated for MMP-1 (DeClerck et al., 1991), MMP-2 (Ward et al., 1991; Itoh et al., 1995), MMP-3 (Benbow et al., 1996), and MMP-9 (Ward et al., 1991). The recombinant MMP-3 intermediate generated by deletion of the first 34 amino acids of the propeptide forms a complex with TIMP-1, but their interaction is looser than that with the fully activated 45-kDa form of MMP-3 (Nagase et al., 1996), suggesting that the MMP-3 activity is tightly regulated by TIMPs during activation.

C. Involvement of the N-Terminal Phenylalanine for the Expression of Full Enzymic Activity

Stromelysins 1 and 2 play a critical role in activation of procollagenases by cleaving the Gln^{80}–Phe^{81} bond of proMMP-1 (interstitial collagenase) (Suzuki et al., 1990) and the Gly^{78}–Phe^{79} bond of proMMP-8 (neutrophil collagenase) (Knäuper et al., 1993, 1996b). When procollagenases are treated with APMA or trypsin in the absence of stromelysin, those particular bonds are not cleaved and only partial (25–40%) collagenolytic activity is detected. This is due to autoprocessing of procollagenases of bonds other than the specific sites mentioned earlier. For example, the treatment of proMMP-1 with APMA generates $[Met^{78}]$MMP-1, $[Val^{82}]$MMP-1, and $[Leu^{83}]$MMP-1 (residues in brackets are the N-terminus), but not $[Phe^{81}]$MMP-1, because Phe at the P_1' site does not fit with the S_1' pocket of MMP-1 (see Nagase and Fields, 1996). Similarly, $[Met^{80}]$MMP-8 and $[Leu^{81}]$MMP-8 with only about 25% of the full collagenolytic activity are generated with $HgCl_2$ (Knäuper et al., 1993). Reduced activity is also observed with MMP-3 when it retains several extra residues before the N-terminal Phe^{83} or when a few residues are trimmed from the N-terminus (Benbow et al., 1996). Treatment of $[Phe^{83}]$MMP-3 with trypsin removes the dipeptide Phe^{83}–Arg^{84} from the N-terminus of MMP-3 (see Fig. 1), which results in reduction of the specific activity and alteration in substrate specificity (Benbow et al., 1996). The reduced activity is due to an increase in K_m and a decrease in k_{cat} on a synthetic substrate (Benbow et al., 1996). The crystal structure of the catalytic domains of $[Phe^{79}]$MMP-8 and $[Met^{80}]$MMP-8 provide important structural insights into the two different forms of the enzyme (Reinemer et al., 1994; Bode et al., 1994). In

[Phe^{79}]MMP-8, the ammonium group of Phe^{79} forms a salt bridge with the side-chain carboxylate group of Asp^{232} in the third helix of the catalytic domain (Reinemer *et al.*, 1994). Without the Phe^{79}, however, the N-terminal hexapeptide MLTPGN (80–85) of MMP-8 is disordered (Bode *et al.*, 1994). However, the geometry of the active sites of the two forms of the enzyme is essentially identical (Reinemer *et al.*, 1994). It is, therefore, postulated that the disruption of this salt bridge may alter stabilization of the active site at the transition state, or that the mobile N-terminal peptide may interfere with the substrate binding to the active site (Reinemer *et al.*, 1994), but the experimental evidence to support these suppositions is not currently available. The residue corresponding to Asp^{232} is conserved in all MMPs, and correctly processed MMPs are thought to possess either Phe or Tyr for their N-termini. Thus, it seems reasonable to consider that the formation of such a salt bridge is critical for the expression of full MMP activities.

D. Activation of Other proMMPs by MMP-3

MMP-3 and MMP-10 cannot activate their own zymogens, but they act directly on the Gly^{78}–Phe^{79} bond of proMMP-8 and remove the entire propeptide (Knäuper *et al.*, 1993, 1996b). MMP-3 can also activate proMMP-9 (progelatinase B) (Ogata *et al.*, 1992; Goldberg *et al.*, 1992; Okada *et al.*, 1992a; Shapiro *et al.*, 1995) and proMMP-13 (collagenase 3) (Knäuper *et al.*, 1996a) in a stepwise manner. MMP-3 initially attacks the Glu^{40}–Met^{41} bond of proMMP-9 (Ogata *et al.*, 1992), and then removes the rest of the propeptide from the zymogen by cleaving the Arg^{87}–Phe^{88} bond (Ogata *et al.*, 1992). In the case of MMP-13, MMP-3 hydrolyzes the Gly^{57}–Leu^{58} bond first, which then allows the removal of the remaining propeptide by autolysis (Knäuper *et al.*, 1996a).

IV. Inhibitors

A. Synthetic Inhibitors

Matrixins are almost completely inhibited by chelating agents such as EDTA, EGTA, or 1,10-phenanthroline. About 50% inhibition of MMP-3 is observed with 0.5 mM cysteine or dithiothreitol, but phosphoramidon, a potent inhibitor of thermolysin and neprilysin, is not effective. Examples of synthetic MMP-3 inhibitors include a phosphonamidate compound, phthaloyl-N-(CH$_2$)$_4$-P(O$_2^-$)-Ile-(β-naphthyl)-Ala-NH-CH$_3$ (K_i = 7 nM; k_{on} = 2.5 × 10^4 M^{-1} s^{-1}; k_{off} = 1.9 × 10^{-4} s^{-1} at pH 5 and 25°C) (Izquierdo-Martin and Stein, 1993), an N-(carboxyalkyl)dipeptide, N-[1(R)-carboxyethyl]α(S)-(2-phenylethyl)-glycyl-L-leucine, N-phenylamide (K$_i$ = 470 nM for human MMP-3;

K_i = 6.5 nM for rabbit MMP-3 at pH 7.5 and 25°C) (Chapman et $al.$, 1993), and pseudopeptide-hydroxamate compounds (Bottomley et $al.$, 1997). Hajduk et $al.$ (1997) reported on a series of nonpeptide hydroxa- mate inhibitors (e.g., 3-[4-(4-cyanophenyl)phenoxy]propanohydroxamic acid; IC_{50} = 25 nM). Selectivity toward MMP-3 is seen with some of these compounds (Chapman et $al.$, 1993; Bottomley et $al.$, 1997), but other MMPs are also inhibited by these synthetic inhibitors.

B. TIMPs

Natural inhibitors include tissue inhibitors of metalloproteinases (TIMPs-1, -2, -3, and -4), α_2-macroglobulins, and ovostatins. These in- hibitors play important roles in regulating the activities of stromelysins and other matrixins in the tissue. TIMPs are about 40% identical to each other in sequence including 12 conserved cysteines (see Douglas et $al.$, 1997). MMP-3 binds to TIMP-1 more readily than to TIMP-2 (k_{on} = 1.9 × 10^6 M^{-1}s^{-1} for TIMP-1 versus k_{on} = 0.3 × 10^6 M^{-1} s^{-1} for TIMP-2) (Nguyen et $al.$, 1994). The disulfide arrangement of human TIMP-1 indicates that the inhibitor consists of the N-terminal domain (126 amino acids) and the C-terminal domain (58 amino acids) with three disulfide bonds in each domain (Williamson et $al.$, 1990), and the N-terminal domain contains the inhibitory site for MMPs (Murphy et $al.$, 1991b). The solution structure of the N-terminal domain of TIMP-2 (N-TIMP-2) by NMR revealed that the protein consists of five-stranded antiparallel β-sheet that forms a closed β-barrel and two short α-helices (Williamson et $al.$, 1994). The recently resolved crystal structure of the complex of the full-length TIMP-1 and the catalytic domain of MMP- 3 has revealed a unique inhibition mechanism of MMPs by TIMP-1 (Gomis-Rüth et $al.$, 1997). The TIMP-1 polypeptide chain folds into a contiguous elongated wedge-shaped molecule with the N- and the C- terminal halves forming two opposing parts. The entire active site cleft of the MMP-3 is blocked with two disulfide-bonded TIMP-1 segments, Cys^1–Val^4 and Ser^{68}–Val^{69}(Cys^{70}), that bind to S_1–S_3' and S_2–S_3 subsites, respectively. The catalytic zinc atom of MMP-3 interacts with a chelat- ing group formed by α-amino and carbonyl groups of the N-terminal Cys^1 of the inhibitor, and the large S_1' pocket occupied by Thr^2. In the complex, however, six sequentially separate polypeptide segments of TIMP-1 interact with MMP-3 and surfaces of about 1300 Å^2 in each molecule are removed from contact with bulk water upon complex for- mation.

The C-terminal part of the TIMPs influences the interaction with MMPs in some cases by interaction with the C-terminal domain of MMPs (Murphy and Willenbrock, 1995). However, the C-terminal domain of MMP-3 does not contribute significantly to the association with TIMP-1 and TIMP-2. The K_i values of the full-length TIMP-1 for the 45-kDa MMP-3 and the catalytic domain of MMP-3 are 0.24 and 0.25 nM, respectively (Huang et al., 1996). The K_i values of N-TIMP-1 for the 45-kDa MMP-3 and the catalytic domain of MMP-3 are 1.4 and 1.9 nM, respectively (Huang et al., 1996).

C. α_2-Macroglobulins

Matrixins in the body fluid are primarily inactivated by a plasma proteinase inhibitor, α_2-macroglobulin (α_2M); 725-kDa α_2M consists of four identical subunits of 180 kDa (Sottrup-Jensen, 1989). The binding of a proteinase and α_2M is initiated by the cleavage of the so-called "bait" region located in the middle of subunit, which in turn traps the enzyme without blocking the active site (Barrett and Starkey, 1973). The reaction of α_2M with MMP-3 is rapid ($k_2/K_i = 5.6 \times 10^4$ M^{-1} s^{-1}), but that with MMP-1 is more favorable ($k_2/K_i = 2.8 \times 10^6$ M^{-1} s^{-1}) (Enghild et al., 1989). Ovostatins (ovomacroglobulins) from avian egg whites also inhibit MMPs (Nagase and Harris, 1983). Other members of the α_2-macroglobulin family such as human pregnancy zone protein, rat α_1M, rat α_2M, and rat α_1-inhibitor 3 are cleaved at their bait regions by human MMP-1 (Sottrup-Jensen and Birkedal-Hansen, 1989), suggesting that they also inhibit MMP-1 and other matrixins.

V. REGULATION OF GENE EXPRESSION

A. Transcriptional Regulation

Matrixins are not readily detected in the normal tissue extracts or cells in culture unless cells are stimulated with an appropriate factor. Early work by Brinckerhoff et al. (1981) showed that the treatment of rabbit synovial fibroblasts with phorbolmyristate acetate (PMA) produced a large amount of collagenase (MMP-1) as a broad triplet band around 50–55 kDa. Later, these bands were proven to be a mixture of procollagenase (proMMP-1) and procollagenase activator (proMMP-3) (Vater et al., 1983). Indeed, numerous factors, including cytokines, growth factors, and chemical and physical stimuli, are found to be stimulatory for both MMP-1 and MMP-3 production. However, the production of these two matrixins is not always co-regulated. For example, although interleukin 1 (IL-1) increases the mRNA level of MMP-3 in human rheumatoid synovial cells more readily than that of MMP-1, tumor necrosis factor α (TNF-α) is more effective for increasing MMP-1 expression than MMP-3 expression (MacNaul et al., 1990). The treatment of the IL-1-stimulated human skin fibroblasts with interferon γ reduces the collagenolytic activity. This was due to the reduction of MMP-3 (collagenase activator) production but not the production of MMP-1 (Unemori et al., 1991). Factors that modulate the production of MMP-3 and MMP-1 are listed in Table III. The effects of these factors, however, appear to be dependent on cell types and species.

Although stromelysin 1 (MMP-3) can be expressed in many cell types such as fibroblasts, chondrocytes, vascular smooth muscle cells, keratinocytes, macrophages, and endothelial cells under certain stimuli, the stromelysin 2 (MMP-10) gene is not transcriptionally active in fibroblasts. It is, however, activated in keratinocytes in culture when treated with transforming growth factor α (TGF-α), epidermal growth factor (EGF), TNF-α, or a phorbol ester, but is not activated by IL-1 or platelet-derived growth factor (Windsor et al., 1993). MMP-10 transcripts were found in T lymphocytes stimulated with PMA and a calcium ionophore A23187 (Conca and Willmroth, 1994).

The stromelysin 1 and stromelysin 2 genes are located on chromosome 11q22–q23 (Formstone et al., 1993), in which other matrixin genes including interstitial collagenase, collagenase 3, macrophage elastase, and matrilysin are also located. Analyses of the promoter region of the MMP-3 gene have revealed the regulatory mechanisms of the gene expression. The 5'-flanking regions of the gene exhibit a TATA box and a tumor promoter-responsive element (TRE) TGAGTCA sequence that can interact with AP-1 proteins (Fos and Jun proteins). These motifs are also formed in the promoter regions of MMP-1, MM-7, MMP-9, MMP-10 (see Gaire et al., 1994), and MMP-12 (Belaaouaj et al., 1995). Stimulation of the expression of these MMPs by PMA, IL-1, or TNF-α is thought to be mediated by TRE in the promoter (Angel et al., 1987; Kerr et al., 1988; Schönthal et al., 1988; Brenner, D.A., et al., 1989; Sato and Seiki, 1993; Gaire et al., 1994). In addition to the AP-1 site, the MMP-3 promoter has two palindromically arranged PEA3 binding sites that mediate activation by c-Ets, Ha-Ras, v-Src, and v-Mos (Wasylyk et al., 1991). The MMP-3 promoter responds to PEA3 and AP-1 independently, and the effect of combining the two is additive (Wasylyk et al., 1991). Buttice et al. (1991), on the other hand, reported that the mutation of the AP-1 site reduced both the basal level and PMA-induced activation of the MMP-3 promoter, but the PMA-induction fold of the mutant promoters was comparable to that of the wild type. These workers suggest that the AP-1 site is required for the basal level expression of the human MMP-3 gene, but is not necessary for the PMA response. It has been reported that the PEA3 site and Ets-2 protein play a major role in PMA induction (Buttice and Kurkinen, 1993). It has also been shown that the AP-1 site is not necessary

TABLE III

FACTORS THAT MODULATE SYNTHESIS OF STROMELYSIN 1

Stimulatory factor	Suppressive factor
Cytokines and Growth Factors	
Interleukin 1 α, β[a,b]	Retinoic acid
Interleukin 6[c]	Glucocorticoid
Interleukin 10	Transforming growth factor β
Tumor necrosis factor α[b]	Interleukin 4[n]
Epidermal growth factor	Tetracycline[o]
Platelet-derived growth factor[d]	Progesterone
Basic fibroblast growth factor[e]	EIA-F[p]
Nerve growth factor	SV40 T-antigen[q]
Transforming growth factor α[f]	Phosphate citrate[r]
Relaxin[g]	cAMP
	Interferon γ
	Calmodulin
	High glucose[s]
Factors Acting at Cell Surface	
Calcium ionophore A23187	
Concanavalin A	
Crystal: urate	
hydroxyapatite	
calcium pyrophosphate	
Anti-$\alpha5\beta1$ integrin antibody	
RGD peptides[h]	
Fibronectin fragments[i]	
SPARC (osteonectin/BM 40)	
EMMPRIN (basigin/M6 antigen)	
Staphylococcal endotoxin A[j]	
Chemical Agents	
Phorbol ester	
Prostaglandin E_2	
Cytochalasin B	
Lipopolysaccharide	
Trifluoperazine	
N-(6-aminohexyl)-5-chloro-1-nephtalene-sulfonamide (W-7)	
Linoleic acid hydroperoxide[k]	
Cycloheximide[l]	
Other Stimuli	
Heat shock	
UV irradiation	
Viral transformation	
Oncogene products	
Cellular aging	
Mechanical injury[m]	

Note: Unless otherwise indicated, factors are referenced in Frisch and Werb (1989), Woessner (1991), and Nagase (1996).

[a] Saus *et al.*, 1988; [b] MacNaul *et al.*, 1990; [c] Ito *et al.*, 1992; [d] Sanz *et al.*, 1994; [e] Chandrasekhar and Harvey, 1992; [f] Hosono *et al.*, 1996b; [g] Qin *et al.*, 1997; [h] Arner and Tortorella, 1995; [i] Xie *et al.*, 1994; [j] Migita *et al.*, 1997; [k] Ohuchida *et al.*, 1991; [l] Otani *et al.*, 1990; [m] James *et al.*, 1993; [n] Prontera *et al.*, 1996; [o] Jonat *et al.*, 1996; [p] Higashino *et al.*, 1995; [q] Logan *et al.*, 1996; [r] Cheung *et al.*, 1996; [s] Kitamura *et al.*, 1992.

for the IL-1-activation of the MMP-3 gene in fibroblasts; the activation occurs through protein interaction with the responsive region immediately surrounding the AP-1 site (Quinones *et al.*, 1994). Nonetheless, the AP-1 site is an absolute requirement to activate the MMP-3 gene by IL-1 in HepG2 cells; it is thus postulated that the regulatory differences depend on a cell-type specific *trans*-acting element or factors (Quinones *et al.*, 1994). On the other hand, Auble and Brinckerhoff (1991) reported that the AP-1 sequence was necessary but not sufficient for phorbol induction of MMP-1 in rabbit synovial fibroblasts and that the enhanced element was located between -182 and -149. DeSouza *et al.* (1995) reported that the nerve growth factor responsive element lay at least in part within a 12-base region between the -241 and -229 region of the MMP-3 promoter, but the AP-1 site was not involved in the growth factor regulation. Induction of the MMP-3 gene in NIH3T3 cells by PDGF was shown to be through the Raf-dependent PEA3-AP-1 unit and the λ/ι protein kinase C-dependent PDGF-responsive element [located in the -1218 to -1202 region (Sanz *et al.*, 1994)] -AP-1 unit (Kirstein *et al.*, 1996).

Several factors have been characterized as suppressors of MMP-3 production (Table III). Inhibition of MMP-3 transcription by retinoic acid was reported to be due to the binding of the retinoic acid receptor–retinoic acid complex to the promoter region from -48 bp to -72 bp, which prevents the interaction of AP-1 and the TRE (Nicholson *et al.*, 1990), and studies by Schüle *et al.* (1991) suggested that the receptor-ligand complex forms a nonproductive complex with c-Jun. Recent studies by Schroen and Brinckerhoff (1996) demonstrate that the retinoic acid receptor and c-Jun form a complex at the AP-1 site in which c-Jun binds directly to the DNA. Formation of such a complex is considered to affect transcriptional activity of AP-1. Suppression of MMP-1 synthesis by glucocorticoids results from the binding of the hormone receptor with c-Jun, which sequesters AP-1 proteins (Jonat *et al.*, 1990; Schüle *et al.*, 1990; Yang-Yen *et al.*, 1990), but, as in the case of retinoic acid receptor, the binding of a glucocorticoid receptor to c-Jun does not disrupt binding of AP-1 to TRE (König *et al.*, 1992). TGF-β down-regulates the MMP-3 transcription in a c-Fos-dependent mechanism that involves the TGF-β inhibitory element (TIE) GnnTTGGtGa sequence at -700 bp in the rat MMP-3 promoter (Kerr *et al.*, 1990).

B. Post-Transcriptional Regulation

The production of MMP-1 and MMP-3 is regulated not only transcriptionally but also post-transcriptionally. When rabbit synovial fibroblasts are stimulated with PMA or EGF mRNA stability of MMP-1 and

MMP-3 increases several-fold (Brinckerhoff *et al.*, 1986; Delany and Brinckerhoff, 1992). Glucocorticoid-treated rat osteoblasts increase the steady-state level of rat collagenase (MMP-13) mRNA. This has been shown to be due to an increase in the half-life of MMP-13 mRNA from 6 to 12 h without altering transcriptional activity (Delany *et al.*, 1995). Hosono *et al.* (1996a, 1996b) reported that the production of MMP-3 is augmented about two to threefold within 1 h after treatment of human uterine cervical fibroblasts with EGF or TGF-α well before any apparent increase of MMP-3 and TIMP-1 mRNAs is observed. Such translational augmentation is not observed with IL-1 or PMA treatment (Hosono *et al.*, 1996b).

VI. BIOLOGICAL AND PATHOLOGICAL ROLES

The remodeling of the extracellular matrix is critical in many biological processes such as embryo development and morphogenesis, tissue resorption, reproduction, and angiogenesis. During these processes, the expression of MMPs and TIMPs is precisely regulated, and timely degradation of matrix components must take place. In the normal steady-state tissue the synthesis and degradation of the matrix are in equilibrium. Disruption of this balance may result in diseases such as arthritis, cancer cell metastasis (Stetler-Stevenson *et al.*, 1993; Coussens and Werb, 1996), glomerulonephritis (Marti *et al.*, 1994), atherosclerosis (Henney *et al.*, 1991; Galis *et al.*, 1994), abdominal aortic aneurysm (Newman *et al.*, 1994; Thompson *et al.*, 1995), dilated cardiomyopathy (Gunja-Smith *et al.*, 1996), periodontal disease (Birkedal-Hansen, 1993), liver fibrosis (Benyon *et al.*, 1996), cornea ulceration (Berman, 1980), gastric ulcers (Saarialho-Kere *et al.*, 1996), skin diseases (Saarialho-Kere *et al.*, 1993a), and encephalomyelitis (Gijbels *et al.*, 1994). A number of MMPs and TIMPs are involved in these biological and pathological processes as being expressed in a tissue- and cell-specific manner. Good examples are seen in mouse uterus during the peri-implantation period (Alexander *et al.*, 1996; Das *et al.*, 1997) and chronic dermal wounds (Saarialho-Kere *et al.*, 1993b, 1994). This section focuses on the biological and pathological events that are accompanied by the expression of stromelysins 1 and 2. Table IV summarizes the tissues and cell types that have been shown to produce MMP-3 and MMP-10.

A. Biological Roles

1. DEVELOPMENT AND MORPHOGENESIS

MMP-3 transcripts are detected at the blastocyst stage of the mouse embryo along with transcripts for MMP-1, -2, and -9 and TIMP-1

TABLE IV

CELLS AND TISSUES ASSOCIATED WITH THE PRODUCTION OF STROMELYSIN 1 (MMP-3) AND
STROMELYSIN 2 (MMP-10)

Normal tissue	Pathological tissue
MMP-3	*MMP-3*
Fibroblast	Osteoarthritic cartilage
Chondrocyte	Rheumatoid synovium (B cell)
Osteoblast	Serum (elevated in patients with
Macrophage	rheumatoid arthritis)
Hepatic lipocyte[a]	Herniated intervertebral disc[h]
Capillary endothelial cell	Cholesteatoma epithelium[i]
Smooth muscle cell (aorta, ciliary[b])	Atherosclerosis lesions
Dermal papilla cell	Aneurysmal abdominal aorta[j]
Cartilage	Proliferating basal keratinocyte
Growth plate	(residing on the basement
Matrix vesicle of growth plate	membrane)
chondrocyte[c]	Respiratory epithelial cell after
Involuting mammary gland	wounding
Developing enamel (bovine)[d]	Gastrointestinal ulcers (Crohn's
Endometrial stroma (menstrual phase)	disease, peptic ulcer, ulcerative
Placenta (aminion, decidua, chorinic	colitis)
villi)	Colorectal cancer (stromal cells)
Trabecular juxtacanalicular cell[e]	Head and neck carcinoma
Retinal pigment epithelium[f]	Basal cell carcinoma
Serum	Bronchial carcinoma (epithelial cells)
	Esophageal squamous carcinoma
	Lung squamous carcinoma
MMP-10	*MMP-10*
Keratinocyte	Head and neck carcinoma
Wound healing skin	Migrating basal keratinocytes (in
T lymphocyte[g]	contact with dermal matrix)

Note: Unless otherwise indicated references are cited in the text.
[a] Vyas *et al.,* 1995; [b] Lindsey *et al.,* 1996; [c] Schmitz *et al.,* 1996; [d] DenBesten *et al.,* 1989;
[e] Samples *et al.,* 1993; [f] Alexander *et al.,* 1990; [g] Conca and Willmroth, 1994; [h] Kanemoto *et al.,* 1996; [i] Schönermark *et al.,* 1996; [j] Newman *et al.,* 1994.

(Brenner, C. A., *et al.,* 1989), and their synthesis, particularly that of
MMP-9, further increases during blastocyst outgrowth and in early
implantation stages *in vivo* (Behrendtsen *et al.,* 1992; Alexander *et al.,*
1996; Das *et al.,* 1997). *In situ* hybridization analyses of later embryonic
mouse tissue revealed the expression of MMP-9 (Reponen *et al.,* 1994)
and MMP-13 (Mattot *et al.,* 1995) in the onset of bone formation, but
the expression of MMP-3 and MMP-10 transcripts during those periods
is not known. However, MMP-3 was immunolocalized in resting, proliferative, hypertrophic zones of growth plates in neonatal rabbits (Brown

et al., 1989). Involvement of matrixins in morphogenesis have been investigated in the branching morphogenesis of salivary gland (Nakanishi *et al.,* 1986), lung rudiments (Ganser *et al.,* 1991), and frog metamorphosis (Patterton *et al.,* 1995), but little information is available for specific roles of MMP-3 and MMP-10 in these processes.

2. TISSUE RESORPTION

MMPs degrade the extracellular matrix in rapidly resorbing tissues. Elevated MMP-3 along with MMP-2, MMP-11, and tissue plasminogen activator was shown in involuting mouse mammary gland (Dickson and Warburton, 1992; Lefebvre *et al.,* 1992; Talhouk *et al.,* 1992). The expression of these proteinase activities is thought to lead to the loss of basement membrane and differentiated lactational phenotype of the gland. This hypothesis was further tested with transgenic mice that express autoactivated MMP-3 (rat MMP-3 containing $Val^{77} \rightarrow Gly^{77}$ or $Pro^{78} \rightarrow Val^{78}$ mutation) under the control of the whey acidic protein gene promoter or the mouse mammary tumor virus promoter (Sympson *et al.,* 1994; Witty *et al.,* 1995b). The transgene was expressed at low levels in the mammary glands of virgin female mice and showed precocious development of alveoli with the similar morphology of a normal 9- to 12-day pregnant gland. High levels of MMP-3 was observed in lacating glands accompanied by the loss of basement membrane integrity and alteration of alveolar morphology. Degradation of the basement membrane components by overexpressed MMP-3 in mammary epithelial cells correlates with the expression of interleukin 1β converting enzyme and the induction of apoptosis of mammary epithelial cells (Boudreau *et al.,* 1995). Those studies suggest that MMP-3 plays a critical role in mammary gland morphogenesis and involution.

Osteoblasts in culture produce MMP-1, MMP-3, and MMP-9 in response to bone resorptive agents (Meikle *et al.,* 1992). These enzymes are thought to be responsible for the removal of nonmineralized collagen, which prevents the adherance and activation of osteoclasts (Chambers *et al.,* 1985). Osteoclasts remove both mineral and organic (collagenous and noncollagenous) constituents by their lysosomal cysteine proteinases and MMPs (Delaissé and Vaes, 1992). The production of MMP-9 (Reponen *et al.,* 1994), MMP-13 (Witty *et al.,* 1996), and membrane type-1 MMP (MT1-MMP/MMP-14) (Sato *et al.,* 1997) in osteoclasts has been reported, but the synthesis of MMP-3 and MMP-10 has not been proven (Witty *et al.,* 1996).

A possible role of MMP-3 in the development of hair follicles has been proposed (Goodman and Ledbetter, 1992). The treatment of rat dermal papilla, the mesenchymal component of the hair follicle, with

PDGF or bFGF potentiates the growth of dermal papilla cells along with increased secretion of proMMP-3. The dermal papilla cells possess inductive properties that regulate hair follicle formation and initiate hair growth cycles (Jahoda et al., 1984). It is thus postulated that the degradation of connective tissue matrix in the dermis is critical as the hair follicle forms.

3. REPRODUCTION

Many MMPs have been implicated in ovulation, blastocyst implantation, uterine cervical dilation, parturition, postpartum uterine involution, and menstruation, but only limited information is available about the expression of MMP-3 and MMP-10 *in vivo*. Das et al. (1997) investigated the expression of MMPs -2, -3, -9, and -13 in mouse uterus during the peri-implantation period from day 1 through 8 of pregnancy. While conspicuous spatial and temporal changes were observed with MMP-2, MMP-9, and TIMP-3, only weak signals for MMP-3 mRNA were detected in the myometrium with little changes in expression throughout the period. In human placenta MMP-3 was found in the amnion, decidua, and chorionic villi at all stages of pregnancy (Vettraino et al., 1996). Elevated mRNA levels for MMP-3 and MMP-10 were detected in stromal cells in the premenstrual and mensural phases of human endometrium (Osteen et al., 1994; Hampton and Salamonsen, 1994; Rodgers et al., 1994). MMP-1 and MMP-9 are also expressed in stromal cells in the luminal region of the tissue, whereas MMP-7 (matrilysin) is expressed in the surface epithelium and glandular epithelial cells (Rodgers et al., 1993, 1994). None of these transcripts was detected during the progesterone-dominated secretory phase. On the other hand, immunolocalization studies of Jeziorska et al. (1996) showed a weak immunostaining for MMP-3 in stromal cells limited to microfocal location at times coincident with stromal edema during proliferative (days 8–10) and secretory (days 21–22) phases as well as wide distribution of MMP-3 during menstruation. Together, those studies emphasize temporal regulation of MMPs in specific cell types and the roles of MMPs in tissue remodeling throughout the normal menstrual cycle.

B. Pathological Roles

1. ARTHRITIS

The erosion of articular cartilage accompanies both rheumatoid arthritis and osteoarthritis. The cartilage matrix is composed of collagens (primarily type II collagen with minor collagen types IX and XI) and proteoglycans. Although a number of proteinases have been found in the arthritic joints (Nagase and Woessner, 1993), matrixins are considered to be key enzymes for the degradation of cartilage matrix.

MMP-3, along with MMP-1 and TIMP-1, was immunolocalized in lining B cells of rheumatoid synovium (Okada *et al.*, 1989, 1990) and the elevated expression of their transcripts was identified by *in situ* hybridization (Firestein *et al.*, 1991; Gravallese *et al.*, 1991; McCachren, 1991). The concentration of MMP-3 in synovial fluid from patients with rheumatoid arthritis was reported to be 40.1 ± 26 μg/mL (n = 19) (Walakovits *et al.*, 1992). Serum levels of MMP-3 were also significantly higher than those in osteoarthritic patients or in normal individuals (Taylor *et al.*, 1994; Yoshihara *et al.*, 1995; Manicourt *et al.*, 1995), and they correlated with the erythrocyte sedimentation rate and C-reactive protein levels (Yoshihara *et al.*, 1995; Manicourt *et al.*, 1995). In osteoarthritis, the destructive enzymes come largely from chondrocytes. The production of MMP-3 in osteoarthritic cartilage is elevated and the immunostaining level of MMP-3 correlates directly with the histological score (Okada *et al.*, 1992b). Nguyen *et al.* (1992) found an abundant amount of MMP-3 mRNA in normal adult cartilage, but a low level in the neonate, suggesting that normal adult cartilage produces a significant amount of proMMP-3. Wolfe *et al.* (1993), on the other hand, reported that the mRNA level for MMP-3 is elevated in arthritic cartilage but not MMP-1 mRNA.

MMP-3 cleaves the core protein of aggrecan and link protein. It also cleaves telopeptides of type II collagen where intermolecular cross-linking occurs and type IX collagen *in vitro*. This suggests that MMP-3 may act as a depolymerizer of cartilage collagens (Wu *et al.*, 1991). The major enzymes that cleave type II collagen in cartilage are thought to be collagenases (MMP-1, MMP-8, and MMP-13); however, MMP-3 and other MMPs cleave the core protein of aggrecan at the Asn[341]–Phe[342] bond located in the region between the G1 and G2 globular domains (Flannery *et al.*, 1992; Fosang *et al.*, 1993). On the other hand, the treatment of cartilage explants with IL-1 generates fragments resulting from the cleavage of the Glu[373]–Ala[374] bond, 32-amino-acid C terminal to the major MMP cleavage site (Sandy *et al.*, 1991). Synovial fluids from patients with osteoarthritis and from knee injury indicated that the cleavage of the Glu[373]–Ala[374] bond increased (Sandy *et al.*, 1992). The enzyme responsible for this cleavage, called *aggrecanase,* is yet to be identified, but inhibition of the Glu[373]–Ala[374] bond cleavage by a peptidyl-hydroxamate MMP inhibitor suggests that it may be a MMP-type metalloproteinase (Buttle *et al.*, 1993). Recent studies by Fosang *et al.* (1996), however, indicated that fragments generated by the cleave of the Asn[34]-Phe[342] bonds were present in synovial fluids in osteoarthritic patients. The G1 fragments generated by MMPs and aggrecanase were immunolocalized in articular cartilage from patients with osteoarthritis and rheumatoid arthritis as well as in normal cartilage (Lark *et al.*, 1997). Those studies suggest that both MMPs and aggrecanase function in releasing aggrecan *in vivo* possibly by two independent pathways under physiological and pathological conditions.

2.   TUMOR INVASION AND METASTASIS

A wealth of information supports the view that the overexpression of MMPs correlates with the invasive behavior of tumor cells (see Stetler-Stevenson *et al.*, 1993; Coussens and Werb, 1996). It is, however, evident that the types of MMP expressed in tumor cells and the surrounding stromal cells depend on tumors. Tumor tissues associated with MMP-3 expression are summarized in Table IV.

In breast cancer MMP-3 is expressed in tumor stroma of invasive cancer in a widespread pattern along with MMP-1, MMP-2, MMP-11, and MT1-MMP (Heppner *et al.*, 1996). In contrast, MMP-9, MMP-12, and MMP-13 are more focal; MMP-9 is expressed at endothelial cells, and MMP-13 is associated with isolated tumor cells and MMP-12 with macrophage-like cells (Heppner *et al.*, 1996). Head and neck carcinoma expresses both MMP-3 and MMP-10 along with MMP-1 and MMP-7 (Muller *et al.*, 1991). Esophageal squamous cell carcinomas express MMP-2 and MMP-3 (Shima *et al.*, 1992). It was reported that colon and gastric carcinomas express MMP-7, but not MMP-3 or MMP-10 (McDonnell *et al.*, 1991). Newell *et al.* (1994) showed that the stromal component of some colorectal carcinomas expressed MMP-3, but not MMP-10. Elevated MMP-3 transcripts were also detected in the foci of normal mucosa adjacent to bronchial lesions, interepithelial lesions, and squamous carcinomas of the lung (Bolon *et al.*, 1996). In microinvasive and invasive bronchial lesions, the expression of MMP-3 was primarily in the stroma. Majmudar *et al.* (1994) reported that all three cutaneous basal cell carcinomas (BCCs) examined from patients with nevoid basal cell carcinoma syndrome (NBCCS), an autosomal dominant disorder characterized by the development of BCCs, overexpressed MMP-3 mRNA, but the expression of MMP-3 mRNA in BCCs from patients without NBCCS was limited. Interestingly, the normal-appearing skin fibroblasts from NBCCS patients in culture constitutively overexpress MMP-3 mRNA, but not the normal skin fibroblasts.

Although the exact role of MMP-3 and MMP-10 in tumor progression is not clearly understood, the chemically induced mouse skin tumor progression model showed that the conversion from a squamous cell carcinoma to an aggressive, metastatic spindle cell carcinoma was associated with the expression of MMP-3 mRNA in the tumor cells as well as in adjacent normal stroma, whereas MMP-3 transcripts were found only in stroma in squamous cell carcinomas (Wright *et al.,* 1994). The transformed rat embryo cell lines with high metastatic potential produced higher levels of MMP-3 and MMP-10 compared with the nonmetastatic lines (Sreenath *et al.,* 1992). These studies suggest a correlation between invasive behavior of tumor cells and the expression of stromelysins. Nonetheless, Witty *et al.* (1995a) reported that when transgenic mice expressing activated MMP-3 under the control of the mouse mammary tumor virus promoter were treated with a carcinogen to induce mammary tumor, the number of developing tumors decreased compared to nontransgenic littermate controls. This was due to an increase in the number of apoptotic cells in MMP-3 transgenic mice. Furthermore, malignant adenocarcinomas developed in the MMP-3 expressing mice showed no detectable alteration in invasion and metastatic potential compared to tumors from control mice. From these studies it was concluded that the expression of a single matrixin, MMP-3, was insufficient for the progression of mammary tumors to an invasive and metastatic phenotype, but the degradation of matrix would alter the basic process of cell proliferation and apoptosis.

3. WOUND HEALING

Distinct expression patterns of MMP-3 and MMP-10 were observed in basal keratinocytes in chronic dermal wounds (Saarialho-Kere *et al.,* 1994). MMP-3 mRNA and protein were detected in proliferating basal keratinocytes adjacent to but distal from the wound edge, whereas MMP-10 mRNA was found only in basal keratinocytes at the migrating front. MMP-3-producing keratinocytes reside on the basement membrane, whereas MMP-10-producing keratinocytes are in contact with dermal matrix without the basement membrane. The latter keratinocytes also express MMP-1 (Saarialho-Kere *et al.,* 1993b). The expression of MMP-3, but not MMP-10 is also prominent in dermal fibroblasts. These findings suggest that two closely related stromelysins may play different roles in tissue repair. The expression of MMP-3 was also detected in the basal keratinocytes surrounding the neutrophilic abscesses in the lesion of dermatitis herpetiformis (Airola *et al.,* 1995), and in migrating human surface respiratory epithelial cells after wounding (Buisson *et al.,* 1996).

In gastrointestinal ulcers such as Crohn's disease, peptic ulcers and ulcerative colitis, MMP-3 mRNA was found in granulation tissues (Saarialho-Kere *et al.,* 1996). MMP-1 mRNA was also detected in granulation tissues and the surface epithelium bordering gastrointestinal area, and MMP-7 in mucosal epithelium bordering the ulcerations

(Saarialho-Kere *et al.*, 1996). Immunolocalization studies of the intestine in Crohn's disease detected MMP-3 in the extracellular matrix in the regions where smooth muscle proliferation and mucosal degradation were seen (Bailey *et al.*, 1994). It is suggested that MMP-1 and MMP-3 are involved in the reparative process in the ulcer bed, whereas MMP-7 participates in epithelial remodeling in gastrointestinal ulcerations.

4. ATHEROSCLEROSIS

In atherosclerotic lesions, MMP-3 was detected in both smooth muscle cells and macrophages in plaques (Henney *et al.*, 1991; Galis *et al.*, 1994). The production of MMP-1 (Nikkari *et al.*, 1995), MMP-2 (Galis *et al.*, 1994; Li *et al.*, 1996), and MMP-9 (Galis *et al.*, 1994; Brown *et al.*, 1995) is also seen primarily in macrophages and smooth muscle cells in the lesion. These enzymes are thought to participate in weakening the connective tissue matrix in the intima, which leads to plaque rupture, acute thrombosis, and smooth muscle cell proliferation and migration. The polymorphism analysis of the MMP-3 gene by Ye *et al.* (1995) identified a common variant in the promoter. This variant gives a rise to one allele with six adenosines (6A) and another with five adenosines (5A) at position −1171. It was shown that the 6A6A genotype was significantly associated with greater progression at coronary atherosclerosis than other genotypes. The promoter with 6A at the polymorphic site was found to be about twofold less active in transcriptional activity than that with 5A (Ye *et al.*, 1996). It is postulated that reduced levels of MMP-3 expression in the 6A homozygotes contribute to a net increase in matrix deposition, leading to more rapid chronic growth of the atherosclerotic plaque (Ye *et al.*, 1996).

VII. CONCLUSIONS AND FUTURE PROSPECTS

Stromelysins 1 and 2 are two closely related matrixins. Biochemical and biophysical studies have revealed the three-dimensional structure of both pro- and catalytic domains of MMP-3, activation mechanisms, and substrate specificity. However, the substrates of these stromelysins *in vivo* have not been well defined. In the tissue, extracellular matrix components are in a solid phase and their concentrations are considerably higher than those in the solution assay. The condensed nature of the matrix may restrict the movement of enzyme while the secreted zymogens may preferentially bind to certain extracellular matrix components. In this regard, it is interesting that proMMP-3, but not proMMP-1, binds to type I collagen (Murphy *et al.*, 1992). Nonetheless, the interaction of these zymogens and active enzymes with other matrix components is not characterized. Such information is essential to gain insights into precise biological and pathological functions of stromelysins and other matrixins.

Further development of specific inhibitors of stromelysins will be useful not only for understanding of the biological functions of MMPs in normal tissue development and matrix turnover, but also for the development of therapeutical strategies for various diseases associated with overexpression of MMPs. Recently resolved three-dimensional structures of various matrixins and the inhibition mechanisms of MMP-3 by TIMP-1 have provided excellent bases to design new, more selective inhibitors for different types of matrixins.

The promoter regions of MMP-3 and of other MMP genes have been analyzed extensively, but little is known about how cell-specific induction of each members of the matrixins occurs. For example, as exemplified in keratinocytes in dermal wounds, expressions of MMP-3 and MMP-10 are distinct and they are cell dependent (Saarialho-Kere *et al.*, 1994). Furthermore, changes of matrix composition resulting from degradation or overproduction also influence the phenotype of cells. The induction of MMP gene expression is also closely linked to signal transduction pathways, but they are poorly understood at present. Elucidation of the cell-specific expression mechanisms and associated signaling pathways may lead to the rational design of inhibitors that perturb the production of MMPs in specific cell types without affecting the others.

ACKNOWLEDGMENTS

This work was supported by National Institutes of Health grants AR39189 and AR40994.

REFERENCES

Airola, K., Vaalamo, M., Reunala, T., and Saarialho-Kere, U.K. (1995). Enhanced expression of interstitial collagenase, stromelysin-1, and urokinase plasminogen activator in lesions of dermatitis herpetiformis. *J. Invest. Dermatol.* **105,** 184–189.

Alexander, C. M., Hansell, E. J., Behrendtsen, O., Flannery, M. L., Kishnani, N. S., Hawkes, S. P., and Werb, Z. (1996). Expression and function of matrix metalloproteinases and their inhibitors at the maternal embryonic boundary during mouse embryo implantation. *Development* **122,** 1723–1736.

Alexander, J. P., Bradley, J. M. B., Gabourel, J. D., and Acott, T. S. (1990). Expression of matrix metalloproteinases and inhibitor by human retinal pigment epithelium. *Invest. Ophthalmol. Vis. Sci.* **31,** 2520–2528.

Angel, P., Imagawa, M., Chiu, R., Stein, B., Imbra, R. J., Rahmsdorf, H. J., Jonat, C., Herrlich, P., and Karin, M. (1987). Phorbol ester-inducible genes contain a common *cis* element recognized by a TPA-modulated *trans*-acting factor. *Cell* **49,** 729–739.

Arner, E. C., and Tortorella, M. D. (1995). Signal transduction through chondrocyte integrin receptors induces matrix metalloproteinase synthesis and synergizes with interleukin-1. *Arthritis Rheum.* **38,** 1304–1314.

Auble, D. T., and Brinckerhoff, C .E. (1991). The AP-1 sequence is necessary but not sufficient for phorbol induction of collagenase in fibroblasts. *Biochemistry* **30,** 4629–4635.

Azzo, W., and Woessner, J. F., Jr. (1986). Purification and characterization of an acid metalloproteinase from human articular cartilage. *J. Biol. Chem.* **261,** 5434–5441.

Bailey, C. J., Hembry, R. M., Alexander, A., Irving, M. H., Grant, M. E., and Shuttleworth, C. A. (1994). Distribution of the matrix metalloproteinases stromelysin, gelatinases A and B, and collagenase in Crohn's disease and normal intestine. *J. Clin. Pathol.* **47,** 113–116.

Barrett, A. J., and Starkey, P.M. (1973). The interaction of α_2-macroglobulin with proteinases. Characteristics and specificity of the reaction and a hypothesis concerning its molecular mechanisms. *Biochem. J.* **133,** 709–724.

Basset, P., Bellocq, J. P., Wolf, C., Stoll, I., Hutin, P., Limacher, J. M., Podhajcer, O. L., Chenard, M. P., Rio, M. C., and Chambon, P. (1990). A novel metalloproteinase gene specificity expressed in stromal cells of breast carcinomas. *Nature* **348,** 699–704.

Becker, J. W., Marcy, A. I., Rokosz, L. L., Axel, M. G., Burbaum, J. J., Fitzgerald, P. M. D., Cameron, M. P., Esser, C. K., Hagmann, W. K., Hermes, J. D., and Springer, J. P. (1995) Stromelysin-1: Three dimensional structure of the inhibited catalytic domain and the C-truncated proenzyme. *Protein Sci.* **4,** 1966–1976.

Behrendtsen, O., Alexander, C. M., and Werb, Z. (1992). Metalloproteinases mediate extracellular matrix degradation by cells from mouse blastocyst outgrowths. *Development* **114,** 447–456.

Belaaouaj, A., Shipley, J. M., Kobayashi, D. K., Zimonjie, D. B., Popescu, N., Silverman, G. A., and Shapiro, S. D. (1995). Human macrophage metalloelastase. Genomic organization, chromosomal location, gene linkage, and tissue-specific expression. *J. Biol. Chem.* **270,** 14568–14575.

Benbow, U., Butticè, G., Nagase, H., and Kurkinen, M. (1996). Characterization of the 46-kDa intermediates of matrix metalloproteinase 3 (stromelysin 1) obtained by site-directed mutation of phenylalanine 83. *J. Biol. Chem.* **271,** 10715–10722.

Benyon, R. C., Iredale, J. P., Goddard, S., Winwood, P. J., and Arthur, M. J. P. (1996). Expression of tissue inhibitor of metalloproteinases 1 and 2 is increased in fibrotic human liver. *Gastroenterology* **110,** 821–831.

Berman, M. B. (1980). Collagenase and corneal ulceration. *In* "Collagenase in Normal and Pathological Connective Tissues" (D. E. Woolley, and J. M. Evanson, eds.), pp. 140–174, Wiley, London.

Bini, A., Itoh, Y., Kudryk, B. J., and Nagase, H. (1996). Degradation of cross-linked fibrin by matrix metalloproteinase 3 (stromelysin 1): Hydrolysis of the γ Gly^{404}–Ala^{405} peptide bond. *Biochemistry* **35,** 13056–13063.

Birkedal-Hansen, H. (1993). Role of matrix metalloproteinases in human periodontal diseases. *J. Periodontol.* **64,** 474–484.

Bode, W., Gomis-Rüth, F.-X., and Stöcker, W. (1993). Astains, serralysins, snake venom and matrix metalloproteinases exhibit identical zinc-binding environments (HEXXHXXGXXH and Met-turn) and topologies and should be grouped into a common family, the "Metzincins." *FEBS Lett.* **331,** 134–140.

Bode, W., Reinemer, P., Huber, R., Kleine, T., Schnierer, S., and Tschesche, H. (1994). The X-ray crystal structure of the catalytic domain of human neutrophil collagenase inhibited by a substrate analogue reveals the essentials for catalysis and specificity. *EMBO J.* **13,** 1263–1269.

Bolon, I., Brambilla, E., Vandenbunder, B., Robert, C., Lantuejoul, S., and Brambilla, C. (1996). Changes in expression of matrix proteases and of the transcription factor c-Ets-1 during progression of precancerous bronchial lesions. *Lab. Invest.* **75**, 1–13.

Bottomley, K. M., Borkakoti, N., Bradshaw, D., Brown, P. A., Broadhurst, M. J., Budd, J. M., Elliott, L., Eyers, P., Hallam, T. J., Handa, B. K., Hill, C. H., James, M., Lahm, H.-W., Lawton, G., Merritt, J. E., Nixon, J. S., Röthlisberger, U., Whittle, A., and Johnson, W. H. (1997). Inhibition of bovine nasal cartilage degradation by selective matrix metalloproteinase inhibitors. *Biochem. J.* **323**, 483–488.

Boudreau, N., Sympson, C. J., Werb, Z., and Bissell, M. J. (1995). Suppression of ICE and apoptosis in mammary epithelial cells by extracellular matrix. *Science* **267**, 891–893.

Breathnach, R., Matrisian, L. M., Gesnel, M.-C., Staub, A., and Leroy, P. (1987). Sequences coding for part of oncogenic-induced transin are highly conserved in a related rat gene. *Nucleic Acids Res.* **15**, 1139–1151.

Brenner, C. A., Adler, R. R., Rappolee, D. A., Pederson, R. A., and Werb, Z. (1989). Genes for extracellular matrix-degrading metalloproteinases and their inhibitor, TIMP, are expressed during early mammalian development. *Genes Develop.* **3**, 848–859.

Brenner, D. A., O'Hara, M., Angel, P., Chojkier, M., and Karin, M. (1989). Prolonged activation of *jun* and collagenase genes by tumor necrosis factor-α. *Nature* **337**, 661–663.

Brinckerhoff, C. E., Vater, C. A., and Harris, E.D. (1981). Effects of retinoids on rabbit synovial fibroblasts and chondrocytes. *In* "Cellular Interactions" (J. T. Dingle and J. L. Gordon, eds.), pp. 215–230, North-Holland Biomedical Press, New York.

Brinckerhoff, C. E., Plucinska, I. M., Sheldon, L. A., and O'Connor, G. T. (1986). Half-life of synovial cell collagenase mRNA is modulated by phorbol myristate acetate but not by all-*trans*-retinoic acid or dexamethasone. *Biochemistry* **25**, 6378–6384.

Brown, C. C., Hembry, R. M., and Reynolds, J. J. (1989). Immunolocalization of metallo-proteinases and their inhibitor in the rabbit growth plate. *J. Bone Joint Surg.* **71-A**, 580–593.

Brown, D. L., Hibbs, M. S., Kearney, M., Loushin, C., and Isner, J.M. (1995). Identification of 92-kD gelatinase in human coronary atherosclerotic lesions. Association of active enzyme synthesis with unstable angina. *Circulation* **91**, 2125–2131.

Browner, M. F., Smith, W. W., and Castelhano, A. L. (1995). Matrilysin-inhibitor complexes: Common themes among metalloproteases. *Biochemistry* **34**, 6602–6610.

Buisson, A.-C., Gilles, C., Polette, M., Zahm, J.-M., Birembaut, P., and Tournier, J.-M. (1996). Wound repair-induced expression of stromelysins is associated with the acquisition of a mesenchymal phenotype in human respiratory epithelial cells. *Lab. Invest.* **74**, 658–669.

Buttice, G., and Kurkinen, M. (1993). A polyomavirus enhancer A-binding protein-3 site and Ets-2 protein have a major role in the 12-O-tetradecanoylphoborl-13-acetate response of the human stromelysin gene. *J. Biol. Chem.* **268**, 7196–7204.

Buttice, G., Quinones, S., and Kurkinen, M. (1991). The AP-1 site is required for basal expression but is not necessary for TPA-response of the human stromelysin gene. *Nucleic. Acids Res.* **19**, 3723–3731.

Buttle, D. J., Handley, C. J., Ilic, M. Z., Saklatvala, J., Murata, M., and Barrett, A. J. (1993). Inhibition of cartilage proteoglycan release by a specific inactivator of cathepsin B and an inhibitor of matrix metalloproteinases. Evidence of two converging pathways of chondrocyte-mediated proteoglycan degradation. *Arthritis Rheum.* **36**, 1709–1717.

Cameron, P. M., Marcy, A. I., Rokosz, L. L., and Hermes, J. D. (1995). Use of an active-site inhibitor of stromelysin to elucidate the mechanism of prostromelysin activation. *Bioorganic Chem.* **23**, 415–426.

Chambers, T. J., Darby, J. A., and Fuller, K. (1985). Mammalian collagenase predisposes bone surfaces to osteoclastic resorption. *Cell Tiss. Res.,* **241**, 671–675.

Chandrasekhar, S., and Harvey, A.K. (1992). Differential regulation of metalloprotease steady-state mRNA levels by IL-1 and FGF in rabbit articular chondrocytes. *FEBS Lett.* **296,** 195–200.

Chapman, K. T., Kopka, I. E., Durette, P. L., Esser, C. K., Lanza, T. J., Izquierdo-Martin, M., Niedzwiecki, L., Chang, B., Harrison, R. K., Kuo, D. W., Lin, T.-Y., Stein, R. L., and Hagmann, W. K. (1993). Inhibition of matrix metalloproteinases by *N*-carboxyalkyl peptides. *J. Med. Chem.* **36,** 4293–4301.

Chen, L.-C., Noelken, M. E., and Nagase, H. (1993). Disruption of the cysteine-75 and zinc ion coordination is not sufficient to activate the precursor of human matrix metalloproteinase 3 (stromelysin 1). *Biochemistry* **32,** 10289–10295.

Cheung, H. S., Sallis, J. D., and Struve, J. A. (1996). Specific inhibition of basic calcium phosphate and calcium pyrophosphate crystal-induction of metalloproteinase synthesis by phosphocitrate. *Biochim. Biophys. Acta* **1315,** 105–111.

Chin, J. R., Murphy, G., and Werb, Z. (1985). Stromelysin, a connective tissue-degrading metalloendopeptidase secreted by stimulated rabbit synovial fibroblasts in parallel with collagenase. Biosynthesis, isolation, characterization, and substrates. *J. Biol Chem.* **260,** 12367–12376.

Conca, W., and Willmroth, F. (1994). Human T lymphocytes express a member of the matrix metalloproteinase gene family. *Arthritis Rheum.* **37,** 951–956.

Coussens, L. M., and Werb, Z. (1996). Matrix metalloproteinases and the development of cancer. *Chem. Biol.* **3,** 895–904.

Das, S. K., Yano, S., Wang, J., Edwards, D. R., Nagase, H., and Dey, S. K. (1997). Expression of matrix metalloproteinases (MMPs) and tissue inhibitors of metalloproteinases (TIMPs) in the mouse uterus during the peri-implantation period. *Develop. Genetics* **21,** 44–54.

DeClerck, Y. A, Yean, T. D., Lu, H. S., Ting, J., and Langley, K. E. (1991). Inhibition of autoproteolytic activation of interstitial procollagenase by recombinant metalloproteinase inhibitor MI/TIMP-2. *J. Biol. Chem.* **266,** 3893–3899.

Delaissé, J. M., and Vaes, G. (1992). Mechanism of mineral solubilization and matrix degradation in osteoclastic bone resorption. In "Biology and Physiology of the Osteoclast" (B. R. Rifkin and C. V. Gay, eds.), pp. 289–314, CRC Press, Boca Raton, FL.

Delany, A. M., and Brinckerhoff, C. E. (1992). Post-transcriptional regulation of collagenase and stromelysin gene expression by epidermal growth factor and dexamethasone in cultured human fibroblasts. *J. Cell. Biochem.* **50,** 400–410.

Delany, A. M., Jeffrey, J. J., Rydziel, S., and Canalis, E. (1995). Cortisol increases interstitial collagenase expression in osteoblasts by post-transcriptional mechanisms. *J. Biol. Chem.* **270,** 26607–26612.

DenBesten, P. K., Heffernan, L. M., Treadwell, B. V., and Awbrey, B. J. (1989). The presence and possible functions of the matrix metalloproteinase collagenase activator protein in developing enamel matrix. *Biochem. J.* **264,** 917–920.

DeSouza, S., Lochner, J., Machida, C. M., Matrisian, L. M., and Ciment, G. (1995). A novel nerve growth factor-responsive element in the stromelysin-1 (transin) gene that is necessary and sufficient for gene expression in PC12 cells. *J. Biol. Chem.* **270,** 9106–9114.

Dickson, S. R., and Warburton, M. J. (1992). Enhanced synthesis of gelatinase and stromelysin by myoepithelial cells during involution of the rat mammary gland. *J. Histochem. Cytochem.* **40,** 697–703.

Douglas, D. A., Shi, Y. E., and Sang, Q. A. (1997). Computational sequence analysis of the tissue inhibitor of metalloproteinase family. *J. Protein Chem.* **16,** 237–255.

Enghild, J. J., Salvesen, G., Brew, K., and Nagase, H. (1989). Interaction of human rheumatoid synovial collagenase (matrix metalloproteinase 1) and stromelysin (matrix metalloproteinase 3) with human α_2-macroglobulin and chicken ovostatin. Binding kinetics and identification of matrix metalloproteinase cleavage sites *J. Biol. Chem.* **264,** 8779–8785.

Fini, M. E., Karmilowicz, M. J., Ruby, P. L., Beeman, A. M., Borges, K. A., and Brinckerhoff, C.E. (1987). Cloning of a complimentary DNA for rabbit proactivator. A metalloproteinase that activates synovial cell collagenase, shares homology with stromelysin and transin and is coordinately regulated with collagenase. *Arthritis Rheum.* **30,** 1254–1264.

Firestein, G.S., Paine, M. M., and Littman, B. H. (1991). Gene expression (collagenase, tissue inhibitor of metalloproteinases, complement and HLA-DR) in rheumatoid arthritis and osteoarthritis synovium. Quantitative analysis and effect of intraarticular corticosteroids. *Arthritis Rheum.* **34,** 1094–1105.

Flannery, C. R., Lark, M. W., and Sandy, J. D. (1992). Identification of a stromelysin cleavage site within the interglobular domain of human aggrecan. Evidence for proteolysis at this site *in vivo* in human articular cartilage. *J. Biol. Chem.* **267,** 1008–1014.

Formstone, C. J., Byrd, P. J., Ambrose, H. J., Riley, J. H., Hernandez, D., McConville, C. M., and Taylor, A. M. R. (1993). The order and orientation of a cluster of metalloproteinase genes, stromelysin-2, collagenase, and stromelysin, together with D11S385, on chromosome-11q22-q23. *Genomics* **16,** 289–291.

Fosang, A. J., Last, K., Knäuper, V., Neame, P. J., Murphy, G., Hardingham, T. E., Tschesche, H., and Hamilton, J. A. (1993). Fibroblast and neutrophil collagenases cleave at two sites in the cartilage aggrecan interglobular domain. *Biochem. J.* **295,** 273–276.

Fosang, A. J., Last, K., and Maciewicz, R. A. (1996). Aggrecan is degraded by matrix metalloproteinases in human arthritis. *J. Clin. Invest.* **98,** 2292–2299.

Fowlkes, J. L., Enghild, J. J., Suzuki, K., and Nagase, H. (1994). Matrix metalloproteinases degrade insulin-like growth factor-binding protein-3 in dermal fibroblast cultures. *J. Biol. Chem.* **269,** 25742–25746.

Freimark, B. D., Feeser, W. S., and Rosenfeld, S. A. (1994). Multiple sites of the propeptide region of human stromelysin-1 are required for maintaining a latent form of the enzyme. *J. Biol. Chem.* **269,** 26982–26987.

Frisch, S. M., and Werb, Z. (1989). Molecular biology of collagen degradation. *In* "Collagen: Molecular Biology" (B. R. Olsen and M. E. Mimni, eds.), Vol. IV, pp. 85–108, CRC Press, Boca Raton, FL.

Gaire, M., Magbanua, Z., McDonnell, S., McNeil, L., Lovett, D. H., and Matrisian, L. M. (1994). Structure and expression of the human gene for the matrix metalloproteinase matrilysin. *J. Biol. Chem.* **269,** 2032–2040.

Galazka, G., Windsor, L. J., Birkedal-Hansen, H., and Engler, J. A. (1996). APMA (4-aminophenylmercuric acetate) activation of stromelysin 1 involves protein interactions in addition to those with cysteine 75 in the propeptide. *Biochemistry* **35,** 11221–11227.

Galis, Z. S., Sukhova, G. K., Lark, M. W., and Libby, P. (1994) Increased expression of matrix metalloproteinases and matrix degrading activity in vulnerable regions of human atherosclerotic plaques. *J. Clin. Invest.* **94,** 2493–2503.

Galloway, W. A., Murphy, G., Sandy, J. D., Gavrilovic, J., Cawston, T. E., and Reynolds, J. J. (1983). Purification and characterization of a rabbit bone metalloproteinase that degrades proteoglycan and other connective-tissue components. *Biochem. J.* **209,** 741–752.

Ganser, G. L., Stricklin, G. P., and Matrisian, L. M. (1991). EGF and TGF α influence *in vitro* lung development by the induction of matrix-degrading metalloproteinases. *Int. J. Dev. Biol.* **35,** 453–461.

Gijbels, K., Galardy, R. E., and Steinman, L. (1994). Reversal of experimental autoimmune encephalomyelitis with a hydroxamate inhibitor of matrix metalloproteinases. *J. Clin. Invest.* **94,** 2177–2182.

Gohlke, U., Gomis-Rüth, F.-X., Crabbe, T., Murphy, G., Docherty, A. J. P., and Bode, W. (1996). The C-terminal (haemopexin-like) domain structure of human gelatinase A (MMP-2): Structural implications for its function. *FEBS Lett.* **378,** 126–130.

Goldberg, G. I., Strongin, A., Collier, I. E., Genrich, L. T., and Marmer, B. L. (1992). Interaction of 92-kDa type IV collagenase with the tissue inhibitor of metalloproteinases prevents dimerization, complex formation with interstitial collagenase, and activation of the proenzyme with stromelysin. *J. Biol. Chem.* **267,** 4583–4591.

Gomis-Rüth, F.-X., Gohlke, Y., Betz, M., Knäuper, V., Murphy, G., López-Otín, C., and Bode, W. (1996). The helping hand of collagenase-3 (MMP-13): 2.7 Å crystal structure of its C-terminal haemopexin-like domain. *J. Mol. Biol.* **264,** 556–566.

Gomis-Rüth, F.-X., Maskos, K., Betz, M. Bergner, A., Huber, R., Suzuki, K., Yoshida, N., Nagase, H., Brew, K., Bourenkow, G. B., Bartunik, H., and Bode, W. (1997). Mechanism of inhibition of the human metalloproteinase stromelysin-1 by TIMP-1. *Nature* **389,** 77–81.

Goodman, L. V., and Ledbetter, S. R. (1992). Secretion of stromelysin by cultured dermal papilla cells: Differential regulation by growth factors and functional role in mitogen-induced cell proliferation. *J. Cell Physiol.* **151,** 41–49.

Gooley, P. R., O'Connell, J. R., Marcy, A. I., Cuca, G. C., Salowe, S. P., Springer, J. B., and Johnson, B. A. (1994). The NMR structure of the inhibited catalytic domain of human stromelysin-1. *Nature Struct. Biol.* **1,** 111–118.

Gravallese, E. M., Darling, J. M., Ladd, A. L., Katz, J. N., and Glimcher, L. H. (1991). *In situ* hybridization studies of stromelysin and collagenase messenger RNA expression in rheumatoid synovium. *Arthritis Rheum.* **34,** 1076–1084.

Gunja-Smith, Z., Morales, A. R., Romanelli, R., and Woessner, J. F., Jr. (1996). Remodeling of human myocardial collagen in idiopathic dilated cardiomyopathy. Role of metalloproteinases and pyridinoline cross-links. *Am. J. Pathol.* **144,** 82–94.

Hajduk, P. J., Sheppard, G., Nettesheim, D. G., Olejniczak, E. T., Shuker, S. B., Meadows, R. P., Steinman, D. H., Carrera, G. M., Jr., Marcotte, P. A., Severin, J., Walter, K., Smith, H., Gubbins, E., Simmer, R., Holzman, T. F., Morgan, D. W., Davidsen, S. K., Summers, J. B., and Fesik, S. W. (1997). Discovery of potent nonpeptide inhibitors of stromelysin using SAR by NMR. *J. Am. Chem. Soc.* **119,** 5818–5821.

Hammani, K., Henriet, P., Silbiger, S. M., and Eeckhout, Y. (1992). Cloning and sequencing a cDNA encoding mouse stromelysin 1. *Gene* **120,** 321–322.

Hampton, A. L., and Salamonsen, L. A. (1994). Expression of messenger ribonucleic acid encoding matrix metalloproteinases and their tissue inhibitors is related to menstruation. *J. Endocrinol.* **141,** R1–R3.

Harrison, R., Teahan, J., and Stein R. (1989). A semicontinuous, high-performance liquid chromatography-based assay for stromelysin. *Anal. Biochem.* **180,** 110–113.

Harrison, R. K., Chang, B., Niedzwiecki, L., and Stein, R. L. (1992). Mechanistic studies on the human matrix metalloproteinase stromelysin. *Biochemistry* **31,** 10757–10762.

Henney, A. M., Wakely, P. R., Davies, M. J., Foster, K., Hembry, R., Murphy, G., and Humphries, S. (1991). Localization of stromelysin gene expression in atherosclerotic plaques by *in situ* hybridization. *Proc. Natl. Acad. Sci. USA* **88,** 8154–8158.

Heppner, K. J., Matrisian, L. M., Jensen, R. A., and Rodgers, W. H. (1996). Expression of most matrix metalloproteinase family members in breast cancer represents a tumor-induced host response. *Am. J. Pathol.* **149,** 273–282.

Higashino, F., Yoshida, K., Noumi, T., Seiki, M., and Fujinaga, K. (1995). Ets-related protein E1A-F can activate three different matrix metalloproteinase gene promoters. *Oncogene* **10,** 1461–1463.

Holz, R. C., Salowe, S. P., Smith, C. K., Cuca, G. C., and Que, L. Jr. (1992). EXAFS evidence for a "cysteine switch" in the activation of prostromelysin. *J. Am. Chem. Soc.* **114,** 9611–9614.

Hosono, T., Ito, A., Sato, T., Nagase, H., and Mori, Y. (1996a). Translational augmentation of promatrix metalloproteinase 3 (prostromelysin 1) and tissue inhibitor of metalloproteinases (TIMP)-1 mRNAs induced by epidermal growth factor in human uterine cervical fibroblasts. *FEBS Lett.* **381,** 115–118.

Hosono, T., Ito, A., Sato, T., Nagase, H., and Mori, Y. (1996b). Translational augmentation of promatrix metalloproteinase 3 (prostromelysin 1) and a tissue inhibitor of metalloproteinases (TIMP)-1 mRNAs by epidermal growth factor and transforming growth factor α, but not by interleukin 1α or 12-*O*-tetradecanoylphobol 13–acetate in human uterine cervical fibroblasts: The possible involvement of an atypical protein kinase C. *Biol. Pharm. Bull.* **19,** 1285–1290.

Housley, T. J., Baumann, A. P., Braun, I. D., David, G., Seperack, P. K., and Wilhelm, S. M. (1993). Recombinant Chinese hamster ovary cell matrix metalloprotease-3 (MMP-3, stromelysin-1). Role of calcium in promatrix metalloprotease-3 (proMMP-3, prostromelysin-1) activation and thermostability of the low mass catalytic domain of MMP-3. *J. Biol. Chem.* **268,** 4481–4487.

Huang, W., Suzuki, K., Nagase, H., Arumugam, S., Van Doren, S. R., and Brew, K. (1996). Folding and characterization of the amino-terminal domain of human tissue inhibitor of metalloproteinases-1 (TIMP-1) expressed at high yield in E. coli. *FEBS Lett.* **384,** 155–161.

Imai, K., Kusakabe, M., Sakakura, T., Nakanishi, I., and Okada, Y. (1994). Susceptibility of tenascin to degradation by matrix metalloproteinases and serine proteinases. *FEBS Lett.* **352,** 216–218.

Imai, K., Shikata, H., and Okada, Y. (1995). Degradation of vitronectin by matrix metalloproteinases-1, -2, -3, -7 and -9. *FEBS Lett.* **369,** 249–251.

Ito, A., Itoh, Y., Sasaguri, Y., Morimatsu, M., and Mori, Y. (1992). Effects of interleukin-6 on the metabolism of connective tissue components in rheumatoid synovial fibroblasts. *Arthritis Rheum.* **35,** 1197–1201.

Ito, A., Mukaiyama, A., Itoh, Y., Nagase, H., Thøgersen, I. B., Enghild, J. J., Sasaguri, Y., and Mori, Y. (1996). Degradation of interleukin 1β by matrix metalloproteinases. *J. Biol. Chem.* **271,** 14657–14660.

Itoh, Y., Binner, S., and Nagase, H. (1995). Steps involved in activation of the complex of promatrix metalloproteinase 2 (progelatinase A) and tissue inhibitor of metalloproteinases (TIMP)-2 by 4-aminophenylmercuric acetate. *Biochem. J.* **308,** 645–651.

Izquierdo-Martin, M., and Stein, R. L. (1993). Mechanistic studies on the inhibition of stromelysin by a peptide phosphonamidate. *Bioorg. Med. Chem.* **1,** 19–26.

Jahoda, C. A., Horne, K. A., and Oliver, R. F. (1984). Induction of hair growth by implantation of cultured dermal papilla cells. *Nature* **311,** 560–562.

James, T. W., Wagner, R., White, L. A., Zwolak, R. M., and Brinckerhoff, C. E. (1993). Induction of collagenase and stromelysin gene expression by mechanical injury in a vascular smooth muscle-derived cell line. *J. Cell. Physiol.* **157,** 426–437.

Jeziorska, M., Nagase, H., Salamonsen, L. A., and Woolley, D. E. (1996). Immunolocalization of the matrix metalloproteinases gelatinase B and stromelysin 1 in human endometrium throughout the menstrual cycle. *J. Reprod. Fertility* **107,** 43–51.

Jonat, C., Rahmsdorf, H. J., Park, K.-K., Cato, A. C.B., Gebel, S., Ponta, H., and Herrlich, P. (1990). Antitumor promotion and antiinflammation: Down-modulation of AP-1 (Fos/Jun) activity by glucocorticoid hormone. *Cell* **62,** 1189–1204.

Jonat, C., Chung, F.-Z., and Baragi, V. M. (1996). Transcriptional down regulation of stromelysin by tetracycline. *J. Cell Biochem.* **60,** 341–347.

Kanemoto, M., Hukuda, S., Komiya, Y., Katsuura, A., and Nishioka, J. (1996). Immunohistochemical study of matrix metalloproteinase-3 and tissue inhibitor of metalloproteinase-1 human intervertebral discs. *Spine* **21,** 1–8.

Kerr, L. D., Holt, J. T., and Matrisian, L. M. (1988). Growth factors regulate transin gene expression by c-*fos*-dependent and c-*fos*-independent pathways. *Science* **242**, 1424–1427.

Kerr, L. D., Miller, D. B., and Matrisian, L. M. (1990). TGF-β1 inhibition of transin/ stromelysin gene expression is mediated through a Fos binding sequence. *Cell* **61**, 267–278.

Kirstein, M., Sanz, L., Quiñones, S., Moscat, J., Diaz-Meco, M. T., and Saus, J. (1996). Cross-talk between different enhancer elements during mitogenic induction of the human stromelysin-1 gene. *J. Biol. Chem.* **271**, 18231–18236.

Kitamura, M., Kitamura, A., Mitarai, T., Maruyama, N., Nagasawa, R., Kawamura, T., Yoshida, H., Takahashi, T., and Sakai, O. (1992). Gene expression of metalloproteinase and its inhibitor in mesangial cells exposed to high glucose. *Biochem. Biophys. Res. Comm.* **185**, 1048–1054.

Knäuper, V., Wilhelm, S. M., Seperack, P. K., DeClerck, Y. A., Langley, K. E., Osthues, A., and Tschesche, H. (1993). Direct activation of human neutrophil procollagenase by recombinant stromelysin. *Biochem. J.* **295**, 581–586.

Knäuper, V., López-Otín, C., Smith, B., Knight, G., and Murphy, G. (1996a). Biochemical characterization of human collagenase-3. *J. Biol. Chem.* **271**, 1544–1550.

Knäuper, V., Murphy, G., and Tschesche, H. (1996b). Activation of human neutrophil procollagenase by stromelysin 2. *Eur. J. Biochem.* **235**, 187–191.

Knight, C. G., Willenbrook, F., and Murphy, G. (1992). A novel coumarin-labeled peptide sensitive continuous assay of the matrix metalloproteinases. *FEBS Lett.* **296**, 263–266.

Koklitis, P., Murphy, G., Sutton, C., and Angal, S. (1991). Purification of recombinant human prostromelysin. Studies on heat activation to give high-M_r and low-M_r active forms, and a comparison of recombinant with natural stromelysin activities. *Biochem. J.* **276**, 217–221.

König, H., Pnota, H., Rahmsdorf, H. J., and Herrlich, P. (1992). Interference between pathway-specific factors: Glucocorticoids antagonize phorbol ester-induced AP-1 activity without altering AP-1 site occupation *in vivo*. *EMBO J.* **11**, 2241–2246.

Lark, M. W., Bayne, E. K., Flanagan, J., Harper, C. F., Hoerrner, L. A., Hutchinson, N. I., Singer, I. I., Donatelli, S. A., Weidner, J. R., Williams, H. R., Mumford, R. A., and Lohmander, L. S. (1997). Aggrecan degradation in human cartilage. Evidence for both matrix metalloproteinase and aggrecanase activity in normal, osteoarthritic, and rheumatoid joints. *J. Clin. Invest.* **100**, 93–106.

Lefebvre, O., Wolf, C., Limacher, J.-M., Hutin, P., Wendling, C., LeMeur, M., and Bassett, P. (1992). The breast cancer-associated stromelysin-3 gene is expressed during mouse mammary gland apoptosis. *J. Cell Biol.* **119**, 997–1002.

Li, J., Brick, P., O'Hare, M. C., Skarzynski, T., Lloyd, L. F., Curry, V. A., Clark, I. M., Bigg, H. F., Hazleman, B. L., Cawston, T. E., and Blow, D. M. (1995). Structure of full-length porcine synovial collagenase reveals a C-terminal domain containing a calcium-linked four-bladed β-propeller. *Structure* **3**, 541–549.

Li, Z., Li, L., Zielke, R., Cheng, L., Xiao, R., Crow, M. T., Stetler-Stevenson, W. G., Froehlich, J., and Lakatta, E. G. (1996). Increased expression of 72–kD type IV collagenase (MMP-2) in human aortic atherosclerotic lesions. *Am. J. Pathol.* **148**, 121–128.

Libson, A. M., Gittis, A. G., Collier, I. E., Marmer, B. L., Goldberg, G. I., and Lattman, E. E. (1995). Crystal structure of the haemopexin-like C-terminal domain of gelatinase A. *Nat. Struct. Biol.* **2**, 938–942.

Lindsey, J. D., Kashiwagi, K., Boyle, D., Kashiwagi, F., Firestein, G. S., and Weinreb, R. N. (1996). Prostaglandins increase proMMP-1 and proMMP-3 secretion by human ciliary smooth muscle cells. *Curr. Eye Res.* **15**, 869–875.

Logan, S. K., Hansell, E. J., Damsky, C. H., and Werb, Z. (1996). T-antigen inhibits metalloproteinase expression and invasion in human placental cells transformed with temperature-sensitive Simian virus 40. *Matrix Biol.* **15,** 81–89.

MacNaul, K. I., Chartrain, N., Lark, M., Tocci, M. J., and Hutchinson, N. I. (1990). Discoordinate expression of stromelysin, collagenase and tissue inhibitor of metalloproteinases-1 in rheumatoid human synovial fibroblasts. Synergistic effects of interleukin-1 and tumor necrosis factor-α on stromelysin expression. *J. Biol. Chem.* **265,** 17238–17245.

Majmudar, G., Nelson, B. R., Jensen, T. C., and Johnson, T. M. (1994). Increased expression of matrix metalloproteinase-3 (stromelysin-1) in cultured fibroblasts and basal cell carcinomas of nevoid basal cell carcinoma syndrome. *Mol. Carcinogenesis* **11,** 29–33.

Manicourt, D.-H., Fujimoto, N., Obata, K., and Thonar, E. J.-M. A. (1995). Levels of circulating collagenase, stromelysin-1, and tissue inhibitor of matrix metalloproteinases 1 in patients with rheumatoid arthritis. Relationship to serum levels of antigenic keratan sulfate and systemic parameters of inflammation. *Arthritis Rheum.* **38,** 1031–1039.

Marti, H. P., Lee, L., Kashgarian, M., and Lovett, D. H. (1994). Transforming growth factor-β 1 stimulates glomerular mesangial cell synthesis of the 72–kD type IV collagenase. *Am. J. Pathol.* **144,** 82–94.

Mast, A. E., Enghild, J. J., Nagase, H., Suzuki, K., Pizzo, S. V., and Salvesen, G. (1991). Kinetics and physiologic relevance of the inactivation of α_1-proteinase inhibitor, α_1-antichymotrypsin, and antithrombin III by matrix metalloproteinases-1 (tissue collagenase), -2 (72-kDa gelatinase/type IV collagenase), and -3 (stromelysin). *J. Biol. Chem.* **266,** 15810–15816.

Matrisian, L. M., Glaichenhaus, N., Gesnel, M. C., and Breathnach, R. (1985). Epidermal growth factor and oncogenes induce transcription of the same cellular mRNA in rat fibroblasts. *EMBO J.* **4,** 1435–1440.

Matrisian, L. M., Bowden, G. T., Krieg, P., Furstenberger, G., Briand, J. P., Leroy, P., and Breathnach, R. (1986). The mRNA coding for the secreted protease transin is expressed more abundantly in malignant than in benign tumors. *Proc. Natl. Acad. Sci. USA,* **83,** 9413–9417.

Mattot, V., Raes, M. B., Henriet, P., Eeckhout, Y., Stehelin, D., Vandenbunder, B., and Desbiens, X. (1995). Expression of interstitial collagenase is restricted to skeletal tissue during mouse embryogenesis. *J. Cell Sci.* **108,** 529–535.

Mayer, U., Mann, K., Timpl, R., and Murphy, G. (1993). Sites of nidogen cleavage by proteases involved in tissue homeostasis and remodelling. *Eur. J. Biochem.* **217,** 877–884.

McCachren, S. S. (1991). Expression of metalloproteinases and metalloproteinase inhibitor in human arthritic synovium. *Arthritis Rheum.* **34,** 1085–1093.

McDonnell, S., Navre, M., Coffey, R. J., Jr., and Matrisian, L. M. (1991). Expression and localization of the matrix metalloproteinase Pump-1 (MMP-7) in human gastric and colon carcinomas. *Mol. Carcinogenesis* **4,** 527–533.

McLaughlin, B., and Weiss, J. B. (1996). Endothelial-cell-stimulating angiogenesis factor (ESAF) activates progelatinase A (72 kDa type IV collagenase), prostromelysin 1 and procollagenase and reactivates their complexes with tissue inhibitors of metalloproteinases: a role for ESAF in non-inflammatory angiogenesis. *Biochem. J.* **317,** 739–745.

Meikle, M. C., Bord, S., Hembry, R. M., Compston, J., Croucher, P. I., and Reynolds, J. J. (1992). Human osteoblasts in culture synthesize collagenase and other matrix metalloproteinases in response to osteotropic hormones and cytokines. *J. Cell Sci.* **103,** 1093–1099.

Migita, K., Eguchi, K., Kawabe, Y., Ichinose, Y. Tsukada, T., Origuchi, T., Aoyagi, T., and Nagataki, S. (1997). Superantigen induced stromelysin production from rheumatoid synovial cells. *Biochem. Biophys. Res. Comm.* **231,** 222–226.

Mott, J. D., Khalifah, R. G., Nagase, H., Sheild, C. F., Hudson, J. K., and Hudson, B. G. (1997). Nonenzymatic glycation of type IV collagen and matrix metalloproteinase susceptibility. *Kidney Int.* **52**, 1302–1312.

Muller, D., Quantin, B., Gesnel, M.-C., Millon-Collard, R., Abecassis, J., and Breathnach, R. (1988). The collagenase gene family in humans consists of at least four members. *Biochem. J.* **253**, 187–192.

Muller, D., Breathnach, R., Engelmann, A., Millon, R., Bronner, G., Flesch, H., Dumont, P., Eber, M., and Abecassis, J. (1991). Expression of collagenase-related metalloproteinase genes in human lung or head and neck tumours. *Int. J. Cancer* **48**, 550–556.

Murphy, G., and Willenbrock, F. (1995). Tissue inhibitors of matrix metalloendopeptidases. *Methods Enzymol.* **248**, 496–510.

Murphy, G., Cockett, M. I., Ward, R. V., and Docherty, A. J. (1991a). Matrix metalloproteinase degradation of elastin, type IV collagen and proteoglycan. A quantitative comparison of the activities of 95 kDa and 72 kDa gelatinases, stromelysins-1 and -2 and punctuated metalloproteinase (PUMP). *Biochem, J.* **277**, 277–279.

Murphy, G., Houbrechts, A., Cockett, M. I., Williamson, R. A., O'Shea, M., and Docherty, A. J. P. (1991b). The N-terminal domain of tissue inhibitor of metalloproteinases retains metalloproteinase inhibitory activity. *Biochemistry* **30**, 8097–8102.

Murphy, G., Allan, J. A., Willenbrock, F., Cockett, M. I., O'Connell, J. P., and Docherty, A. J. P. (1992). The role of the C-terminal domain in collagenase and stromelysin specificity. *J. Biol. Chem.* **267**, 9612–9618.

Murphy, G. J. P., Murphy, G., and Reynolds, J. J. (1991). The origin of matrix metalloproteinases and their familial relationships. *FEBS Lett.* **289**, 4–7.

Nagase, H. (1995). Human stromelysins 1 and 2. *Methods Enzymol.* **248**, 449–470.

Nagase, H. (1996). Matrix metalloproteinases. *In* "Zinc Metalloproteases in Health and Disease" (N.M. Hooper, ed.), pp. 153–205, Taylor & Francis, London.

Nagase, H. (1997). Activation mechanisms of matrix metalloproteinases. *Biol. Chem.* **378**, 151–160.

Nagase, H., and Fields, G. B. (1996). Human matrix metalloproteinase specificity studies using collagen sequence-based synthetic peptides. *Biopolymers* **40**, 399–416.

Nagase, H., and Harris, E. D., Jr. (1983). Ovostatin: A novel proteinase inhibitor from chicken egg white. II. Mechanism of inhibition studied with collagenase and thermolysin. *J. Biol. Chem.* **258**, 7490–7498.

Nagase, H., and Woessner, J. F., Jr. (1993). Role of endogenous proteinases in the degradation of cartilage matrix. *In* "Joint Cartilage Degradation. Basic and Clinical Aspects" (J.F. Woessner, Jr., and D. S. Howell, eds.), pp. 159–185, Marcel Dekker, New York.

Nagase, H., Enghild, J. J., Suzuki, K., and Salvesen, G. (1990). Stepwise activation mechanisms of the precursor of matrix metalloproteinase 3 (stromelysin) by proteinases and (4-aminophenyl) mercuric acetate. *Biochemistry* **29**, 5783–5789.

Nagase, H., Barrett, A. J., and Woessner, J. F., Jr. (1992). Nomenclature and glossary of the matrix metalloproteinases. *In* "Matrix Metalloproteinases and Inhibitors" (H. Birkedal-Hansen, Z. Werb, H. Welgus, and H. Van Wart, eds.), pp. 421–424, Gustav Fischer, Stuttgart.

Nagase, H., Fields, C. G., and Fields, G. B. (1994). Design and characterization of a fluorogenic substrate selectively hydrolyzed by stromelysin 1 (matrix metalloproteinase-3). *J. Biol. Chem.* **269**, 20952–20957.

Nagase, H., Suzuki, K., Itoh, Y., Kan, C.-C., Gehring, M. R., Huang, W., and Brew, K. (1996). Involvement of tissue inhibitors of metalloproteinases (TIMPs) during matrix metalloproteinase activation. *Adv. Exp. Med. Biol.* **389**, 23–31.

Nakanishi, Y., Sugiura, F., Kishi, J., and Hayakawa, T. (1986). Collagenase inhibitor stimulates cleft formation during early morphogenesis of mouse salivary gland. *Dev. Biol.* **113**, 201–206.

Netzel-Arnett, S., Mallya, S. K., Nagase, H., Birkedal-Hansen, H., and Van Wart, H. E. (1991). Continuously recording fluorescent assays optimized for five human matrix metalloproteinases. *Anal. Biochem.* **195**, 86–92.

Newell, K. J., Witty, J. P., Rodgers, W. H., and Matrisian, L. M. (1994). Expression and localization of matrix-degrading metalloproteinases during colorectal tumorigenesis. *Mol. Carcinogenesis* **10**, 199–206.

Newman, K. M., Ogata, Y., Malon, A. M., Irizarry, E., Gandhi, R. H., Nagase, H., and Tilson, M. D. (1994). Identification of matrix metalloproteinases 3 (stromelysin-1) and 9 (gelatinase B) in abdominal aortic aneurysm. *Arterioscler. Thromb.* **14**, 1315–1320.

Nguyen, Q., Murphy, G., Roughley, P. J., and Mort, J. S. (1989). Degradation of proteoglycan aggregate by a cartilage metalloproteinase. Evidence for the involvement of stromelysin in the degradation of link protein heterogeneity *in situ*. *Biochem. J.* **259**, 61–67.

Nguyen, Q., Mort, J. S., and Roughley, P. J. (1992). Preferential mRNA expression of prostromelysin relative to procollagenase and *in situ* localization in human articular cartilage. *J. Clin. Invest.* **89**, 1189–1197.

Nguyen, Q., Murphy, G., Hughes, C., Mort, J. S., and Roughley, P. (1993). Matrix metalloproteinases cleave at two distinct sites on human cartilage link protein. *Biochem. J.* **295**, 595–598.

Nguyen, Q., Willenbrock, F., Cockett, M. I., O'Shea, M., Docherty, A. J. P., and Murphy, G. (1994). Different domain interactions are involved in the binding of tissue inhibitors of metalloproteinases to stromelysin-1 and gelatinase A. *Biochemistry* **33**, 2089–2095.

Nicholson, R., Murphy, G., and Breathnach, R. (1989). Human and rat malignant-tumor-associated mRNAs encode stromelysin-like metalloproteinases. *Biochemistry*, **28**, 5195–5203.

Nicholson, R. C, Mader, S., Nagpal, S., Leid, M., Rochette-Egly, C., and Chambon, P. (1990). Negative regulation of the rat stromelysin gene promoter by retinoic acid is mediated by an AP1 binding site. *EMBO J.* **9**, 4443–4454.

Niedzwiecki, L., Teahan, J., Harrison, R. K., and Stein, R. L. (1992). Substrate specificity of the human matrix metalloproteinase stromelysin and the development of continuous fluorometric assays. *Biochemistry* **31**, 12618–12623.

Nikkari, S. T., O'Brien, K. D., Ferguson, M., Hatsukami, T., Welgus, H. G., Alpers, C. E., and Clowes, A. W. (1995). Interstitial collagenase (MMP-1) expression in human carotid atherosclerosis. *Circulation* **92**, 1393–1398.

Ogata, Y., Enghild, J. J., and Nagase, H. (1992). Matrix metalloproteinase 3 (stromelysin) activates the precursor for the human matrix metalloproteinase 9. *J. Biol. Chem.* **267**, 3581–3584.

Ohuchida, M., Sasaguri, Y., Morimatsu, M., Nagase, H., and Yagi, K. (1991). Effect of linoleic acid hydroperoxide on production of matrix metalloproteinases by human skin fibroblasts. *Biochem. Int.* **25**, 447–452.

Okada, Y., Nagase, H., and Harris, E. D., Jr. (1986). A metalloproteinase from human rheumatoid synovial fibroblasts that digests connective tissue matrix components. Purification and characterization. *J. Biol. Chem.* **261**, 14245–14255.

Okada, Y., Harris, E. D., Jr., and Nagase, H. (1988). The precursor of a metalloendopeptidase from human rheumatoid synovial fibroblasts. Purification and mechanisms of activation by endopeptidases and 4-aminophenylmercuric acetate. *Biochem. J.* **254**, 731–741.

Okada, Y., Takeuchi, N., Tomita, K., Nakanishi, I., and Nagase, H. (1989). Immunolocalisation of matrix metalloproteinase 3 (stromelysin) in rheumatoid synovioblasts (B cells): Correlation with rheumatoid arthritis. *Ann. Rheum. Dis.* **48**, 645–653.

Okada, Y., Gonoji, Y., Nakanishi, I., Nagase, H., and Hayakawa, T. (1990). Immunohistochemical demonstration of collagenase and tissue inhibitor of metalloproteinases

(TIMP) in synovial lining cells of rheumatoid synovium. *Virchows Arch. B.* **59,** 305–312.

Okada, Y., Gonoji, Y., Naka, K., Tomita, K., Nakanishi, I., Iwata, K., Yamashita, K., and Hayakawa, T. (1992a). Matrix metalloproteinase 9 (92–kDa gelatinase/type IV collagenase) from HT 1080 human fibrosarcoma cells. Purification and activation of the precursor and enzymic properties. *J. Biol. Chem.* **267,** 21712–21719.

Okada, Y., Shinmei, M., Tanaka, O., Naka, K., Kimura, A., Nakanishi, I., Bayliss, M. T., Iwata, K., and Nagase, H. (1992b). Localization of matrix metalloproteinase 3 (stromelysin) in osteoarthritic cartilage and synovium. *Lab. Invest.* **66,** 680–690.

Osteen, K. G., Rodgers, W. H., Gaire, M., Hargrove, J. T., Gorstein, F., and Matrisian, L. M. (1994). Stromal-epithelial interaction mediates steroidal regulation of metalloproteinase expression in human endometrium. *Proc. Natl. Acad. Sci. USA* **91,** 10129–10133.

Otani, Y., Quinones, S., Saus, J., Kurkinen, M., and Harris, E. D., Jr. (1990). Cycloheximide induces stromelysin mRNA in cultured human fibroblasts. *Eur. J. Biochem.* **192,** 75–79.

Patterton, D., Hayes, W. P., and Shi, Y. B. (1995). Transcriptional activation of the matrix metalloproteinase gene stromelysin-3 coincides with thyroid hormone-induced cell death during frog metamorphosis. *Dev. Biol.* **167,** 252–262.

Perides, G., Asher, R. A., Lark, M. W., Lane, W. S., Robinson, R. A., and Bignami, A. (1995). Glial hyaluronate-binding protein: A product of metalloproteinase digestion of versican? *Biochem. J.* **312,** 377–384.

Prontera, C., Crescenzi, G., and Rotilio, D. (1996). Inhibition by interleukin-4 of stromelysin expression in human skin fibroblasts: role of PKC. *Exp. Cell Res.* **224,** 183–188.

Qin, X., Chua, P. K., Ohira, R. H., and Bryant-Greenwood, G. D. (1997). An autocrine/paracrine role of human decidual relaxin. II. Stromelysin-1 (MMP-3) and tissue inhibitor matrix metalloproteinase-1 (TIMP-1). *Biol. Reprod.* **56,** 812–820.

Quinones, S., Buttice, G., and Kurkinen, M. (1994). Promoter elements in the transcriptional activation of the human stromelysin-1 gene by the inflammatory cytokine, interleukin 1. *Biochem. J.* **302,** 471–477.

Reinemer, P., Grams, F., Huber, R., Kleine, T., Schnierer, S., Piper, M., Tschesche, H., and Bode, W. (1994). Structural implications for the role of the N terminus in the 'superactivation' of collagenases. A crystallographic study. *FEBS Lett.* **338,** 227–233.

Reponen, P., Sahlberg, C., Munaut, C., Thesleff, I., and Tryggvason, K. (1994). High expression of 92–kD type IV collagenase (gelatinase B) in the osteoclast lineage during mouse development. *J. Cell. Biol.* **124,** 1091–1102.

Rodgers, W. H., Osteen, K. G., Matrisian, L. M., Naure, M., Guidice, L. C., and Gorstein, F. (1993). Expression and localization of matrilysin, a matrix metalloproteinase, in human endometrium during the reproductive cycle. *Am. J. Obstet. Gynecol.* **168,** 253–260.

Rodgers, W. H., Matrisian, L. M., Giudice, L. C., Dsupin, B., Cannon, P., Svitek, C., Gorstein, F., and Osteen, K. G. (1994). Patterns of matrix metalloproteinase expression in cycling endometrium imply differential functions and regulation by steroid hormones. *J. Clin. Invest.* **94,** 946–953.

Saarialho-Kere, U. K., Chang, E. S., Welgus, H. G., and Parks, W. C. (1993a). Expression of interstitial collagenase, 92–kDa gelatinase, and tissue inhibitor of metalloproteinases-1 in granuloma annulare and necrobiosis lipoidica diabeticorum. *J. Invest. Dermatol.* **100,** 335–342.

Saarialho-Kere, U. K., Kovacs, S. O., Pentland, A. P., Olerud, J. E., Welgus, H. G., and Parks, W. G. (1993b). Cell-matrix interactions modulate interstitial collagenase expression by human keratinocytes actively involved in wound healing. *J. Clin. Invest.* **92,** 2858–2866.

Saarialho-Kere, U. K., Pentland, A. P., Birkedal-Hansen, H., Parks, W. C., and Welgus, H. G. (1994). Distinct populations of basal keratinocytes express stromelysin-1 and stromelysin-2 in chronic wounds. *J. Clin. Invest.* **94,** 79–88.

Saarialho-Kere, U. K., Vaalamo, M., Puolakkainen, P., Airola, K., Parks, W. G., and Karjalainen-Lindsberg, M.-L. (1996). Enhanced expression of matrilysin, collagenase, and stromelysin-1 in gastrointestinal ulcers. *Am. J. Pathol.* **148,** 519–526.

Salowe, S. P., Marcy, A. I., Cuca, G. C., Smith, C. K., Kopka, I. E., Hagmann, W. K., and Hermes, J. D. (1992). Characterization of zinc-binding sites in human stromelysin-1: Stoichiometry of the catalytic domain and identification of a cysteine ligand in the proenzyme. *Biochemistry* **31,** 4535–4540.

Samples, J. R., Alexander, J. P., and Acott, T. S. (1993). Regulation of the levels of human trabecular matrix metalloproteinases and inhibitor by interleukin-1 and dexamethasone. *Invest. Ophthalmol. Vis. Sci.* **34,** 3386–3395.

Sandy, J. D., Neame, P. J., Boynton, R. E., and Flannery, C. R. (1991). Catabolism of aggrecan in cartilage explants. Identification of a major cleavage site within the interglobular domain. *J. Biol. Chem.* **266,** 8683–8685.

Sandy, J. D., Flannery, C. R., Neame, P. J., and Lohmander, L. S. (1992). The structure of aggrecan fragments in human synovial fluid. Evidence for the involvement in osteoarthritis of a novel proteinase which cleaves the Glu^{373}–Ala^{374} bond of the interglobular domain. *J. Clin. Invest.* **89,** 1512–1516.

Sanz, L., Berra, E., Municio, M. M., Dominguez, I., Lozano, J., Johansen, T., Mosca, J., and Diaz-Meco, M. T. (1994). ζ PKC plays a critical role during stromelysin promoter activation by platelet-derived growth factor through a novel palindromic element. *J. Biol. Chem.* **269,** 10044–10049.

Sapolsky, A. I., Howell, D. S., and Woessner, J. F., Jr. (1974). Natural proteases and cathepsin D in human articular cartilage. *J. Clin. Invest.* **53,** 1044–1053.

Sasaki, T., Mann, K., Murphy, G., Chu, M.-L., and Timpl, R. (1996). Different susceptibilities of fibulin-1 and fibulin-2 to cleavage by matrix metalloproteinases and other tissue proteases. *Eur. J. Biochem.* **240,** 427–434.

Sasaki, T., Göhring, W., Mann, K., Maurer, P., Hohenester, E., Knäuper, V., Murphy, G., and Timpl, R. (1997). Limited cleavage of extracellular matrix protein BM-40 by matrix metalloproteinases increases its affinity for collagens. *J. Biol. Chem.* **272,** 9237–9243.

Sato, H., and Seiki, M. (1993). Regulatory mechanism of 92 kDa type IV collagenase gene expression which is associated with invasiveness of tumor cells. *Oncogene* **8,** 395–405.

Sato, T., del Carmen Ovejero, M., Hou, P., Heegaard, A.-M., Kumegawa, M., Foged, N. T., and Delaissé, J.-M. (1997). Identification of the membrane-type matrix metalloproteinase MT1-MMP in osteoclasts. *J. Cell Sci.* **110,** 589–596.

Saus, J., Quinones, S., Otani, Y., Nagase, H., Harris, E. D., Jr., and Kurkinen, M. (1988). The complete primary structure of human matrix metalloproteinase-3. Identity with stromelysin. *J. Biol Chem.* **263,** 6742–6745.

Schmitz, J. P., Schwartz, Z., Sylvia, V. L., Dean, D. D., Calderon, F., and Boyan, B. D. (1996). Vitamin D_3 regulation of stromelysin-1 (MMP-3) in chondrocyte cultures is mediated by protein kinase C. *J. Cell. Physiol.* **168,** 570–579.

Schönthal, A., Herrlich, P., Rahmsdorf, H. J., and Ponta, H. (1988). Requirement for *fos* gene expression in the transcriptional activation of collagenase by other oncogenes and phorbol esters. *Cell* **54,** 325–334.

Schönermark, M., Mester, B., Kempf, H.-G., Bläser, J., Tschesche, H., and Lenarz, T. (1996). Expression of matrix-metalloproteinases and their inhibitors in human cholesteatomas. *Acta Otolaryngol.* **116,** 451–456.

Schroen, D. J., and Brinckerhoff, C. E. (1996). Inhibition of rabbit collagenase (matrix metalloproteinase-1; MMP-1) transcription by retinoid receptors: Evidence for binding of RARs/RXRs to the -77 AP-1 site through interactions with c-Jun. *J. Cell. Physiol.* **169**, 320–332.

Schüle, R., Rangarajan, P., Kliewer, S., Ransome, L. J., Bolado, J., Yang, N., Verma, I. M., and Evans, R. M. (1990). Functional antagonism between oncoprotein c-Jun and the glucocorticoid receptor. *Cell* **62**, 1217–1226.

Schüle, R., Rangarajan, P., Yang, N., Kliewer, S., Ransone, L. J., Bolado, J., Verma, I. M., and Evans, R. M. (1991). Retinoic acid is a negative regulator of AP-1-responsive genes. *Proc. Natl. Acad. Sci. USA.* **88**, 6092–6096.

Shapiro, S. D., Fliszar, C. J., Broekelmann, T. J., Mecham, R. P., Senior, R. B., and Welgus, H. G. (1995). Activation of the 92–kDa gelatinase by stromelysin and 4-aminophenylmercuric acetate. *J. Biol. Chem.* **270**, 6351–6356.

Shima, I., Sasaguri, Y., Kusukawa, J., Yamana, H., Fujita, H., Kakegawa, T., and Morimatsu, M. (1992). Production of matrix metalloproteinase-2 and metalloproteinase-3 related to malignant behavior of esophageal carcinoma. A clinicopathologic study. *Cancer* **70**, 2747–2753.

Siri, A., Knäuper, V., Veirana, N., Caocci, F., Murphy, G., and Zardi, L. (1995). Different susceptibility of small and large human tenascin-C isoforms to degradation by matrix metalloproteinases. *J. Biol. Chem.* **270**, 8650–8654.

Smith, M. M., Shi, L., and Navre, M. (1995). Rapid identification of highly active and selective substrates for stromelysin and matrilysin using bacteriophage peptide display libraries. *J. Biol. Chem.* **270**, 6440–6449.

Sottrup-Jensen, L. (1989). α-macroglobulins: Structure, shape and mechanism of proteinase complex formation. *J. Biol. Chem.* **264**, 11539–11542.

Sottrup-Jensen, L., and Birkedal-Hansen, H. (1989). Human fibroblast collagenase-α-macroglobulin interactions. Localization of cleavage sites in the bait regions of five mammalian α-macroglobulins. *J. Biol. Chem.* **264**, 393–401.

Springman, E. B., Angleton, E. L., Birkedal-Hansen, H., and Van Wart, H. E. (1990). Multiple modes of activation of latent human fibroblast collagenase: Evidence for the role of a Cys73 active- site zinc complex in latency and a "cysteine-switch" mechanism for activation. *Proc. Natl. Acad. Sci. USA* **87**, 364–368.

Sreenath, T., Matrisian, L. M., Stetler-Stevenson, W., Gattoni-Celli, S., and Pozzatti, R. O. (1992). Expression of matrix metalloproteinase genes in transformed rat cell lines of high and low metastatic potential. *Cancer Res.* **52**, 4942–4947.

Stetler-Stevenson, W. G., Aznavoorian, S., and Liotta, L. A. (1993). Tumor cell interactions with the extracellular matrix during invasion and metastasis. *Ann. Rev. Cell Biol.* **9**, 541–573.

Stöcker, W., Grams, F., Baumann, U., Reinemer, P., Gomis-Rüth, F. -X., McKay, D. B., and Bode, W. (1995). The metzincins—Topological and sequential relations between the astacins, adamalysins, serralysins, and matrixins (collagenases) define a superfamily of zinc-peptidases. *Protein Sci.* **4**, 823–840.

Suzuki, K., Enghild, J. J., Morodomi, T., Salvesen, G., and Nagase, H. (1990). Mechanisms of activation of tissue procollagenase by matrix metalloproteinase 3 (stromelysin). *Biochemistry* **29**, 10261–10270.

Suzuki, K., Nagase, H., Murphy, G., and Docherty, A. J. P. (1997). Human stromelysin 2 (matrix metalloproteinase 10). Expression and activation of the recombinant precursor and enzymic properties. In preparation.

Sympson, C. J., Talhouk, R. S., Alexander, C. M., Chin, J. R., Clift, S. M., Bissell, M. J., and Werb, Z. (1994). Targeted expression of stromelysin-1 in mammary gland provides evidence for a role of proteinases in branching morphogenesis and the requirement for an intact basement membrane for tissue-specific gene expression. *J. Cell Biol.* **125**, 681–693.

Talhouk, R. S., Bissell, M. J., and Werb, Z. (1992). Coordinated expression of extracellular matrix-degrading proteinases and their inhibitors regulates mammary epithelial function during involution. *J. Cell Biol.* **118,** 1271–1282.

Taylor, D. J., Cheung, N. T., and Dawes, P. T. (1994). Increased serum proMMP-3 in inflammatory arthritides: A potential indicator of synovial inflammatory monokine activity. *Ann. Rheum. Dis.* **53,** 768–772.

Thompson, R. W., Holmes, D. R., Mertens, R. A., Liao, S., Botney, M. D., Mecham, R. P., Welgus, H. G., and Parks, W. C. (1995). Production and localization of 92–kilodalton gelatinase in abdominal aortic aneurysms. An elastolytic metalloproteinase expressed by aneurysm-infiltrating macrophages. *J. Clin. Invest.* **96,** 318–326.

Treadwell, B. V., Neidel, J., Pavia, M., Towle, C. A., Trice, M. E., and Mankin, H. J. (1986). Purification and characterization of collagenase activator protein synthesized by articular cartilage. *Arch. Biochem. Biophys.* **251,** 715–723.

Unemori, E. N., Bair, M. J., Bauer, E. A., and Amento, E. P. (1991). Stromelysin expression regulates collagenase activation in human fibroblasts. Dissociable control of two metalloproteinases by interferon-γ. *J. Biol. Chem.* **266,** 23477–23482.

Van Doren, S. R., Kurochkin, A. V., Hu, W., Ye, Q. Z., Johnson, L. L., Hupe, D. J., and Zuiderweg, E. R. (1995). Solution structure of the catalytic domain of human stromelysin complexed with a hydrophobic inhibitor. *Protein Sci.* **4,** 2487–2498.

Van Wart, H. E., and Birkedal-Hansen, H. (1990). The cysteine switch: A principle of regulation of metalloproteinase activity with potential applicability to the entire matrix metalloproteinase gene family. *Proc. Natl. Acad. Sci. U. S.A.* **87,** 5578–5582.

Vater, C. A., Nagase, H., and Harris, E. D., Jr. (1983). Purification of an endogenous activator of procollagenase from rabbit synovial fibroblast culture medium. *J. Biol. Chem.* **258,** 9374–9382.

Vettraino, I. M., Roby, J., Tolley, T., and Parks, W. C. (1996). Collagenase-1, stromelysin-1, and matrilysin are expressed within the placenta during multiple stages of human pregnancy. *Placenta* **17,** 557–563.

Vyas, S. K., Leyland, H., Gentry, J., and Arthur, M. J. P. (1995). Rat hepatic lipocytes synthesize and secrete transin (stromelysin) in early primary culture. *Gastroenterology* **109,** 889–898.

Walakovits, L. A., Moore, V. L., Bhardwaj, N., Gallick, G. S., and Lark, M. W. (1992). Detection of stromelysin and collagenase in synovial fluid from patients with rheumatoid arthritis and posttraumatic knee injury. *Arthritis Rheum.* **35,** 35–42.

Ward, R. V., Hembry, R. H., Reynolds, J. J., and Murphy, G. (1991). The purification of tissue inhibitor of metalloproteinases-2 from its 72 kDa progelatinase complex. *Biochem. J.* **278,** 179–187.

Wasylyk, C., Gutman, A., Nicholson, R., and Wasylyk, B. (1991). The c-Ets oncoprotein activates the stromelysin promoter through the same elements as several non-nuclear oncoproteins. *EMBO. J.* **10,** 1127–1134.

Werb, Z., and Reynolds, J. J. (1974). Stimulation of endocytosis of the secretion of collagenase and neutral proteinase from rabbit synovial fibroblasts. *J. Exp. Med.* **140,** 1482–1497.

Wetmore, D. R., and Hardman, K. D. (1996). Roles of the propeptide and metal ions in the folding and stability of the catalytic domain of stromelysin (matrix metalloproteinase 3). *Biochemistry* **35,** 6549–6558.

Whitelock, J. M., Murdoch, A. D., Iozzo, R. V., and Underwood, P. A. (1996). The degradation of human endothelial cell-derived perlecan and release of bound basic fibroblast growth factor by stromelysin, collagenase, plasmin, and heparanases. *J. Biol. Chem.* **271,** 10079–10086.

Whitham, S. E., Murphy, G., Angel, P., Rahmsdorf, H.-J., Smith, B. J., Lyons, A., Harris, T. J. R., Reynolds, J. J., Herrlich, P., and Docherty, A. J. P. (1986). Comparison of human stromelysin and collagenase by cloning and sequence analysis. *Biochem. J.* **240,** 913–916.

Wilhelm, S. M., Collier, I. E., Kronberger, A., Eisen, A. Z., Marmer, B. L., Grant, G. A., Bauer, E. A., and Goldberg, G. I. (1987). Human skin fibroblast stromelysin: Structure, glycosylation, substrate specificity, and differential expression in normal and tumorigenic cells. *Proc. Natl. Acad. Sci. USA.,* **84,** 6725–6729.

Wilhelm, S. M., Shao, Z.-H., Housley, T. J., Seperack, P. K., Baumann, A. P., Gunja-Smith, Z., and Woessner, J. F., Jr. (1993). Matrix metalloproteinase-3 (stromelysin-1). Identification as the cartilage acid metalloprotease and effect of pH on catalytic properties and calcium affinity. *J. Biol. Chem.* **268,** 21906–21913.

Williamson, R. A., Marston, F. A., Angal, S., Koklitis, P., Panico, M., Morris, H. R., Carne, A. F., Smith, B. J., Harris, T. J., and Freedman, R. B. (1990). Disulphide bond assignment in human tissue inhibitor of metalloproteinases (TIMP). *Biochem. J.* **268,** 267–274.

Williamson, R. A., Martorell, G., Carr, M. D., Murphy, G., Docherty, A. J. P., Freedman, R. B., and Feeney, J. (1994). Solution structure of the active domain of tissue inhibitor of metalloproteinase-2. A new member of the OB fold protein family. *Biochemistry* **33,** 11745–11759.

Windsor, L. J., Grenett, H., Birkedal-Hansen, B., Bodden, M. K., Engler, J. A., and Birkedal-Hansen, H. (1993). Cell type-specific regulation of SL-1 and SL-2 genes. Induction of the SL-2 gene but not the SL-1 gene by human keratinocytes in response to cytokines and phorbolesters. *J. Biol. Chem.* **268,** 17341–17347.

Witty, J. P., Lempka, T., Coffey, R. J., Jr., and Matrisian, L. M. (1995a). Decreased tumor formation in 7,12-dimethylbenzanthracene-treated stromelysin-transgenic mice is associated with alterations in mammary epithelial cell apoptosis. *Cancer Res.* **55,** 1401–1406.

Witty, J. P., Wright, J. H., and Matrisian, L. M. (1995b). Matrix metalloproteinases are expressed during ductal and alveolar mammary morphogenesis, and misregulation of stromelysin-1 in transgenic mice induces unscheduled alveolar development. *Mol. Biol. Cell* **6,** 1287–1303.

Witty, J. P., Foster, S. A., Stricklin, G. P., Martisian, L. M., and Stern, P. H. (1996). Parathyroid hormone-induced resorption in fetal rat limb bones is associated with production of the metalloproteinases collagenase and gelatinase B. *J. Bone Mineral Res.* **11,** 72–78.

Woessner, J. F., Jr. (1991). Matrix metalloproteinases and their inhibitors in connective tissue remodeling. *FASEB J.* **5,** 2145–2154.

Wolfe, G. C., MacNaul, K. L., Buechel, F. F., McDonnell, J., Noerrner, L. A., Lark, M. W., Moore, V. L., and Hutchinson, N. I., (1993). Differential *in vivo* expression of collagenase messenger RNA in synovium and cartilage. Quantitative comparison with stromelysin messenger RNA levels in human rheumatoid arthritis and osteoarthritis patients and in two animal models of acute inflammatory arthritis. *Arthritis Rheum.* **36,** 1540–1547.

Wright, J. H., McDonnell, S., Portella, G., Bowden, G. T., Balmain, A., and Matrisian, L. M. (1994). A switch from stromal to tumor cell expression of stromelysin-1 mRNA associated with the conversion of squamous to spindle carcinomas during mouse skin tumor progression. *Mol. Carcinog.* **10,** 207–215.

Wu, J. J., Lark, M. W., Chun, L. E., and Eyre, D. R. (1991). Sites of stromelysin cleavage in collagen types II, IX, X and XI of cartilage. *J. Biol. Chem.* **266,** 5625–5628.

Xie, D.-l., Hui, F., Meyers, R., and Homandberg, G. A. (1994). Cartilage chondrolysis by fibronectin fragments is associated with release of several proteinases: Stromelysin plays a major role in chondrolysis. *Arch. Biochem. Biophys.* **311,** 205–212.

Yang-Yen, H.-F., Chambard, J. C., Sun, Y. L., Smeal, T., Schmidt, T. J., Drouin, J., and Karin, M. (1990). Transcriptional interference between c-Jun and the gluctocorticoid receptor: Mutual inhibition of DNA binding due to direct protein-protein interaction. *Cell* **62,** 1205–1215.

Ye, S., Watts, G. F., Mandalia, S., Humphries, S. E., and Henney, A. M. (1995). Preliminary report: Genetic variation in the human stromelysin promoter is associated with progression of coronary atherosclerosis. *Br. Heart J.* **73,** 209–215.

Ye, S., Eriksson, P., Hamsten, A., Kurkinen, M., Humphries, S. E., and Henney, A. M. (1996). Progression of coronary atherosclerosis is associated with a common genetic variant of the human stromelysin-1 promoter which results in reduced gene expression. *J. Biol. Chem.* **271,** 13055–13060.

Yoshihara, Y., Obata, K., Fujimoto, N., Yamashita, K., Hayakawa, T., and Shimmei, M. (1995). Increased levels of stromelysin-1 and tissue inhibitor of metalloproteinases-1 in sera from patients with rheumatoid arthritis. *Arthritis Rheum.* **38,** 969–975.

72-kDa Gelatinase (Gelatinase A): Structure, Activation, Regulation, and Substrate Specificity

Anita E. Yu, Anne N. Murphy,*
and William G. Stetler-Stevenson

*Extracellular Matrix Pathology Section, Laboratory of Pathology, National Cancer Institute, National Institutes of Health, Bethesda, Maryland 20892; and *Department of Biochemistry, George Washington University Medical School, Washington, DC 20037*

I. Introduction

The 72-kDa gelatinase (also known as matrix metalloproteinase-2, MMP-2, gelatinase A, or 72-kDa type IV collagenase; EC 3.4.24.24) was first described by Liotta *et al.* (1979) who found that an enzyme secreted by a metastatic murine tumor degraded soluble type IV collagen. This enzyme apparently did not attack types I, II, and III collagen and cleaved type IV collagen to produce 1/4 N-terminal and

Matrix Metalloproteinases

3/4 C-terminal fragments consistent with a single cleavage site. In 1988 (Collier *et al.,* 1988) a cDNA clone for this type IV collagenase was reported by Greg Goldberg and colleagues, and the zymogen form was found to have a molecular weight of 72 kDa. Now known as gelatinase A or MMP-2, this protease was the second member of a growing family of metalloenzymes known as *matrix metalloproteinases* (MMPs) or *matrixins* (see Chapter 1 for a detailed description of the matrixin family). This family currently consists of at least 20 members. Like most MMPs, gelatinase A is secreted from cells as a zymogen (progelatinase A) and its activity in the extracellular environment is controlled by various activators and the endogeneous tisssue inhibitors of metalloproteinases (TIMPs).

Gelatinase A is unique in that it is constitutively expressed by many cells, has a ubiquitous tissue distribution, and has a cell surface mode of activation that differs from other members of the family. Furthermore, unlike other members in this family, progelatinase A and progelatinase B (92-kDa gelatinase) are usually isolated in complex with TIMP-2 and TIMP-1, respectively. Progelatinase A selectively binds to TIMP-2 but not to TIMP-1. This progelatinase A–TIMP-2 interaction may mediate the cell surface activation of gelatinase A. Also, recent findings suggest that MMP-2 activity is involved in initiating or regulating cell responses such as proliferation, adhesion, and migration. This extends our concept of MMP functions beyond simple removal of extracellular matrix (ECM) barriers that limit cell movement. MMP modification of the ECM may initiate a host of cellular events that occur in response to specific proteolytic fragments that are generated and/or the selective alteration of cell–ECM interactions. In this respect the action of MMPs on the ECM may be viewed as the initiation of a signaling cascade, similar to intracellular proteolytic events involved in regulation of a variety of cellular processes, such as programmed cell death [caspases, interleukin-1 converting enzyme (ICE)-like proteases (Chinnaiyan *et al.,* 1997)].

II. Matrix Metalloproteinase Family

A. *Overview*

These enzymes are collectively able to cleave most classes of ECM proteins in their native forms. There are currently 20 members (see Chapter 1) that have been classified into six subgroups based on their putative substrate specificity and internal homologies: interstitial collagenases; gelatinases (type IV collagenases); stromelysins; metalloelastases; secreted RXKR containing MMPs (stromelysin-3-like); and the membrane-type MMPs (MT-MMPs). These latter two groups are distinguished by the insertion of a furin cleavage site, RXKR, between the classic prodomain and the catalytic domain (Pei and Weiss, 1995). This cleavage site has been implicated in the intracellular activation of stromelysin-3, and may also be important in the activation of MT-MMPs (Pei and Weiss, 1995).

B. Domain Structure of Gelatinase A

The members of the MMP family possess different substrate specificities, but all show sequence homology that defines the following shared domains: a signal sequence, a pro-domain, a zinc-binding domain containing the catalytic site, and a hemopexin-like carboxyl-terminal domain (Fig. 1). This last domain is not required for catalytic activity, as demonstrated by the activity of matrilysin, which is the smallest member of the MMP family and does not contain a hemopexin-like domain. In addition to these shared domains, gelatinase A, like gelatinase B, has an additional fibronectin-like gelatin-binding domain that

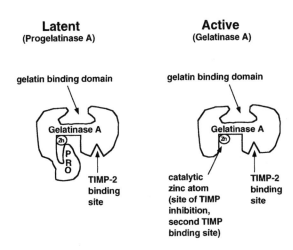

FIG. 1. The domain structure of progelatinase A and activated enzyme is illustrated in this schematic. Progelatinase A has three essential domains: the pro-domain with a conserved Cys residue, which coordinates to Zn atom in the active site to maintain enzyme latency; the gelatin-binding domain, which is homologous to the gelatin-binding domain of fibronectin; and the TIMP-2 binding site in the C terminus of the enzyme. Through this site the progelatinase A forms a tight complex with TIMP-2. This complex, as well as TIMP-2 free progelatinase A, can be activated through removal of the profragment to generate active enzyme or active enzyme–TIMP-2 complex. However, the gelatinase A–TIMP-2 complex retains 10% of the proteolytic activity of the TIMP-2 free enzyme. A second TIMP-2 binding site becomes available upon cleavage of the profragment when the enzyme becomes activated. This site is not specific for TIMP-2 and the active enzyme is inhibited by all TIMPs.

contributes to the substrate specificity of these enzymes (Banyai *et al.,* 1994; Strongin *et al.,* 1993) (Fig. 1). The recently described MT-MMPs have been shown to be membrane associated via a transmembrane domain at the carboxy-terminal end of the molecule (Sato *et al.,* 1994). Further details regarding the structural domains and substrate preferences of the MMP family members, including gelatinase A, can be found in other chapters of this volume and several excellent reviews (Birkedal-Hansen *et al.,* 1993; MacDougall and Matrisian, 1995; Matrisian, 1992).

III. MULTILEVEL REGULATION OF GELATINASE A

Synthesis of gelatinase A appears to be regulated at both the transcriptional and post-transcriptional levels in fibroblasts studied *in vitro.* However, the constitutive, high levels of gelatinase A mRNA expression seen in many tissues suggest that regulation of proenzyme activation is more relevant for the control and induction of gelatinase A action than other members of the matrixin family in which transcriptional control is more important. Following activation, enzyme activity is then down-regulated by TIMPs.

A. *Transcriptional Regulation*

Unlike other MMPs, progelatinase A is constitutively expressed by many cell types, and its transcription is not readily induced by agents such as 12-*O*-tetradecanoylphorbol-13-acetate (TPA) or interleukin 1α (IL-1a) which have been shown to increase transcription of other MMPs (i.e., interstitial collagenase and gelatinase B) by as much as 100-fold (Brown *et al.,* 1990; Collier *et al.,* 1988; Overall and Sodek, 1990). Inducers of MMP expression are generally thought to act at the level of transcriptional activation of the gene and promoter elements. Analysis of the gelatinase A promoter regions shows marked differences between this protease and other MMPs (Matrisian, 1994) (Fig. 2). The gelatinase A promotor is distinguished by the lack of known transactivator sequences (AP-1, PEA-3), the presence of a unique noncanonical TATA box, as well as a putative enhancer element located at nucleotides −223 to −422 relative to the translational start site (Huhtala *et al.,* 1990; Templeton and Stetler-Stevenson, 1991). The TATA site is usually located at ∼−30 bp from the transcriptional start site and plays a pivotal role in the transcriptional regulation of most promotors. Other distinctive features of the gelatinase A promotor include an SP-1 consensus sequence located 120 bp from the start site and an adenovirus E1A repressor element (Frisch *et al.,* 1990). Unlike other MMPs, the gelatinase A promoter also lacks an upstream TIE (Matrisian, 1994).

a) Gelatinase A promoter

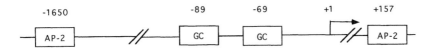

b) Generalized MMP promoter

Fɪɢ. 2. Comparison of the promoters of 72-kDa gelatinase and other MMPs. TATA box, activator protein 1 (AP-1), AP-2 sites, polyomavirus enhancer A-binding protein 3 (PEA-3) elements and transforming growth factor β (TGF-β) inhibitory elements (TIE) are indicated. Transcription initiation sites are indicated by bent arrows. Nucleotide positions are shown above each element. *TIE-like sequences are present for gelatinase B (-474), collagenase (-245), and matrilysin (-475, -500, and -820). **PEA-3 elements are present for collagenase (-89), stromelysin-1 (-208, -216), stromelysin-2 (-181), and matrilysin (-475, -500, and -820), but not for gelatinase B.

Consistent with this sequence information is the observation that gelatinase A transcription is not suppressed but slightly upregulated by TGF-β (Brown *et al.*, 1990; Overall *et al.*, 1991). The presence of the AP-2 element (5'-GCCCCAGGC-3') contributes to cell-type specific expression of at least two MMPs: gelatinase A and gelatinase B (Benbow and Brinckerhoff, 1997). An AP-2 enhancer element (-1650 bp) is indispensable for transcription of the human gelatinase A gene in HT1080 cells (Frisch *et al.*, 1990). Similarly, in the rabbit gelatinase B promotor, the AP-1 site at -56 bp and an AP-2 like element located between -432 and -330 bp have been shown to be essential for PMA inducibility in primary epithelial cells (Fini *et al.*, 1994). The unique transcriptional regulation of gelatinase A, and the fact that gelatinase A is constitutively expressed and not well regulated, has led to the suggestion that gelatinase A acts as a "housekeeping" gene (Overall *et al.*, 1991). However, this incorrectly implies that transcriptional activation of the gelatinase A gene results in a direct increase in gelatinase A activity. The proenzyme form of gelatinase A is stable and does not undergo spontaneous activation under physiologic conditions

(Kleiner *et al.*, 1993). The cellular regulation of progelatinase A activation plays a pivotal role in the initiation of ECM turnover.

B. Post-Transcriptional Regulation

There is evidence for post-transcriptional regulation of gelatinase A expression (Overall *et al.*, 1991). While TGF-β_1 suppresses overall proteolytic activity through reduced proteinase synthesis and by increased TIMP expresssion, it increases the levels of gelatinase A secreted by human fibroblasts and rat bone cells (Overall *et al.*, 1991). The total cellular abundance of gelatinase A mRNA was found to increase from between 1.5- to 2.2-fold, an amount similar to the increase seen in the synthesis of gelatinase A protein. In human fibroblasts treated with TGF-β_1, the half-life of gelatinase A mRNA increased from 46 to 150 h, but the stability of TIMP-1 mRNA was unaffected (Overall *et al.*, 1991). This approximately 3-fold increase in the stability of the gelatinase A mRNA induced by TGF-β_1 would augment the elevated levels of gelatinase A mRNA produced by increased gelatinase A transcription. These data indicate that TGF-β1 acts both transcriptionally and post-transcriptionally to elevate gelatinase A expression; however, the mechanism of these effects has not been studied in detail.

C. Regulation of Activity by TIMPs

Like many members of the matrixin family, gelatinase A is secreted as latent proenzyme and must be activated extracellularly. However, unlike most other members of the MMP family, progelatinase A is usually found complexed with the endogenous inhibitor TIMP-2 (Fig. 3). While, in general, active MMPs are inhibited by TIMPs binding to the active site, progelatinase A interacts selectively to TIMP-2, but not TIMP-1 (Goldberg *et al.*, 1989, 1992; Stetler-Stevenson *et al.*, 1989a; Wilhelm *et al.*, 1989). By deletion mutagenesis, it has been shown that interactions between the C-terminal domains of TIMPs and MMPs influence the rate of complex formation. Ionic interactions involving the negatively charged C terminus of TIMP-2, are thought to be important in the rapid binding to the C terminus of progelatinase A (Willenbrock *et al.*, 1993). Evidence from studies involving truncated enzymes and cross-linking methods suggests the following model: The N-terminal domain of TIMP is inhibitory, while the C-terminal domain confers binding specificity. So, while the N-terminal domain of TIMP-1 and TIMP-2 bind to the N-terminal domain of active gelatinase A, only the C-terminal domain of TIMP-2 will bind specifically to the C-terminal domain of progelatinase A (Fig. 1). As evidence to support this model, C-terminally truncated forms of progelatinase A do not

a   **Free ProgelA/TIMP-2 complex**

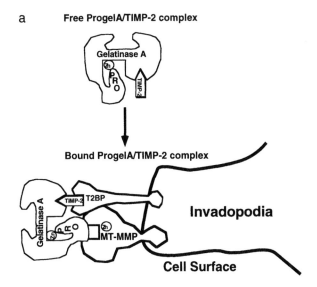

FIG. 3. Cellular mechanism for progelatinase A activation: (a) Attachment of progelatinase A–TIMP-2 complex to a cell surface receptor. Activation of progelatinase A occurs at the cell surface. This activity is thought to be localized at the tip of projections on invading cells called invadopodia, and it is an MT-MMP-mediated event. For activation to occur, progelatinase A must be bound in proximity to the active site of MT-MMP. This binding is thought to be mediated by TIMP-2 complexed to progelatinase A. The TIMP-2 component of the complex may attach to a binding site on MT-MMP. Alternatively, it may attach to another TIMP-2-binding protein (T2BP) as shown in the diagram, which is closely associated with MT-MMP. Upon binding TIMP-2, this receptor may help orient the progelatinase A–TIMP-2 complex for activation by MT-MMP. (b) Subsequent detachment of gelatinase A–TIMP-2 complex from the cell receptor. Once activated, gelatinase A–TIMP-2 complex would then dissociate from the cell surface and be free to carry out its biological functions. However, this active complex is short lived, and upon molecular rearrangement, will become inhibited by the TIMP-2 to form an inhibited gelatinase A–TIMP-2 complex.

bind TIMP-2, but activated C-terminally truncated gelatinase A is inhibited by TIMP-2 (Fridman *et al.*, 1992; Murphy *et al.*, 1992; Nguyen *et al.*, 1994; Willenbrock *et al.*, 1993). In addition, cross-linked progelatinase A/TIMP-2 complex can still be activated and show gelatinase activity, which suggests that the TIMP-2 is not entirely bound to the active site (Kleiner *et al.*, 1992). However, this gelatinase–TIMP-2 complex retains 10% of the proteolytic activity of the enzyme that is free of TIMP-2. Upon cleavage of the profragment, the active site of the gelatinase A–TIMP-2 complex becomes available for binding the second domain of TIMP-2. This site is not specific for TIMP-2 and the active

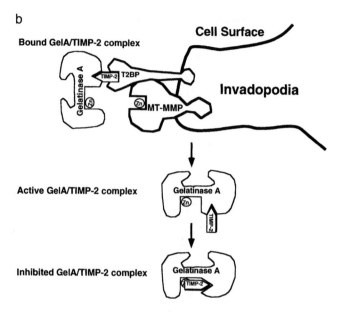

FIG. 3. continued

enzyme is inhibited by all TIMPs. It has also been found that the latent progelatinase A–TIMP-2 complex is capable of binding activated gelatinase B to form a ternary complex devoid of gelatinase activity (Curry *et al.,* 1992; Kolkenbrock *et al.,* 1991). This evidence suggests that when TIMP-2 is complexed to progelatinase A, the inhibitory N-terminal domain of TIMP-2 is available for interaction with gelatinase B. Recently, the newest member of the TIMP family, TIMP-4, was found to bind to the C terminus of gelatinase A with high affinity (apparent $K_d = 1.7 \times 10^{-7}$ M) (Bigg *et al.,* 1997). The interaction of progelatinase A with any TIMP-4 has not been observed in cell cultures or tissues to date, so the physiological significance of these *in vitro* interactions remains to be clarified.

Recent data suggest that TIMP-2 may mediate the cell surface activation of progelatinase A by binding to a MT1-MMP containing complex on the cell surface (see Section III,D,2 and Figs. 3 and 4). Thus, TIMP-2 may play a dual role in its interaction with gelatinase A. Interaction of TIMP-2 with progelatinase A may facilitate cell surface-mediated activation, whereas interaction with active gelatinase A results in inhibition. In summary, much progress has been made recently in explaining the interactions between TIMPs and the gelatinases. However, the question remains whether the two TIMP domains interact with the two gelatinase-binding sites simultaneously, or whether the TIMP molecule can only interact with one binding site at a time.

D. Regulation of Proenzyme Activation

1. *In Vitro* ACTIVATION BY APMA AND CHAOTROPIC AGENTS

The latency of MMPs is thought to be maintained by coordination between the unpaired cysteine residue in the pro-domain and the Zn atom at the active site. Disruption of this Zn atom interaction with the sulfhydryl functionality of the cysteine residue and coordination with H_2O are necessary to obtain catalytic activity. The unpaired cysteine residue in progelatinase A remains cryptic in the holo-proenzyme but is exposed upon removal of divalent cations by chelating agents (Stetler-Stevenson *et al.*, 1989). Upon the dissociation of this Cys–Zn interaction, gelatinase A undergoes an autolytic cleavage, which removes an 8-kDa peptide by hydrolysis at the Asn^{90}–Tyr^{91} bond, at a site homologous (Phe>Tyr) to that cleaved in other MMPs, eight residues downstream from the conserved Cys residue in the propeptide activation locus (Stetler-Stevenson *et al.*, 1989b). This processing appears to occur in two steps with an initial cleavage at Asn^{37}–Leu^{38} that generates an obligatory 65-kDa intermediate similar to that observed in the cell surface activation of progelatinase A (see Section III,D,2) (Kleiner and Stetler-Stevenson, unpublished observations; Itoh *et al.*, 1995).

2. CELL SURFACE LOCALIZATION AND ACTIVATION OF GELATINASE A: ROLE OF MT1-MMP AND $\alpha_v\beta_3$ INTEGRIN

Physiologic activation of progelatinase A is regulated at the level of the cell surface by a membrane-associated activator as first proposed in 1990 by Brown *et al.* (1990). Activation of progelatinase A involves interaction of progelatinase A, TIMP-2, and the activator protein, and requires an intact progelatinase A C-terminal domain. A membrane-associated specific activator of gelatinase A can be induced in a range of cell types by agents such as concanavalin A, cytochalasin D, phorbol esters, and TGF-β_1 (Brown *et al.*, 1993c; Overall and Sodek, 1990; Ward *et al.*, 1991). Activation of progelatinase A, by membrane preparations from cells treated with these agents, is blocked by MMP inhibitors, including 1,10-phenanthroline, synthetic peptide inhibitors, and the TIMPs (Murphy *et al.*, 1994; Ward *et al.*, 1991). This membrane-dependent activation of progelatinase A results in initial cleavage of the propeptide at Asn^{37}–Leu^{38} to give a 64-kDa intermediate, followed by cleavage at Asn^{80}–Phe^{81} to give the final 62-kDa active enzyme

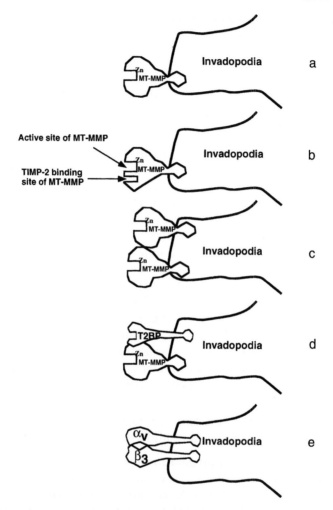

FIG. 4. Summary of potential cell surface receptors for progelatinase A or progelatinase A–TIMP-2 complex. Several possibilities exist for localization and activation of progelatinase A or progelatinase A–TIMP-2 complex on the cell surface. (a) As described in the text recent experiments have identified MT-MMP as a cell suface activator for progelatinase A–TIMP-2 complex. (b) Controversy still exists over the possible role of TIMP-2 in cell surface-mediated activation of progelatinase A. Currently, there are two schools of thought on this subject: Some investigators feel that TIMP-2 inhibits cell surface activation of progelatinase A, and others feel that TIMP-2 mediates activation through formation of a progelatinase A–TIMP-2 complex. It is possible that MT-MMP might contain a TIMP-2 specific binding site as shown in the diagram (much like that of progelatinase A), but this has not been demonstrated experimentally. (c) Another possibility is the existence of a dimeric form of MT-MMP on the cell surface. Since active MT-MMP is inhibited by the N terminus of TIMP-2, it will provide a mechanism for the localization of the progelatinase A–TIMP-2 complex. Once recruited, the progelatinase A–TIMP-2

(Strongin *et al.*, 1993). Cleavage at Asn^{80}–Phe^{81} also occurs during autoactivation in the presence of an organomercurial reagent (Stetler-Stevenson *et al.*, 1989b), and it has been suggested that the association of gelatinase A with membrane component(s) may initiate autoactivation of the enzyme (Bergmann *et al.*, 1995; Strongin *et al.*, 1993).

Sato *et al.* (1994) have cloned a novel MMP with a transmembrane domain (MT1-MMP) which they propose represents the cell membrane-associated activator of progelatinase A. MT-MMP is also proposed to act as a receptor for TIMP-2. The MT1-MMP–TIMP-2 complex can bind progelatinase A, to form a trimeric complex, through the carboxyl terminus of progelatinase A (Strongin *et al.*, 1995) (Fig. 3a). Formation of this complex may lead to subsequent activation of progelatinase A (Fig. 3b). MT1-MMP has a transmembrane (TM) domain, and truncation of this domain was initially reported to abolish gelatinase A activation (Cao *et al.*, 1995). This suggests that the TM domain serves to anchor and localize MT1-MMP at a position on the cell surface that facilitates progelatinase A–TIMP-2 complex activation. Binding, localization, and activation of progelatinase A on the cell surface depends on cooperative interactions between a number of different molecules, including MT1-MMP and TIMP-2. The role of TIMP-2 in these interactions is a controversial matter, and while some investigators have shown that TIMP-2 inhibits the cell surface activation of progelatinase A, others suggest it may enhance activation through formation of progelatinase A–TIMP-2 complex, which facilitates binding of the progelatinase A to the cell surface (Fig. 3a). TIMP-2 may achieve this effect either by binding to MT1-MMP or to another T2BP, thus localizing progelatinase A in proximity to MT1-MMP (Fig. 4).

Interestingly, some tumor cell lines have been shown to bind progelatinase A without activating the enzyme (Emonard *et al.*, 1992), which suggests that there is a receptor for progelatinase A that may be distinct from MT1-MMP. There is evidence that the integrin $\alpha_v\beta_3$ is involved in binding progelatinase A to the cell surface (Brooks *et al.*, 1996) (Fig. 4). These molecules were shown to form an SDS-stable complex *in vitro*

---

complex can then be activated by the second molecule of MT-MMP. (d) Also presented in this scheme is the possibility of a cell surface receptor for TIMP-2 that is distinct from MT-MMP. Upon binding TIMP-2, this receptor may help orient the progelatinase A–TIMP-2 complex for activation by MT-MMP. (e) Recent experimental evidence demonstrated that a proteolytically active form of gelatinase A can be localized on the surface of invasive cells, based on its ability to bind directly integrin $\alpha_v\beta_3$. *In vitro*, these proteins form an SDS-stable complex that depended on the noncatalytic C terminus of gelatinase A, since a truncation mutant lost the ability to bind $\alpha_v\beta_3$. It is not clear how $\alpha_v\beta_3$ interacts with gelatinase A but this interaction seems to be specific for $\alpha_v\beta_3$ because gelatinase A bound poorly to other integrins such as $\alpha_v\beta_5$ and $\alpha_5\beta_1$.

that involved binding of $\alpha_v\beta_3$ to the noncatalytic C-terminal domain of gelatinase A. Expression of $\alpha_v\beta_3$ on the surface of cultured melanoma cells enabled them to bind gelatinase A in a proteolytically active form, which allowed cell-mediated degradation of collagen. It has not yet been reported what effect TIMP-2 may have in this interaction or if progelatinase A binding to $\alpha_v\beta_3$ results in autoproteolytic activation. It would also be interesting to know if this integrin receptor is proteolytically modified by gelatinase A activity and if this may influence cell attachment (see later discussion).

Employing sensitive and specific sandwich ELISAs, Murphy and Stetler-Stevenson (1997) determined the levels of total TIMP-2 and gelatinase A–TIMP-2 complex in conditioned media samples from nine different tumor cell lines established from human melanoma, fibrosarcoma, breast carcinoma, lung carcinoma, and pancreatic carcinoma. The results of this study showed that TIMP-2 may be present in significant excess over the levels of gelatinase A in the culture fluid of many human tumor cells (Table I). In most cell lines examined (8/10) the excess free TIMP-2 concentration was greater than 10 pM/10^5 cells/ 24 h and in one cell line was as great as 250 pM/10^5 cells/24 h (Table I). These results indicate that the ratio of secreted TIMP-2 to gelatinase A in culture varies greatly from one tumor cell type to another, and as such may have a variable influence on the invasive behavior of tumor cells. Thus, any model for the cellular mechanism of progelatinase A activation should also account for this excess TIMP-2 relative to gelatinase A that is observed in the conditioned media from different cell lines. These results suggest that most of the progelatinase A produced by cells is immediately complexed with TIMP-2 and that the species that must be activated by the cell is the progelatinase A–TIMP-2 complex.

It has also been reported that thrombin and urokinase facilitate the cell surface activation of progelatinase A (Ginestra *et al.*, 1997; Mazzieri *et al.*, 1997; Zucker *et al.*, 1995). This suggests that multiple mechanisms may be evoked for the cellular control of progelatinase A activation. The exact details of these multiple activation mechanisms and their relative significance in various physiologic processes and pathologic states remain to be elucidated. However, the redundancy of possible mechanisms for the cellular control of progelatinase A activation hints at the significance of this level of regulation in controlling matrix turnover.

The fact that processing and activation of gelatinase A appears to be restricted to the cell surface is likely to have important implications *in vivo*. Cell surface activation allows migrating and invading cells to focus gelatinase A activity immediately adjacent to areas of the ECM

TABLE I

DETERMINATION OF FREE TIMP-2 IN CONDITIONED MEDIA FROM VARIOUS TUMOR
CELL LINES[a]

	$[\text{TIMP-2}]_{total}$	[GEL A/TIMP-2 complex]$_{total}$	Free calculated $[\text{TIMP-2}]_f$	Measured [TIMP-2]
Melanoma				
A2058	107.3 ± 8.5	53.4 ± 2.9	53.9	60.6 ± 4.8
HT-144	293.8 ± 4.4	46.0 ± 8.1	247.8	251.1 ± 1.4
LOX	8.3 ± 4.7	5.4 ± 1.6	2.9	0.8 ± 0.3
RPMI-7951	144.0 ± 1.7	177.1 ± 17.0	0.0	N.D.[b]
Fibrosarcoma				
HT1080	124.1 ± 2.6	113.6 ± 6.1	10.5	35.9 ± 0.03
Breast carcinoma				
BT20	32.0 ± 1.4	N.D.	32.0	18.2 ± 2.7
ZR-75-1	15.1 ± 2.3	1.7 ± 0.7	13.4	10.1 ± 1.4
Lung carcinoma				
A549	45.5 ± 4.9	N.D.	45.5	34.9 ± 2.3
Pancreatic carcinoma				
CAPAN-2	79.5 ± 3.2	N.D.	79.5	66.6 ± 0.6

[a] Concentration (pM per 105 cells per 24 h ± S.D.) The ELISA assays for TIMP-2 and for the gelatinase A–TIMP-2 complex can be used to estimate the concentration of free TIMP-2 in conditioned media. The values for TIMP-2 and complex represent total concentrations determined using the ELISAs for TIMP-2 and complex, respectively. The calculated uncomplexed TIMP-2 is the numerical difference between TIMP-2 and complex, and estimates the quantity of TIMP-2 in the conditioned media that is not complexed to gelatinase A. To confirm this method, complex was immunoprecipitated from conditioned media samples, and the resulting complex-depleted media was assayed, again using the TIMP-2 ELISA. These values are shown in the final column. Calculated and measured TIMP-2 concentrations are in close agreement, providing evidence that the use of the ELISA assays for TIMP-2 and complex to estimate the quantity of uncomplexed TIMP-2 is valid.

[b] N.D., not detected.

that are in the cell path and thus limits the amount of matrix that needs to be turned over (Brown *et al.*, 1990). Consistent with this idea, active gelatinase A has been shown to be localized to the invadopodia of malignant cells (Monsky *et al.*, 1993).

IV. SUBSTRATE SPECIFICITY

Gelatinase A cleaves a number of peptide bonds such as Gly–Val, Gly–Leu, Gly–Glu, Gly–Asn, and Gly–Ser in denatured collagen to produce small peptides (Seltzer *et al.*, 1981, 1990). Besides gelatin and types IV and V collagen, gelatinase A also cleaves type VII collagen, found in anchoring fibrils (Seltzer *et al.*, 1989), cartilage type X collagen (Gadher *et al.*, 1989; Welgus *et al.*, 1990), and elastin (Senior *et al.*, 1991). It is interesting to note that pepsin-solubilized type IV collagen is degraded by both gelatinase A and gelatinase B (Mainardi *et al.*, 1980), whereas the full-length type IV collagen apparently is resistant to proteolysis by both gelatinases (Mackay *et al.*, 1990). These findings raised concerns as to whether gelatinases actually function as true "type IV collagenases" *in vivo*. While the initial cleavage site of soluble type IV collagen by gelatinase A has been found to be located 1/4 of the distance from the NH_2 terminus (Fessler *et al.*, 1984; Murphy *et al.*, 1989), the cleavage site of the full-length native type IV collagen remains to be determined.

Although both gelatinase A and B bind to type I collagen, only gelatinase A has been shown to cleave soluble, triple helical type I collagen generating 3/4 N-terminal and 1/4 C-terminal fragments characteristic of vertebrate interstitial collagenases (Aimes and Quigley, 1995). Gelatinase A cleaves type I collagen at the same Gly–Ile/Leu bond in the collagen α-chains as interstitial collagenases. Gelatinase A was also capable of degrading reconstituted type I collagen fibrils. The closely related gelatinase B was unable to cleave soluble or fribrillar collagen under identical conditions, indicating that the type I collagenolytic activity of gelatinase A is not a general property of the gelatinase subgroup of MMPs. In addition to fibrillar and nonfibrillar collagen, gelatinase A has been shown to degrade elastin (Murphy *et al.*, 1991; Senior *et al.*, 1991). These findings suggest that the name *gelatinase A* is a misnomer and that this protease may play a much greater role in turnover of a variety of ECM components than previously appreciated. This should emphasize the importance of determining the critical substrates for these proteases *in vivo*.

To date, very little is known about the *in vivo* substrate specificity of gelatinase A. Aside from its ability to degrade various components of ECM, gelatinase A has been shown to possess a β-secretase activity. In HeLa cells transfected with an amyloid protein precursor (APP)695 expression plasmid, gelatinase A cleaved the 110-kDa cell surface APP, releasing a 100-kDa form of the protein. While a peptide homologous to the β-secretase site was cleaved by gelatinase A, a peptide homologous to the α-secretase site was not cleaved (LePage *et al.*, 1995; Roher *et al.*, 1994). This recent finding may represent one of many in vivo substrates of gelatinase A. Additonal potential substrates indetified *in vitro* include vitronectin and laminin-5 (Giannelli *et al.*, 1997; Imai *et al.*, 1995).

V. Role of Gelatinase A in Cellular Processes

A. *Regulation of Cell Proliferation and Differentiation*

A novel role for gelatinase A was reported by Lovett and coworkers (Turck *et al.,* 1996). Using glomerular mesangial cells, which are centrally involved in both the acute inflammatory and chronic renal sclerotic processes, they were able to demonstrate that the gelatinase A is synthesized as a consequence of mesangial cell activation. Secreted gelatinase A was shown to be concentrated at sites immediately involved in the limited proteolysis of critical ECM proteins and subsequent transition to an inflammatory phenotype. To determine the significance of gelatinase A secretion for the acquisition of the inflammatory phenotype, the authors reduced the constitutive level of gelatinase A by cultured mesangial cells with antisense RNA expressed by an episomally replicating vector or with specific anti-gelatinase A ribozymes expressed by a retroviral transducing vector. The phenotype of the transfected, or retrovirally infected, cells exhibited pronounced differences including a change in the synthesis and organization of the ECM loss of activation markers, and a nearly complete exit from the cell cycle. The features of the inflammatory phenotype could be restored by exposure to active, but not latent, gelatinase A. This observation indicates that proteolytic modification of the ECM by gelatinase A regulates the mesangial cell phenotype, inflammatory versus quiescent. Although they were unable to demonstrate the presence of MT-MMP transcripts or enzymatic activity or the specific target substrate responsible for this activity, they were able to show that gelatinase A in isolated mesangial cell plasma membrane preparations can be activated by an autocatalytic mechanism.

Notably, the other member of the gelatinase family, gelatinase B, was not effective in reconstituting the inflammatory phenotype in these experiments (Turck *et al.,* 1996). Although the two gelatinases are similar in many respects, they have distinctively different substrate specificities. For example, gelatinase B has a much higher degree of activity than gelatinase A for types IV and V collagen. On the other hand, only gelatinase A can degrade fibronectin and laminin. This difference in substrate specificity is significant considering a direct relationship between the presence of fibronectin-degrading surface proteases and the ability of cells to proliferate and migrate (Chen and Chen, 1987). On the other hand, actions that facilitate fibronectin pericellular matrix assembly, including protease inhibition or overexpression of fibronectin integrin receptors, result in a loss of cell motility and invasive properties (Giancotti and Ruoslahti, 1990; Juliano and Haskill,

1993). These observations, along with the specific requirement for proteolytic activity, suggest that the specific action of gelatinase A in mesangial cells is due to the cleavage of certain ECM proteins, which are not cleaved by gelatinase B.

B. *Modulation of Cell Adhesion and Migration*

Certain structural changes of the ECM accompany cell migration during physiologic tissue remodeling and tumor cell invasion, and gelatinase A has been shown to play a key role in promoting invasiveness of both normal and neoplastic cells. At the interface between histologically distinct tissues, the ECM forms highly organized structures such as basement membranes that modulate cell adhesion, differentiation and morphogenesis, mitogenesis, cell motility, and the selective exchange of molecules between cells and interstitial fluids. Basement membranes form physical barriers that separate and define epithelial compartments from the mesenchymal tissue compartments. This barrier is routinely crossed by inflamatory cells in mediating immune surveillance and response to immune triggers. It also is crossed and modified during glandular morphogenesis, angiogenesis, and tumor cell invasion. Modifications or degradation of the ECM by tumor-derived proteases was originally thought to destroy physical barriers to cell migration. However, more recently modification of the ECM by proteases has been shown to have profound effects on cell adhesion and migration.

For instance, Ray and Stetler-Stevenson (1995) found that adhesion of A2058 human melanoma cells to various tissue culture substrates including fibronection, gelatin, and vitronectin was decreased by treatment of cells with exogenous gelatinase A. In contrast, inhibiting the endogenous gelatinase A with either neutralizing antibody or TIMP-2 resulted in enhanced attachment to these substrates when compared to untreated cells. To study this phenomenon, retroviral infection vectors were used to control the amount of the TIMP-2 relative to endogenous gelatinase A in human melanoma A2058 cells. Altering the production of TIMP-2 modulted not only proteolysis of the ECM, but also the adhesive and spreading properties of the cells and resulted in altered cell morphology. These effects of TIMP-2 were shown to be mediated by inhibition of gelatinase A activity.

Recently, Giannelli *et al.* (1997) have shown that gelatinase A activity induces the migration of breast epithelial cells by cleaving and regulating the function of a specific ECM component, laminin-5 (Ln-5). Ln-5 is a component of epithelial basement membranes, which also contain collagen type IV, laminin-1 (Ln-1), and fibronectin (Fn) (Burgeson *et al.*, 1994; Laurie *et al.*, 1982). Cells adhere or migrate on these ECM substrates using integrin receptors. For example, Ln-5 interaction with integrins is essential for adhesion of epithelial cells to basement membranes (Baker *et al.*, 1996) and to promote migration (Miyazaki *et al.*, 1993). To test whether the effect of proteolytic modification of these basement membrane components altered cell migration, these authors studied the effects of gelatinase A, gelatinase B, and plasmin on select basement membrane components. Subsequent experiments showed that the active form of gelatinase A cleaves the Ln-5γ2 subunit at residue 587, exposing a putative cryptic pro-migratory site that triggers cell motility but not cell adhesion. This effect was specific for gelatinase A cleavage of Ln-5. Cleavage of Ln-1 did not have this effect, and plasmin or gelatinase B did not produce active fragments of Ln-5. Alternatively, this cleavage may mask a site that suppresses cell motility. The existence of pro-motility cryptic sites and suppressor sites on laminins has been hypothesized, and evidence for stimulatory sites has been demonstrated for other ECM molecules (Calof *et al.*, 1994; Stefansson and Lawrence, 1996). The fact that this pro-motility cryptic site does not support adhesion suggests that the proteolytic activity of gelatinase A may provide a signaling mechanism for cells to begin migration during mammary gland morphogenesis. The neoplastic breast epithelial cell line, MCF-7, also demonstrates an enhanced migratory response to gelainase A modification of Ln-5.

The findings of Ray and Stetler-Stevenson (1995), as well as Gianelli *et al.* (1997), suggest that cell adhesion and migration are markedly influenced by local levels of gelatinase A activity. According to the "three-step" hypothesis of tumor cell invasion (Liotta, 1986; Stetler-Stevenson *et al.*, 1993), key events in the process of tumor invasion are tumor cell adhesion to ECM structures, ECM degradation by proteolysis, and then tumor cell migration into the degraded area. In this model, proteolytic degradation of the ECM was principally viewed as removal of ECM barriers to the migration of invading tumor cells. In light of the recent findings described earlier, this view must be modified to include the findings that protease activity may generate proteolytic fragments that stimulate cell movement. Thus, vectorial secretion of gelatinase A activity may create a gradient of chemotactic or haptotactic stimuli that promotes and directs cell migration.

While it is well established that proteolysis is required for tumor invasion, these recent findings also suggest that excessive proteolysis may inhibit this process by impairing tumor cell adhesion or disrupting and degrading the cell–matrix interactions or matrix signals required for migration and invasion. It may therefore be necessary for the level of proteolytic activity to fall within a certain critical range in order for both tumor invasion and angiogenesis to occur. Precise regulation of

the balance between gelatinase A and TIMP-2 expression is therefore
likely to be essential.

C.   Overexpression of Active Gelatinase A in Cell Invasion: Tumor Invasion and Angiogenesis

A common feature in the invasive processes associated with tumor
invasion and angiogenesis is the degradation of ECM, basement membranes, basal laminae, and interstitial stroma. As described previously,
this activity is thought to remove a physical barrier that prevents
the cell migration required for invasive cells to advance into adjacent
tissues. Benign (noninvasive, nonmetastatic) epithelial tumors are always characterized by an intact basement membrane that separates
the neoplastic epithelium from the connective tissue stroma, whereas
malignant epithelial tumors (carcinomas) have a defective basement
membrane that forms an incomplete barrier to spread of the neoplastic
cells (Barsky *et al.*, 1983). Defective basement membrane organization
or loss could be either due to decreased synthesis or increased degradation of its components. Neoplastic cells that invade surrounding tissues
and go on to metastasize through the bloodstream must penetrate the
basement membranes of blood vessel walls in order to enter and to leave
the vascular compartment. The ability of neoplastic cells to degrade
basement membranes is therefore one important determinant of metastatic potential. It is this requirement for tumor cells to degrade basement membranes that led to the discovery of gelatinase A (Liotta *et
al.*, 1979). It is now well established that gelatinase A is not tumor cell
specific, and is involved in ECM remodeling in a wide range of non-
neoplastic processes, including embryonic development, trophoblast invasion, angiogenesis, T-cell transmigration, and wound healing (Agren,
1994; Casasco *et al.*, 1995; Lelongt *et al.*, 1997; McLaughlin and Weiss,
1996; Romanic and Madri, 1994; Romanic *et al.*, 1997; Sharkey *et al.*,
1996; Shimonovitz *et al.*, 1994; Stetler-Stevenson and Corcoran, 1997).
Evidence for the enhanced expression of gelatinase A in human tumors comes from many experimental studies correlating enzyme expression and tumor grade (Liotta *et al.*, 1980; Nakajima *et al.*, 1987,
1989; Turpeenniemi-Hujanen *et al.*, 1985). For example, immunocyto-
chemical and *in situ* hybridization (ISH) studies have shown increased
expression of gelatinase A in many human tumors, including carcinomas of the colon (Poulsom *et al.*, 1992; Pyke *et al.*, 1993), pancreas
(Gress *et al.*, 1995), prostate (Boag and Young, 1994), bladder (Davies
et al., 1993b), skin (squamous and basal cell carcinomas) (Pyke *et al.*,
1992), breast (Davies 1993; Polette *et al.*, 1994), and ovary (Autio-
Harmainen *et al.*, 1993). In contrast, benign proliferative disorders of

these tissues usually show negative immunoreactivity for gelatinase A (Hoyhtya *et al.*, 1994). In addition to its role in metastasis, gelatinase A also plays a key role in angiogenesis. When endothelial cells were cultured on matrigel (Schnaper *et al.*, 1993), the formation of tube networks was increased by the addition of recombinant gelatinase A, and decreased both by neutralizing antibody and TIMP-2, suggesting that network formation was limited by enzyme activity. However, further addition of gelatinase A above a critical level resulted in decreased tube formation, suggesting that excess enzyme activity was deleterious to this complex morphological change. This effect was reversed by addition of exogenous recombinant TIMP-2. These results suggest that while gelatinase A activity has an important role in angiogenesis, excessive levels of activity will inhibit the process. Furthermore, inhibitors such as TIMP-2 may both block or facilitate an angiogenic response depending on the level of protease expression. This is similar to reports in which serine protease inhibitors have been shown to enhance endothelial cell invasion and alter the morphology of endothelial cell outgrowths (Pepper *et al.*, 1990; Tsuboi and Rifkin, 1990).

Cellular localization of gelatinase A by ISH in carcinomas indicates that gelatinase A mRNA appears to be localized to the stromal fibroblasts adjacent to the sites of tumor invasion (Poulsom *et al.*, 1993; Pyke *et al.*, 1992). Exceptions to this observation include prostate cancer (Boag and Young, 1994) and carcinomas of the lung (Urbanski *et al.*, 1992), where both mRNA and protein were detected mainly in the neoplastic epithelium, as well as carcinomas of the hypopharynx (Gress *et al.*, 1995) and pancreas (Miyajima *et al.*, 1995), in which gelatinase A mRNA was localized to both neoplastic cells and tumor stroma. For carcinomas of the colon, breast, and ovary, ISH studies show that gelatinase A mRNA can only be detected in the connective tissue stroma (Autio-Harmainen *et al.*, 1993; Polette *et al.*, 1994; Poulsom *et al.*, 1992, 1993; Pyke *et al.*, 1993). This stromal localization of the gelatinase A mRNA contrasts with the predominant localization of gelatinase A in the malignant epithelium that is seen with immunohistochemical techniques. Gelatinase A protein is detected mainly in the neoplastic epithelium of these tumors by immunocytochemistry (Autio-Harmainen *et al.*, 1993; Levy *et al.*, 1991; Polette *et al.*, 1994). One explanation for this discrepancy could be that the half-life of the gelatinase A mRNA, and /or the efficiency of translation, differs between the stromal cells and neoplastic epithelial cells. Another explanation is that the enzyme is synthesized predominantly in the tumor stroma, but that on secretion it is recruited by receptors on the surface of neoplastic epithelial cells.

Increasingly, it is becoming clear that ECM degradation by gelatinase A is regulated by the net balance of proenzyme expression, secretion, and activation, and by the local concentration of TIMPs. All of the correlative studies reviewed earlier used techniques that measure total level of gelatinase A immunoreactive protein or mRNA transcripts. The fact that gelatinase A overexpression in tumors can precede acquisition of the invasive phenotype suggests that it is the activation of gelatinase A that makes a significant contribution to the acquisition of the invasive phenotype. Thus, the measurement of active enzyme levels may be more informative with respect to the stage and aggressive behavior of the tumor, and possibly of more diagnotic or prognostic utility to clinicians. One such technique is quantitative gelatin zymography, which has so far been used in a small number of studies to assess the contribution of active gelatinase species to the invasive phenotype of human breast, non-small-lung cancer, and colon cancer (Brown *et al.*, 1993a, 1993b; Davies, 1993a; Emmert-Buck *et al.*, 1994). For example, Nomura *et al.* (1996) used gelatin zymography to show that active forms of gelatinase A and gelatinase B were more frequently detected in human gastric carcinoma tissue, and that the activation of the zymogen form of gelatinase A, but not that of gelatinase B, correlated well with the degree of local invasion and lymphatic permeation. These results support the idea that activation of gelatinase A is a key step in the acquisition of malignant phenotype.

TIMP-2 selectively binds to progelatinase A and inhibits the protease activity of gelatinase A. Consequently, TIMP-2 can suppress invasion, metastasis, neovascularization, and growth of some rodent and human tumors (DeClerck *et al.*, 1992; Liotta *et al.*, 1991; Montgomery *et al.*, 1994; Moses *et al.*, 1990; Stetler-Stevenson *et al.*, 1989a). In a study on gastric cancer patients, TIMP-2 expression was found to be lower in advanced cases, while gelatinase A expression was either increased or maintained (Grigioni *et al.*, 1994). In addition, patients who died from their primary gastric tumor had higher a percentage of gelatinase A positive cells and a lower percentage of TIMP-2 positive cells, compared to survivors. In urothelial cancer patients with muscular invasion or with lymph node metastasis, the mean gelatinase A : TIMP-2 ratio in patients with recurrence was significantly higher than that in patients without recurrence ($P < 0.05$) (Gohji *et al.*, 1996). This same study showed that disease-free survival of patients with high gelatinase A : TIMP-2 ratios was extremely poor compared with that of patients with lower ratios ($P < 0.01$). In another study, the imbalance of the gelatinase A : TIMP-2 ratio was examined in lymph-node-positive breast carcinomas as measured by RT-PCR (Onisto *et al.*, 1995). The ratio between gelatinase $A_{tumor/normal}$ and TIMP-2$_{tumor/normal}$ was usually

lower in carcinomas without lymph-node involvement (LN⁻). In contrast, five out of six patients with lymph-node metastasis (LN⁺) showed
much higher ratios, suggesting enhanced gelatinase A expression relative to TIMP-2. The ratio ranged between 2 and 4. However, the study
found that the magnitude of this ratio was not related to the frequency
of positive lymph nodes, nor to relapse status at follow-up. These studies
indicate that evaluation of gelatinase A : TIMP-2 mRNA balance and/
or serum level may constitute an early prognostic indicator in various
types of human cancer.

VI. ROLE OF GELATINASE A IN PLATELET AGGREGATION

Recently, Sawicki et al. (1997) reported a novel involvement of gelatinase A in a new pathway of platelet aggregation that is distinct
from the two known pathways involving the release of endoperoxide/
thromboxane A_2 and ADP (Colman, 1990; Sargeant and Sage, 1994;
Siess, 1989). The authors investigated whether human platelets express gelatinase A and if its release affects platelet aggregation. The
study found that gelatinase A is released during aggregation induced
by collagen and thrombin, but not by PMA, and its release can be
positively correlated with platelet aggregation. Nitric oxide (NO) and
prostacyclin, which are inhibitors of aggregation, inhibited the secretion of gelatinase A induced by collagen and thrombin in intact platelets, indicating that enzyme release is controlled by these agents. Both
monoclonal anti-gelatinase A antibodies, which neutralize gelatinase
A activity, and TIMP-2 inhibited aggregation. Selected synthetic inhibitors of MMPs also decreased aggregation at concentrations similar to
those required for inhibition of gelatinase A. Also, low concentrations
of recombinant activated (but not latent) gelatinase A stimulated aggregation, indicating that enzymatic activity is required for its role in
platelet aggregation. These results show that the release of gelatinase
A can mediate platelet aggregation in vitro and suggest that TIMP-2
acts as an endogenous regulator of the pathway of platelet aggregation
operated by gelatinase A. This suggests that gelatinase A specific,
synthetic inhibitors may be useful not only in anti-cancer and anti-
angiogenic therapies, but could also be useful in expanding anti-
thrombotic therapies as well.

VII. PERSPECTIVES

Gelatinase A was the second member of the MMP family to be identified. The overexpression of this protease in tumor cells was helpful in
the identification and isolation of gelatinase A. The close association

of gelatinase A with the process of tumor cell invasion has led to the tendency to neglect the role of this protease in physiologic processes and other pathologic conditions. However, this is changing with the identification of a role for gelatinase A in such diverse processes as T-cell transmigation, platelet activation, and mammary epithelial cell migration. The ubiquitous distribution and lack of significant transcriptional regulation suggest that gelatinase A is unique among matrixin family members. Identification of a cellular mechanism for the control of gelatinase A activation is also unique. However, the details of this mechanism, as well as the possibility of multiple pathways for cell surface binding and activation, need to be worked out. Identification of new substrates is also changing our concept of the function of gelatinase A and of matrix metalloproteinases in general. As described, data now demonstrate that gelatinase A activity can generate biologically active products and alter cell–ECM interactions. Research must now examine the effects of these products of gelatinase A proteolytic activity on cellular functions such as proliferation, attachment, and migration. The identification of these "new" roles for an "old" protease suggest that the development of specific and selective inhibitors for gelatinase A may be useful not only in cancer therapy, but in a variety of clinical settings such as anti-thrombotic therapy and possibly autoimmune diseases as well.

REFERENCES

Agren, M. S. (1994). Gelatinase activity during wound healing. *Br. J. Dermatol.* **131**, 634–640.

Aimes, R. T., and Quigley, J. P. (1995). Matrix metalloproteinase-2 is an interstitial collagenase. Inhibitor-free enzyme catalyzes the cleavage of collagen fibrils and soluble native type I collagen generating the specific 3/4- and 1/4-length fragments. *J. Biol. Chem.* **270**, 5872–5876.

Autio-Harmainen, H., Karttunen, T., Hurskainen, T., Hoyhtya, M., Kauppila, A., and Tryggvason, K. (1993). Expression of 72 kilodalton type IV collagenase (gelatinase A) in benign and malignant ovarian tumors. *Lab. Invest.* **69**, 312–321.

Baker, S. E., Hopkinson, S. B., Fitchmun, M., Andreason, G. L., Frasier, F., Plopper, G., Quaranta, V., and Jones, J. C. (1996). Laminin-5 and hemidesmosomes: Role of the alpha 3 chain subunit in hemidesmosome stability and assembly. *J. Cell Sci.* **109**, 2509–2520.

Banyai, L., Tordai, H., and Patthy, L. (1994). The gelatin-binding site of human 72 kDa type IV collagenase (gelatinase A). *Biochem. J.* **298**, 403–407.

Barsky, S. H., Siegal, G. P., Jannotta, F., and Liotta, L. A. (1983). Loss of basement membrane components by invasive tumors but not by their benign counterparts. *Lab. Invest.* **49**, 140–147.

Benbow, U., and Brinckerhoff, C. E. (1997). The AP-1 site and MMP gene regulation: What is all the fuss about? *Matrix Biol.* **15**, 519–526.

Bergmann, U., Tuuttila, A., Stetler-Stevenson, W. G., and Tryggvason, K. (1995). Autolytic activation of recombinant human 72 kilodalton type IV collagenase. *Biochemistry* **34,** 2819–2825.

Bigg, H. F., Shi, Y. E., Liu, Y. E., Steffensen, B., and Overall, C. M. (1997). Specific, high affinity binding of tissue inhibitor of metalloproteinase-4 (TIMP-4) to the COOH-terminal hemopexin-like domain of human gelatinase A. *J. Biol. Chem.* **272,** 15496–15500.

Birkedal-Hansen, H., Moore, W. G. I., Bodden, M, K., Windsor, L. J., Birkedal-Hansen, B., DeCarlo, A., and Engler, J. A. (1993). Matrix metalloproteinases: A review. *Crit. Rev. Oral Biol. Med.* **4,** 197–250.

Boag, A. H., and Young, I. D. (1994). Increased expression of the 72–kd type IV collagenase in prostatic adenocarcinoma. Demonstration by immunohistochemistry and *in situ* hybridization. *Am. J. Pathol.* **144,** 585–591.

Brooks, P. C., Stromblad, S., Sanders, L. C., von Schalscha, T. L., Aimes, R. T., Stetler-Stevenson, W. G., Quigley, J. P., and Cheresh, D. A. (1996). Localization of matrix metalloproteinase MMP-2 to the surface of invasive cells by interaction with integrin alpha v beta 3. *Cell* **85,** 683–693.

Brown, P. D., Levy, A. T., Margulies, I. M., Liotta, L. A., and Stetler-Stevenson, W. G. (1990). Independent expression and cellular processing of Mr 72,000 type IV collagenase and interstitial collagenase in human tumorigenic cell lines. *Cancer Res.* **50,** 6184–6191.

Brown, P. D., Bloxidge, R. E., Anderson, E., and Howell, A. (1993a). Expression of activated gelatinase in human invasive breast carcinoma. *Clin. Exp. Metastasis* **11,** 183–189.

Brown, P. D., Bloxidge, R. E., Stuart, N. S., Gatter, K. C., and Carmichael, J. (1993b). Association between expression of activated 72–kilodalton gelatinase and tumor spread in non-small-cell lung carcinoma. *J. Natl. Cancer Inst.* **85,** 574–578.

Brown, P. D., Kleiner, D. E., Unsworth, E. J., and Stetler-Stevenson, W. G. (1993c). Cellular activation of the 72 kDa type IV procollagenase/TIMP-2 complex. *Kidney Int.* **43,** 163–170.

Burgeson, R. E., Chiquet, M., Deutzmann, R., Ekblom, P., Engel, J., Kleinman, H., Martin, G. R., Meneguzzi, G., Paulsson, M., Sanes, J., *et al.* (1994). A new nomenclature for the laminins. *Matrix Biol.* **14,** 209–211.

Calof, A. L., Campanero, M. R., O'Rear, J. J., Yurchenco, P. D., and Lander, A. D. (1994). Domain-specific activation of neuronal migration and neurite outgrowth-promoting activities of laminin. *Neuron.* **13,** 117–130.

Cao, J., Sato, H., Takino, T., and Seiki, M. (1995). The C-terminal region of membrane type matrix metalloproteinase is a functional transmembrane domain required for pro-gelatinase A activation. *J. Biol. Chem.* **270,** 801–805.

Casasco, A., Casasco, M., Reguzzoni, M., Calligaro, A., Tateo, S., Stetler-Stevenson, W. G., and Liotta, L. A. (1995). Occurrence and distribution of matrix metalloproteinase-2-immunoreactivity in human embryonic tissues. *Eur. J. Histochem.* **39,** 31–38.

Chen, J. M., and Chen, W. T. (1987). Fibronectin-degrading proteases from the membranes of transformed cells. *Cell* **48,** 193–203.

Chinnaiyan, A. M., O'Rourke, K., Lane, B. R., and Dixit, V. M. (1997). Interaction of CED-4 with CED-3 and CED-9: A molecular framework for cell death [see comments]. *Science* **275,** 1122–1126.

Collier, I. E., Wilhelm, S. M., Eisen, A. Z., Marmer, B. L., Grant, G. A., Seltzer, J. L., Kronberger, A., He, C. S., Bauer, E. A., and Goldberg, G. I. (1988). H-ras oncogene-transformed human bronchial epithelial cells (TBE-1) secrete a single metalloprotease capable of degrading basement membrane collagen. *J. Biol. Chem.* **263,** 6579–6587.

Colman, R. W. (1990). Aggregin: A platelet ADP receptor that mediates activation. *FASEB J.* **4,** 1425–1435.

Curry, V. A., Clark, I. M., Bigg, H., and Cawston, T. E. (1992). Large inhibitor of metalloproteinases (LIMP) contains tissue inhibitor of metalloproteinases (TIMP)-2 bound to 72,000–M(r) progelatinase. *Biochem. J.* **285,** 143–147.

Davies, B., Miles, D. W., Happerfield, L. C., Naylor, M. S., Bobrow, L. G., Rubens, R. D., and Balkwill, F. R. (1993a). Activity of type IV collagenases in benign and malignant breast disease. *Br. J. Cancer* **67,** 1126–1131.

Davies, B., Waxman, J., Wasan, H., Abel, P., Williams, G., Krausz, T., Neal, D., Thomas, D., Hanby, A., and Balkwill, F. (1993b). Levels of matrix metalloproteases in bladder cancer correlate with tumor grade and invasion. *Cancer Res.* **53,** 5365–5369.

DeClerck, Y. A., Perez, N., Shimada, H., Boone, T. D., Langley, K. E., and Taylor, S. M. (1992). Inhibition of invasion and metastasis in cells transfected with an inhibitor of metalloproteinases. *Cancer Res.* **52,** 701–708.

Emmert-Buck, M. R., Roth, M. J., Zhuang, Z., Campo, E., Rozhin, J., Sloane, B. F., Liotta, L. A., and Stetler-Stevenson, W. G. (1994). Increased gelatinase A (MMP-2) and cathepsin B activity in invasive tumor regions of human colon cancer samples. *Am. J. Pathol.* **145,** 1285–1290.

Emonard, H. P., Remacle, A. G., Noel, A. C., Grimaud, J. A., Stetler-Stevenson, W. G., and Foidart, J. M. (1992). Tumor cell surface-associated binding site for the M(r) 72,000 type IV collagenase. *Cancer Res.* **52,** 5845–5848.

Fessler, L. I., Duncan, K. G., Fessler, J. H., Salo, T., and Tryggvason, K. (1984). Characterization of procollagen IV cleavage products produced by a specific tumor collagenase. *J. Biol. Chem.* **259,** 9783–9789.

Fini, M. E., Bartlett, J. D., Matsubara, M., Rinehart, W. B., Mody, M. K., Girard, M. T., and Rainville, M. (1994). The rabbit gene for 92-kDa matrix metalloproteinase. Role of AP1 and AP2 in cell type-specific transcription. *J. Biol. Chem* **269,** 28620–28628.

Fridman, R., Fuerst, T. R., Bird, R. E., Hoyhtya, M., Oelkuct, M., Kraust, S., Komarek, D., Liotta, L. A., Berman, M. L., and Stetler-Stevenson, W.G. (1992). Domain structure of human 72-kDa gelatinase/type IV collagenase. *J. Biol. Chem.* **267,** 15398–15405.

Frisch, S. M., Reich, R., Collier, I. E., Genrich, L. T., Martin, G., and Goldberg, G. I. (1990). Adenovirus E1A represses protease gene expression and inhibits metastasis of human tumor cells. *Oncogene* **5,** 75–83.

Gadher, S. J., Schmid, T. M., Heck, L. W., and Woolley, D. E. (1989). Cleavage of collagen type X by human synovial collagenase and neutrophil elastase. *Matrix* **9,** 109–115.

Giancotti, F. G., and Ruoslahti, E. (1990). Elevated levels of the alpha 5 beta 1 fibronectin receptor suppress the transformed phenotype of Chinese hamster ovary cells. *Cell* **60,** 849–859.

Giannelli, G., Falk-Marzillier, J., Schiraldi, O., Stetler-Stevenson, W. G., and Quaranta, V. (1997). Induction of cell migration by matrix metalloproteinase-2 cleavage of laminin-5. *Science* **277,** 225–228.

Ginestra, A., Monea, S., Seghezzi, G., Dolo, V., Nagase, H., Mignatti, P., and Vittorelli, M. L. (1997). Urokinase plasminogen activator and gelatinases are associated with membrane vesicles shed by human HT1080 fibrosarcoma cells. *J. Biol. Chem.* **272,** 17216–17222.

Gohji, K., Fujimoto, N., Fujii, A., Komiyama, T., Okawa, J., and Nakajima, M. (1996). Prognostic significance of circulating matrix metalloproteinase-2 to tissue inhibitor of metalloproteinase-2 ratio in recurrence of urothelial cancer after complete resection. *Cancer Res.* **56,** 3196–3198.

Goldberg, G. I., Marmer, B. L., Grant, G. A., Eisen, A. Z., Wilhelm, S., and He, C. S. (1989). Human 72-kilodalton type IV collagenase forms a complex with a tissue inhibitor of metalloproteases designated TIMP-2. *Proc. Natl. Acad. Sci. USA* **86,** 8207–11.

Goldberg, G. I., Strongin, A., Collier, I. E., Genrich, L. T., and Marmer, B. L. (1992). Interaction of 92–kDa type IV collagenase with the tissue inhibitor of metallopro-teinases prevents dimerization, complex formation with interstitial collagenase, and activation of the proenzyme with stromelysin. *J. Biol. Chem.* **267**, 4583–4591.

Gress, T. M., Muller-Pillasch, F., Lerch, M. M., Friess, H., Buchler, M., and Adler, G. (1995). Expression and *in-situ* localization of genes coding for extracellular matrix proteins and extracellular matrix degrading proteases in pancreatic cancer. *Int. J. Cancer* **62**, 407–413.

Grigioni, W. F., D'Errico, A., Fortunato, C., Fiorentino, M., Mancini, A. M., Stetler-Stevenson, W. G., Sobel, M. E., Liotta, L. A., Onisto, M., and Garbisa, S. (1994). Prognosis of gastric carcinoma revealed by interactions between tumor cells and base-ment membrane. *Mod. Pathol.* **7**, 220–225.

Hoyhtya, M., Fridman, R., Komarek, D., Porter-Jordan, K., Stetler-Stevenson, W. G., Liotta, L. A., and Liang, C. M. (1994). Immunohistochemical localization of matrix metalloproteinase 2 and its specific inhibitor TIMP-2 in neoplastic tissues with mono-clonal antibodies. *Int. J. Cancer* **56**, 500–505.

Huhtala, P., Chow, L. T., and Tryggvason, K. (1990). Structure of the human type IV collagenase gene. *J. Biol. Chem.* **265**, 11077–11082.

Imai, K., Shikata, H., and Okada, Y. (1995). Degradation of vitronectin by matrix metallo-proteinases-1, -2, -3, -7 and -9. *FEBS Lett.* **369**, 249–251.

Itoh, Y., Binner, S., and Nagase, H. (1995). Steps involved in activation of the complex of pro-matrix metalloproteinase 2 (progelatinase A) and tissue inhibitor of metallopro-teinases (TIMP)-2 by 4-aminophenylmercuric acetate. *Biochem. J.* **308**, 645–651.

Juliano, R. L., and Haskill, S. (1993). Signal transduction from the extracellular matrix. *J. Cell Biol.* **120**, 577–585.

Kleiner, D. E., Jr., Unsworth, E. J., Krutzsch, H. C., and Stetler-Stevenson, W. G. (1992). Higher-order complex formation between the 72-kilodalton type IV collagenase and tissue inhibitor of metalloproteinases-2. *Biochemistry* **31**, 1665–1672.

Kleiner, D. E., Jr., Tuuttila, A., Tryggvason, K., and Stetler-Stevenson, W. G. (1993). Stability analysis of latent and active 72-kDa type IV collagenase: the role of tissue inhibitor of metalloproteinases-2 (TIMP-2). *Biochemistry* **32**, 1583–1592.

Kolkenbrock, H., Orgel, D., Hecker-Kia, A., Noack, W., and Ulbrich, N. (1991). The complex between a tissue inhibitor of metalloproteinases (TIMP-2) and 72kDa progela-tinase is a metalloproteinase inhibitor. *Eur. J. Biochem.* **198**, 775–781.

Laurie, G. W., Leblond, C. P., and Martin, G. R. (1982). Localization of type IV collagen, laminin, heparan sulfate proteoglycan, and fibronectin to the basal lamina of basement membranes. *J. Cell. Biol.* **95**, 340–344.

Lelongt, B., Trugnan, G., Murphy, G., and Ronco, P. M. (1997). Matrix metalloproteinases MMP2 and MMP9 are produced in early stages of kidney morphogenesis but only MMP9 is required for renal organogenesis in vitro. *J. Cell Biol.* **136**, 1363–1373.

LePage, R. N., Fosang, A. J., Fuller, S. J., Murphy, G., Evin, G., Beyreuther, K., Masters, C. L., and Small, D. H. (1995). Gelatinase A possesses a beta-secretase-like activity in cleaving the amyloid protein precursor of Alzheimer's disease. *FEBS Lett.* **377**, 267–270.

Levy, A. T., Cioce, V., Sobel, M. E., Garbisa, S., Grigioni, W. F., Liotta, L. A., and Stetler-Stevenson, W. G. (1991). Increased expression of the Mr 72,000 type IV collagenase in human colonic adenocarcinoma. *Cancer Res.* **51**, 439–444.

Liotta, L. A. (1986). Tumor invasion and metastasis—Role of the extracellular matrix: Rhoads memorial award lecture. *Cancer Res.* **46**, 1–7.

Liotta, L. A., Abe, S., Robey, P. G., and Martin, G. R. (1979). Preferential digestion of basement membrane collagen by an enzyme derived from a metastatic murine tumor. *Proc. Natl. Acad. Sci. USA* **76**, 2268–2272.

Liotta, L. A., Tryggvason, K., Garbisa, S., Hart, I., Foltz, C. M., and Shafie, S. (1980). Metastatic potential correlates with enzymatic degradation of basement membrane collagen. *Nature* **284,** 67–68.

Liotta, L. A., Steeg, P. S., and Stetler-Stevenson, W. G. (1991). Cancer metastsis and angiogenesis: An imbalance of positive and negative regulation. *Cell* **64,** 327–336.

MacDougall, J. R., and Matrisian, L. M. (1995). Contributions of tumor and stromal matrix metalloproteinases to tumor progression, invasion and metastasis. *Cancer Metast. Rev.* **14,** 351–362.

Mackay, A. R., Hartzler, J. L., Pelina, M. D., and Thorgeirsson, U. P. (1990). Studies on the ability of 65-kDa and 92-kDa tumor cell gelatinases to degrade type IV collagen. *J. Biol. Chem.* **265,** 21929–21934.

Mainardi, C. L., Dixit, S. N., and Kang, A. H. (1980). Degradation of type IV (basement membrane) collagen by a proteinase isolated from human polymorphonuclear leukocyte granules. *J. Biol. Chem.* **255,** 5435–5441.

Matrisian, L. M. (1992). The matrix-degrading metalloproteinases. *BioEssays* **14,** 455–462.

Matrisian, L. M. (1994). Matrix metalloproteinase gene expression. *Ann. NY Acad. Sci.* **732,** 42–50.

Mazzieri, R., Masiero, L., Zanetta, L., Monea, S., Onisto, M., Garbisa, S., and Mignatti, P. (1997). Control of type IV collagenase activity by components of the urokinase-plasmin system: A regulatory mechanism with cell-bound reactants. *EMBO J.* **16,** 2319–2332.

McLaughlin, B., and Weiss, J. B. (1996). Endothelial-cell-stimulating angiogenesis factor (ESAF) activates progelatinase A (72 kDa type IV collagenase), prostromelysin 1 and procollagenase and reactivates their complexes with tissue inhibitors of metalloproteinases: A role for ESAF in non-inflammatory angiogenesis. *Biochem. J.* **317,** 739–745.

Miyajima, Y., Nakano, R., and Morimatsu, M. (1995). Analysis of expression of matrix metalloproteinases-2 and -9 in hypopharyngeal squamous cell carcinoma by *in situ* hybridization. *Ann. Otol. Rhinol. Laryngol.* **104,** 678–684.

Miyazaki, K., Kikkawa, Y., Nakamura, A., Yasumitsu, H., and Umeda, M. (1993). A large cell-adhesive scatter factor secreted by human gastric carcinoma cells. *Proc. Natl. Acad. Sci. USA* **90,** 11767–11771.

Monsky, W. L., Kelly, T., Lin, C. Y., Yeh, Y., Stetler-Stevenson, W. G., Mueller, S. C., and Chen, W. T. (1993). Binding and localization of M(r) 72,000 matrix metalloproteinase at cell surface invadopodia. *Cancer Res.* **53,** 3159–3164.

Montgomery, A. M. P., Mueller, B. M., Reisfeld, R. A., Taylor, S. M., and DeClerck, Y. A. (1994). Effect of tissue inhibitor of the matrix metalloproteinase-2 expression on the growth and spontaneous metastasis of a human melonama cell line. *Cancer Res.* **54,** 5467–5473.

Moses, M. A., Sudhalter, J., and Langer, R. (1990). Identification of an inhibitor of neovascularization from cartilage. *Science* **248,** 1408–1410.

Murphy, A. N., and Stetler-Stevenson, W. G. (1997). Determination of total and free TIMP-2 in tumor cell conditioned media. Unpublished data.

Murphy, G., Ward, R., Hembry, R. M., Reynolds, J. J., Kuhn, K., and Tryggvason, K. (1989). Characterization of gelatinase from pig polymorphonuclear leukocytes. *Biochem. J.* **258,** 463–472.

Murphy, G., Cockett, M. I., Ward, R. V., and Docherty, A. J. (1991). Matrix metalloproteinase degradation of elastin, type IV collagen and proteoglycan. A quantitative comparison of the activities of 95 kDa and 72 kDa gelatinases, stromelysins-1 and -2 and punctuated metalloproteinase (PUMP). *Biochem. J.* **277,** 277–279.

Murphy, G., Willenbrock, F., Ward, R. V., Cockett, M. I., Eaton, D., and Docherty, A. J. (1992). The C-terminal domain of 72 kDa gelatinase A is not required for catalysis, but is essential for membrane activation and modulates interactions with tissue inhibitors of metalloproteinases. *Biochem. J.* **283,** 637–41. [Erratum. *Biochem. J.* **284,** 935.]

Murphy, G., Willenbrock, F., Crabbe, T., O'Shea, M., Ward, R., Atkinson, S., O'Connell, J., and Docherty, A. (1994). Regulation of matrix metalloproteinase activity. *Ann. NY Acad. Sci.* **732,** 31–41.

Nakajima, M., Welch, D. R., Belloni, P. N., and Nicolson, G. L. (1987). Degradation of basement membrane type IV collagen and lung subendothelial matrix by rat mammary adenocarcinoma cell clones of differing metastatic potentials. *Cancer Res.* **47,** 4869–4876.

Nakajima, M., Lotan, D., Baig, M. M., Carralero, R. M., Wood, W. R., Hendrix, M. J., and Lotan, R. (1989). Inhibition by retinoic acid of type IV collagenolysis and invasion through reconstituted basement membrane by metastatic rat mammary adenocarcinoma cells. *Cancer Res.* **49,** 1698–1706.

Nguyen, Q., Willenbrock, F., Cockett, M. I., O'Shea, M., Docherty, A. J., and Murphy, G. (1994). Different domain interactions are involved in the binding of tissue inhibitors of metalloproteinases to stromelysin-1 and gelatinase A. *Biochemistry* **33,** 2089–2095.

Nomura, H., Fujimoto, N., Seiki, M., Mai, M., and Okada, Y. (1996). Enhanced production of matrix metalloproteinases and activation of matrix metalloproteinase 2 (gelatinase A) in human gastric carcinomas. *Int. J. Cancer* **69,** 9–16.

Onisto, M., Riccio, M. P., Scannapieco, P., Caenazzo, C., Griggio, L., Spina, M., Stetler-Stevenson, W. G., and Garbisa, S. (1995). Gelatinase A/TIMP-2 imbalance in lymph-node-positive breast carcinomas, as measured by RT-PCR. *Int. J. Cancer* **63,** 621–626.

Overall, C. M., and Sodek, J. (1990). Concanavalin A produces a matrix-degradative phenotype in human fibroblasts. Induction and endogenous activation of collagenase, 72–kDa gelatinase, and Pump-1 is accompanied by the suppression of the tissue inhibitor of matrix metalloproteinases. *J. Biol. Chem.* **265,** 21141–21151.

Overall, C. M., Wrana, J. L., and Sodek, J. (1991). Transcriptional and post-transcriptional regulation of 72-Da gelatinase/type IV collagenase by transforming growth factor-beta 1 in human fibroblasts. Comparisons with collagenase and tissue inhibitor of matrix metalloproteinase gene expression. *J. Biol. Chem.* **266,** 14064–14071.

Pei, D., and Weiss, S. J. (1995). Furin-dependent intracellular activation of the human stromelysin-3 zymogen. *Nature* **375,** 244–247.

Pepper, M. S., Belin, D., Montesano, R., Orci, L., and Vassalli, J. D. (1990). Transforming growth factor-beta 1 modulates basic fibroblast growth factor-induced proteolytic and angiogenic properties of endothelial cells in vitro. *J. Cell Biol.* **111,** 743–755.

Polette, M., Gilbert, N., Stas, I., Nawrocki, B., Noel, A., Remacle, A., Stetler-Stevenson, W. G., Birembaut, P., and Foidart, M. (1994). Gelatinase A expression and localization in human breast cancers. An *in situ* hybridization study and immunohistochemical detection using confocal microscopy. *Virchows Arch.* **424,** 641–645.

Poulsom, R., Pignatelli, M., Stetler-Stevenson, W. G., Liotta, L. A., Wright, P. A., Jeffery, R. E., Longcroft, J. M., Rogers, L., and Stamp, G. W. (1992). Stromal expression of 72 kda type IV collagenase (MMP-2) and TIMP-2 mRNAs in colorectal neoplasia. *Am. J. Pathol.* **141,** 389–396.

Poulsom, R., Hanby, A. M., Pignatelli, M., Jeffery, R. E., Longcroft, J. M., Rogers, L., and Stamp, G. W. (1993). Expression of gelatinase A and TIMP-2 mRNAs in desmoplastic fibroblasts in both mammary carcinomas and basal cell carcinomas of the skin. *J. Clin. Pathol.* **46,** 429–436.

Pyke, C., Ralfkiaer, E., Huhtala, P., Hurskainen, T., Dano, K., and Tryggvason, K. (1992). Localization of messenger RNA for Mr 72,000 and 92,000 type IV collagenases in human skin cancers by *in situ* hybridization. *Cancer Res.* **52,** 1336–1341.

Pyke, C., Ralfkiaer, E., Tryggvason, K., and Dano, K. (1993). Messenger RNA for two type IV collagenases is located in stromal cells in human colon cancer. *Am. J. Pathol.* **142,** 359–365.

Ray, J. M., and Stetler-Stevenson, W. G. (1995). Gelatinase A activity directly modulates melanoma cell adhesion and spreading. *EMBO J.* **14,** 908–917.

Roher, A. E., Kasunic, T. C., Woods, A. S., Cotter, R. J., Ball, M. J., and Fridman, R. (1994). Proteolysis of A beta peptide from Alzheimer disease brain by gelatinase A. *Biochem. Biophys. Res. Commun.* **205,** 1755–1761.

Romanic, A. M., and Madri, J. A. (1994). Extracellular matrix-degrading proteinases in the nervous system. *Brain Pathol.* **4,** 145–156.

Romanic, A. M., Graesser, D., Baron, J. L., Visintin, I., Janeway, C. A., Jr., and Madri, J. A. (1997). T cell adhesion to endothelial cells and extracellular matrix is modulated upon transendothelial cell migration. *Lab. Invest.* **76,** 11–23.

Sargeant, P., and Sage, S. O. (1994). Calcium signaling in platelet and other non-excitable cells. *Pharmac. Ther.* **64,** 395–443.

Sato, H., Takino, T., Okada, Y., Cao, J., Shinagawa, A., Yamamoto, E., and Seiki, M. (1994). A matrix metalloproteinase expressed on the surface of invasive tumour cells. *Nature* **370,** 61–65.

Sawicki, G., Salas, E., Murat, J., Miszta-Lane, H., and Radomski, M. W. (1997). Release of gelatinase A during platelet activation mediates aggregation. *Nature* **386,** 616–619.

Schnaper, H. W., Grant, D. S., Stetler-Stevenson, W. G., Fridman, R., D'Orazi, G., Murphy, A. N., Bird, R. E., Hoythya, M., Fuerst, T. R., French, D. L., *et al.* (1993). Type IV collagenase(s) and TIMPs modulate endothelial cell morphogenesis in vitro. *J. Cell Physiol.* **156,** 235–246.

Seltzer, J. L., Adams, S. A., Grant, G. A., and Eisen, A.Z. (1981). Purification and properties of a gelatin-specific neutral protease from human skin. *J. Biol. Chem.* **256,** 4662–4668.

Seltzer, J. L., Eisen, A. Z., Bauer, E. A., Morris, N. P., Glanville, R. W., and Burgeson, R. E. (1989). Cleavage of type VII collagen by interstitial collagenase and type IV collagenase (gelatinase) derived from human skin. *J. Biol.Chem.* **264,** 3822–3826.

Seltzer, J. L., Akers, K. T., Weingarten, H., Grant, G. A., McCourt, D. W., and Eisen, A.Z. (1990). Cleavage specificity of human skin type IV collagenase (gelatinase). *J. Biol.Chem.* **265,** 20409–20413.

Senior, R. M., Griffin, G. L., Fliszar, C. J., Shapiro, S. D., Goldberg, G. I., and Welgus, H. G. (1991). Human 92– and 72–kilodalton type IV collagenases are elastases. *J. Biol.Chem.* **266,** 7870–7875.

Sharkey, M. E., Adler, R. R., Brenner, C. A., and Nieder, G. L. (1996). Matrix metalloproteinase expression during mouse peri-implantation development. *Am. J. Reprod. Immunol.* **36,** 72–80.

Shimonovitz, S., Hurwitz, A., Dushnik, M., Anteby, E., Geva-Eldar, T., and Yagel, S. (1994). Developmental regulation of the expression of 72 and 92 kd type IV collagenases in human trophoblasts: A possible mechanism for control of trophoblast invasion [see comments]. *Am. J. Obstet. Gynecol.* **171,** 832–838.

Siess, W. (1989). Molecular mechanisms of platelet activation. *Pharmacol. Rev.* **69,** 58–178.

Stefansson, S., and Lawrence, D. A. (1996). The serpin PAI-1 inhibits cell migration by blocking integrin alpha V beta 3 binding to vitronectin [see comments]. *Nature* **383,** 441–443.

Stetler-Stevenson, W. G., and Corcoran, M. L. (1997). Tumor angiogenesis: functional similarities with tumor invasion. *In* "Regulation of Angiogenesis" (I. D. Goldberg and E. M. Rosen, eds.), pp. 413–418, Birkhäuser Verlag, Basel.

Stetler-Stevenson, W. G., Krutzsch, H. C., and Liotta, L. A. (1989a). Tissue inhibitor of metalloproteinase (TIMP-2). A new member of the metalloproteinase inhibitor family. *J. Biol. Chem.* **264,** 17374–17378.

Stetler-Stevenson, W. G., Krutzsch, H. C., Wacher, M. P., Margulies, I. M., and Liotta, L. A. (1989b). The activation of human type IV collagenase proenzyme. Sequence identification of the major conversion product following organomercurial activation. *J. Biol. Chem.* **264,** 1353–1356.

Stetler-Stevenson, W. G., Aznavoorian, S., and Liotta, L. A. (1993). Tumor cell interactions with the extracellular matrix during invasion and metastasis. *Annu. Rev. Cell Biol.* **9,** 541–573.

Strongin, A. Y., Collier, I. E., Krasnov, P. A., Genrich, L. T., Marmer, B. L., and Goldberg, G. I. (1993). Human 92 kDa type IV collagenase: Functional analysis of fibronectin and carboxyl-end domains. *Kidney Int.* **43,** 158–162.

Strongin, A. Y., Collier, I., Bannikov, G., Marmer, B. L., Grant, G. A., and Goldberg, G. I. (1995). Mechanism of cell surface activation of 72-kDa type IV collagenase. Isolation of the activated form of the membrane metalloprotease. *J. Biol. Chem.* **270,** 5331–5338.

Templeton, N. S., and Stetler-Stevenson, W. G. (1991). Identification of a basal promoter for the human Mr 72,000 type IV collagenase gene and enhanced expression in a highly metastatic cell line. *Cancer Res.* **51,** 6190–6193.

Tsuboi, R., and Rifkin, D. B. (1990). Bimodal relationship between invasion of the amniotic membrane and plasminogen activator activity. *Int. J. Cancer* **46,** 56–60.

Turck, J., Pollock, A. S., Lee, L. K., Marti, H. P., and Lovett, D. H. (1996). Matrix metalloproteinase 2 (gelatinase A) regulates glomerular mesangial cell proliferation and differentiation. *J. Biol. Chem.* **271,** 15074–15083.

Turpeenniemi-Hujanen, T., Thorgeirsson, U. P., Hart, I. R., Grant, S. S., and Liotta, L. A. (1985). Expression of collagenase IV (basement membrane collagenase) activity in murine tumor cell hybrids that differ in metastatic potential. *J. Natl. Cancer Inst.* **75,** 99–103.

Urbanski, S. J., Edwards, D. R., Maitland, A., Leco, K. J., Watson, A., and Kossakowska, A. E. (1992). Expression of metalloproteinases and their inhibitors in primary pulmonary carcinomas. *Br. J. Cancer* **66,** 1188–1194.

Ward, R. V., Atkinson, S. J., Slocombe, P. M., Docherty, A. J., Reynolds, J. J., and Murphy, G. (1991). Tissue inhibitor of metalloproteinases-2 inhibits the activation of 72 kDa progelatinase by fibroblast membranes. *Biochim. Biophys. Acta* **1079,** 242–246.

Welgus, H. G., Fliszar, C. J., Seltzer, J. L., Schmid, T. M., and Jeffrey, J. J. (1990). Differential susceptibility of type X collagen to cleavage by two mammalian interstitial collagenases and 72-kDa type IV collagenase. *J. Biol. Chem.* **265,** 13521–13527.

Wilhelm, S. M., Collier, I. E., Marmer, B. L., Eisen, A. Z., Grant, G. A., and Goldberg, G. I. (1989). SV40-transformed human lung fibroblasts secrete a 92–kDa type IV collagenase which is identical to that secreted by normal human macrophages. *J. Biol. Chem.* **264,** 17213–17221. [Erratum. *J. Biol. Chem.* **265,** 22570.]

Willenbrock, F., Crabbe, T., Slocombe, P. M., Sutton, C. W., Docherty, A. J., Cockett, M. I., O'Shea, M., Brocklehurst, K., Phillips, I. R., and Murphy, G. (1993). The activity of the tissue inhibitors of metalloproteinases is regulated by C-terminal domain interactions: A kinetic analysis of the inhibition of gelatinase A. *Biochemistry* **32,** 4330–4337.

Zucker, S., Conner, C., DiMassmo, B. I., Ende, H., Drews, M., Seiki, M., and Bahou, W. F. (1995). Thrombin induces the activation of progelatinase A in vascular endothelial cells. Physiologic regulation of angiogenesis. *J. Biol. Chem.* **270,** 23730–23738.

Gelatinase B: Structure, Regulation, and Function

Thiennu H. Vu* and Zena Werb†

*Department of Medicine, Division of Pulmonary and Critical Care Medicine, and
†Department of Anatomy, University of California, San Francisco, California 94143

I. INTRODUCTION

In 1962 Gross and Lapiere demonstrated the existence of a specific enzyme, collagenase, that cleaves native collagens into discrete fragments. Because this enzyme had no activity against denatured collagens or collagen fragments, it was hypothesized that another enzyme must accompany collagenase in order to completely degrade collagens in the extracellular matrix (ECM). In 1974 Sopata and Dancewicz purified a neutral protease from human neutrophils that could degrade

Matrix Metalloproteinases

denatured collagens (gelatins) (Sopata and Dancewicz, 1974). Proteolytic activities against gelatins and type IV and V collagens were subsequently identified from rabbit bone culture medium (Murphy *et al.*,
1981) and from human neutrophils (Murphy *et al.*, 1982) and characterized as metalloproteinases of ~90–110 kDa. Similar activities of 82–97
kDa were also detected in pig neutrophils (Murphy *et al.*, 1989), rabbit
and human alveolar macrophages (Hibbs *et al.*, 1987; Mainardi *et al.*,
1984), and in a variety of tumors and transformed cell lines (Collier *et
al.*, 1988; Moll *et al.*, 1990; Yamagata *et al.*, 1989). Wilhelm *et al.*
(1989) purified a gelatinolytic activity from simian virus 40 (SV40)-
transformed human lung fibroblasts and showed that it was identical
to a 92-kDa enzyme secreted by the monocytic U937 cells, fibrosarcoma
HT1080 cells, alveolar macrophages, and neutrophils. The cDNA sequence (Wilhelm *et al.*, 1989) shows that this enzyme belongs to the
matrix metalloproteinase (MMP) family, and is homologous to the previously identified 72-kDa type IV collagenase (Fig. 1). Both the 72-kDa
enzyme and the 92-kDa enzyme degrade native type IV and V collagens
and gelatins, but have no activity against native type I collagen, proteoglycans, or laminins. Because of its size and substrate preference, it
has been referred to as 92-kDa type IV collagenase, 92-kDa gelatinase,
or gelatinase B. Following the convention proposed by the Destin Beach
matrix metalloproteinases meeting in 1989 (Nagase *et al.*, 1992), it is
designated MMP-9. Since its discovery, gelatinase B has been implicated in a variety of physiological and pathological processes. This
article summarizes what has been learned regarding its structure,
expression, regulation of activities, substrate specificities, and functions.

II. GENE STRUCTURE

The human gelatinase B gene is 7.7 kb and contains 13 exons, ranges
in size from 104 bp (exon 12) to 280 bp (exon 9) (Huhtala *et al.*, 1991).
The introns range from 96 bp (intron 11) to 1800 bp (intron 12). The
mouse gene shows a similar organization, also with 13 exons (Masure
et al., 1993). All correspond in size to those in the human genes except
for exons 9 and 13, which contain 54 and 15 more bases, respectively,
than their counterparts in the human gene.

Analyses of the 5'-flanking region of both the human and mouse
genes show several putative control elements (Fig. 2) (Gum *et al.*, 1996;
Huhtala *et al.*, 1991; Munaut *et al.*, 1994; Sato *et al.*, 1993; Sato and
Seiki, 1993). There is a TATA-like box at position −29 in both genes.
There is no CCAAT box. There are several GC boxes that can serve as
binding sites for Sp1 transcription factors. There are also several 12-

```
MMP-9    ..........  MSLWQPLVLV LLVLGCCFAA  PRQRQSTLVL  FPGDLRTNLT  DRQLAEEYLY
MMP-2    .......MEA  LMARGALTGP LRALCLLGCL  LSHAAAPSP   IIKFPGDVAP  KTDKELAVQY
MMP-1    ................MH  SFPPLLLLLF WGVVSHSFPA  TLETEQDVDL  VQKYLEKYYN


51
RYGYTRVAEM  RGESKSLGPA  LLLLQKQLSL  PETGELDSAT  LKAMRTPRCG  VDPLGRFQTF
LNTFYGCPKE  SCNLFVLKDT  LKKMQKFFGL  PQTGDLDQNT  IETMRKPRCG  NPDVANYNFF
LKNDGRQVEK  RRNSGPVVEK  LKQMQEFFGL  KVTGKPDAET  LKVMKQPRCG  VPDVAQFVLT


111
EGDLKWHHHN  ITYWIQNYSE  DLPRAVIDDA  FARAFALWSA  VTPLTFTRVY  SRDADIVIQF
PRKPRWDKNQ  ITYRIIGYTP  DLDPETVDDA  FARAFQVWSD  VTPLRFSRIH  DGEADIMINF
EGNPRWEQTH  LTYRIENYTP  DLPRADVDHA  IEKAFQLWSN  VTPLTFTKVS  EGQADIMISF


171
GVAEHGDGYP  FDGKDGLLAH  AFPPGPGIQG  DAHFDDDELW  SLGKGVVVPT  RFGNADGAAC
GRWEHGDGYP  FDGKDGLLAH  AFAPGTGVGG  DSHFDDDELW  TLGEGQVVRV  KYGNADGEYC
VRGDHRDNSP  FDGPGGNLAH  AFQPGPGIGG  DAHFDEDERW  TNNFREYN..  ..........


231
HFPFIFEGRS  YSACTTDGRS  DGLPWCSTTA  NYDTDDRFGF  CPSERLYTRD  GNADGKPCQF
KFPFLFNGKE  YNSCTDTGRS  DGFLWCSTTY  NFEKDGKYGF  CPHEALFTMG  GNAEGQPCKF
..........  ..........  ..........  ..........  ..........  ..........


291
PFIFQGQSYS  ACTTDGRSDG  YRWCATTANY  DRDKLFGFCP  TRADSTVMGG  NSAGELCVFP
PFRFQGTSYD  SCTTEGRTDG  YRWCGTTEDY  DRDKKYGFCP  ETAMSTV.GG  NSEGAPCVFP
..........  ..........  ..........  ..........  ..........  ..........


351
FTFLGKEYST  CTSEGEGDGR  LWCATTSNFD  SDKKWGFCPD  QGYSLFLVAA  HEFGHALGLD
FTFLGNKYES  CTSAGRSDGK  MWCATTANYD  DDRKWGFCPD  QGYSLFLVAA  HEFGHAMGLE
..........  ..........  ..........  ..........  ...LHRVAA   HELGHSLGLS


411
HSSVPEALMY  PMYRFTEGPP  LHKDDVNGIR  HLYGPRPEPE  PRPPTTTTPQ  PTAPPTVCPT
HSQDPGALMA  PIYTYTKNFR  LSQDDIKGIQ  ELYG......  ..........  ..........
HSTDIGALMY  PSYTFSGDVQ  LAQDDIDGIQ  AIYG......  ..........  ..........


471
GPPTVHPSER  PTAGPTGPPS  AGPTGPPTAG  PSTATTVPLS  PVDDACNVNI  FDAIAEIGNQ
..........  ..........  ......ASP   DIDLGTGPTP  TLGPVTPEIC  KQDIVFDGIA
.....:....  ..........  ......RSQ   NPVQPIGPQT  PKACDSKLTF  DAITTIRGEV


531
LYLFKDGKYW  RFSEGRGSRP  QGPFLIADKW  PALPRKLDSV  FEEPLSKKLF  FFSGRQVWVY
QIRGEIFFFK  DRFIWRTVTP  RDKPMGPLLV  ATFWPELPEK  IDAVYEAPQE  EKAVFFAGNE
MFFKDRFYMR  TNPFYPEVEL  NFISVFWPQL  PNGLEAAYEF  ADRDEVRFFK  GNKYWAVQGQ


591
TGASVLGPRR  LDKLGLGADV  AQVTGALRSG  RGKMLLFSGR  RLWRFDVKAQ  MVDPRSASEV
YWIYSASTLE  RGYPKPLTSL  GLPPDVQRVD  AAFNWSKNKK  TYIFAGDKFW  RYNEVKKKMD
NVLHGYPKDI  YSSFGFPRTV  KHIDAALSEE  NTGKTYFFVA  NKYWRYDEYK  RSMDPGYPKM


651
DRMFPGVPLD  THDVFQYREK  AYFCQDRFYW  RVSSRSELNQ  VDQVGYVTYDILQCPED
PGFPKLIADA  WNAIPDNLDA  VVDLQGGGHS  YFFKGAYYLK  LENQSLKSVKFGSIKSDWLGC
IAHDFPGIGH  KVDAVFMKDG  FFYFFHGTRQ  YKFDPKTKRI  LTLQKANSWFNCRKN
```

FIG. 1. Amino acid sequences of human gelatinase B (MMP-9), gelatinase A (MMP-2), and interstitial collagenase (MMP-1). The shadowed sequences denote the catalytic domains, which are interrupted by the fibronectin repeats in MMP-2 and MMP-9. The single-underlined sequences denote the propeptides. The boxed sequence denotes the collagen V-like domain in MMP-9. Sequences were obtained from the GenBank database. MMP-1: Acc. No. P03956; MMP-2: Acc. No. P08253; MMP-9: Acc. No. P14780.

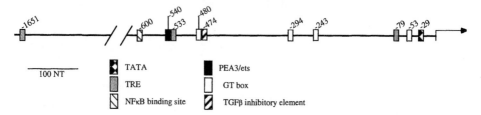

FIG. 2. The promoter region of the human gelatinase B gene.

O-tetradecanyol phorbol 4-acetate (TPA) responsive elements (TRE) (located at −79 , −533, and −1651 in the human gene) that can bind AP1 transcription factors. In the human gene there is a NFκB binding site at −600 and a PEA3/Ets binding site at −540. There are also several GT boxes that are similar to the retinoblastoma control elements (RCEs) at −53, −243, −294, and −480. There is a consensus sequence for a transforming growth factor β (TGF-β) inhibitory element at position −474 in the human gene, but there is no counterpart in the mouse gene. In the human promoter, mutations of the AP1, NFκB, and Sp1 sites located at −79, −600, and −558, respectively, reduce or abolish the induction by TPA or tumor necrosis factor α (TNF-α) in the human OST osteosarcoma or HepG2 hepatoma cell lines (Sato and Seiki, 1993). In OST cells a nuclear factor that binds to Sp1 is constitutively expressed, whereas factors binding to AP1 and NFκB sites are rapidly induced with TNF-α treatment (Sato *et al.*, 1993). The AP1 site is indispensible, but alone is not sufficient for the induction by TPA and TNF-α and requires, in addition, either the NFκB or the Sp1 site. The AP1 site at −79 is also essential, along with a GT box at −52 for activation of the promoter by v-src, whereas the GT box is not essential for activation by TPA. In the fibrosarcoma cell line HT1080, gel shift assays show that RCE-binding proteins, including Sp1 family members, can bind to GT boxes (He, 1996). The GT box sequence can also compete with a c-fos RCE for binding to Sp1 transcription factors. Recently it was shown that activated Ha-ras can induce gelatinase B expression in the ovarian adenocarcinoma cell line OVCAR-3 (Gum *et al.*, 1996). Optimal induction requires a region between −634 to −331, which contains a PEA3/Ets motif, AP1 site, NFκB site, and a GT box. Deletion or mutation of any of these sites reduces activation of the promoter by Ras. Several growth factors, including TGF-α and epidermal growth factor (EGF), may mediate their effect through this GTP-binding protein.

A 2-kb 5′-flanking region of the gene is sufficient for developmental and tissue-specific expression; thus, a β-galactosidase reporter transgene ligated to this region has an identical pattern of developmental expression to that of gelatinase B *in vivo* (Munaut *et al.*, 1994).

III. Protein Structure and Activation

Gelatinase B contains several structural domains, each of which is encoded by one or more exons. Similar to other enzymes of the MMP family, it has a propeptide, a catalytic domain, and a C-terminal hemopexin domain (Wilhelm *et al.*, 1989) (Fig. 3). The gelatinases A and B also have a fibronectin-like domain inserted in the middle of the catalytic domain. Unique to gelatinase B is a collagen V-like domain located between the catalytic domain and the hemopexin domain.

The structure–function relationships of the various domains of the MMP molecule have been extensively investigated. The propeptide maintains the enzyme in an inactive state through a cysteine switch mechanism (Van Wart and Birkedal-Hansen, 1990). Activation of the enzyme occurs when the interaction between the active site zinc molecule and a cysteine in the pro-domain is disrupted, rendering the active site accessible. This can be achieved by proteolytic removal of the propeptide, or by disruption of the cysteine–zinc interaction by organomercurials and chaotropic agents, leading to an active enzyme which then cleaves the propeptide autocatalytically.

The activation of gelatinase B is a complex process. During activation the molecule is processed in different ways depending on whether it is free or in a complex with other molecules. In most cells, gelatinase B is secreted as a complex with TIMP-1, a tissue inhibitor of metallopro-

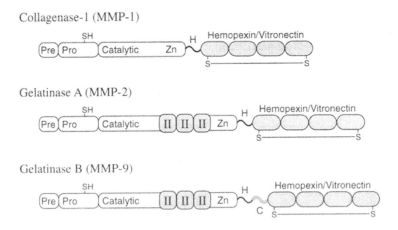

FIG. 3. Functional domains of matrix metalloproteinases.

teinase that can inhibit gelatinase B activity (Moll *et al.*, 1990; Murphy *et al.*, 1989). The enzyme secreted by neutrophils is complexed with a 25-kDa protein of the lipocalin family (Kjeldsen *et al.*, 1993). The function of this interaction is not known. Activation of TIMP-1-free gelatinase B with 4-aminophenylmercuric acetate (APMA) results in cleavage first of the NH_2-terminal peptide at the Ala^{74}–Met^{75} bond, yielding an 83-kDa intermediate, then of the COOH terminus, resulting in a 67-kDa enzyme (Okada *et al.*, 1992); both have similar activity. TIMP-1 binding to gelatinase B prohibits processing to the 67-kDa form. Gelatinase B is also activated by cathepsin G, trypsin, α-chymotrypsin, and stromelysin-1 (MMP-3), but not by plasmin, thrombin, or interstitial collagense (MMP-1) (Okada *et al.*, 1992). Leukocyte elastase and tissue kallikrein were reported to both activate (Ferry *et al.*, 1997; Menashi *et al.*, 1994) and not activate gelatinase B (Okada *et al.*, 1992). Activation by stromelysin-1 results in cleavage initially of the Glu^{40}–Met^{41} bond, then the Arg^{87}–Phe^{88} bond, resulting in an 82-kDa form (Ogata *et al.*, 1992). If gelatinase is not complexed with TIMP-1, it may be further processed by stromelysin-1 to either an active 67-kDa form with a COOH-terminal truncation or an inactive 50-kDa form lacking the catalytic domain (Shapiro *et al.*, 1995). Collagenase-1, matrilysin, mast cell chymase, gelatinase A, and trypsin also activate gelatinase B (Fang *et al.*, 1996; Fridman *et al.*, 1995; Sang *et al.*, 1995). Activation by all of these proteinases results in cleavage of the Arg^{87}–Phe^{88} bond, but no COOH-terminal truncation has been noted, even with TIMP-1-free recombinant enzymes. The propeptide removed by the proteases contains the cysteine switch, Cys^{50}, however, this cysteine remains with the enzyme during organomercurial activation. It has been suggested that the continuing interaction of this cysteine with organomercurials removes the normal inhibitory interaction with the zinc moeity (Sang *et al.*, 1995). However, Triebel *et al.* (1992) observed C-terminal cleavages of gelatinase B on activation with $HgCl_2$, with loss of three cysteine residues at positions 497, 653, and 683. The authors suggested a mechanism of activation by rearrangement of intramolecular disulfide bridges that leads to active site accessibility.

The activation of gelatinase B complexed with TIMP-1 either by trypsin or APMA results in the processing to lower molecular weight forms with decreased activity. However, if the progelatinase B-TIMP-1 complex is allowed to interact with another MMP, such as stromelysin-1 or collagenase-1, before activation, the activated form has a higher activity (Kolkenbrock *et al.*, 1995; Ogata *et al.*, 1995). Thus the activity of gelatinase B is regulated not only by interactions with TIMP-1 but also with other MMPs. When the monocytic leukemic cell line U937 is treated with TPA and LPS, it secretes large amounts of gelatinase B and collagenase-1, partly complexed with each other free of TIMP-1 (Goldberg *et al.*, 1992). The complex of gelatinase B with collagenase-1 can be activated by stromelysin-1.

Activation of recombinant unglycosylated 57.5-kDa gelatinase B with APMA in the absence of Ca^{2+} generates a 49-kDa inactive intermediate, whereas activation in the presence of Ca^{2+} results in an active 41.5-kDa molecule that requires Ca^{2+} for activity (Bu and Pourmotabbed, 1995). The 41.5-kDa form has the same NH_2 terminus, Met^{75}, as the 49-kDa one, indicating that it must have a COOH-terminal truncation. When gelatinase B is activated by trypsin in the absence of Ca^{2+}, it is degraded. However, when it is activated by trypsin in the presence of Ca^{2+}, a 40-kDa active form is generated that does not require Ca^{2+} for activity. The NH_2 terminus of this form is $Phe^{88.}$ When the 41.5-kDa form obtained by activation with APMA is treated with trypsin, the same Phe^{88} NH_2 terminus is generated and this also abolished the Ca^{2+} dependence of enzyme activity. Phe^{88} probably forms a salt linkage with the active site Asp^{432}, an interaction necessary for activity, and Ca^{2+} may substitute for Phe^{88} in this interaction (Bu and Pourmotabbed, 1996).

The catalytic domain of all MMPs contains a conserved sequence HExGHxxGxxHS/T that binds the catalytic zinc (Nagase *et al.*, 1992). In gelatinase B, this sequence begins with His^{400}. Substitution of this His residue results in loss of enzymatic activity (Pourmotabbed *et al.*, 1995). Besides the zinc-binding sequence, the catalytic domain of gelatinase B contains another highly conserved sequence, ^{429}LxxDDxxGI, which has similiarity to the calcium-binding sites of other calcium-dependent proteins (Pourmotabbed *et al.*, 1995). Mutations of Asp^{432} and Asp^{433} in this sequence result in a three- to fivefold reduction in K_{cat} with no change in K_m for substrate or K_i for the inhibitor GM6001, suggesting that these Asp residues are involved in the stabilization of the active site.

The fibronectin-like domain in gelatinase B consists of three contiguous fibronectin type II homology units (T2HU-1, T2HU-2, and T2HU-3). FN type II homology units can bind gelatin, as shown by the ability to confer gelatin binding when fused to an exogenous protein (Collier *et al.*, 1992; Strongin *et al.*, 1993). T2HU-2 binds several-fold more efficiently than T2HU-3, but T2HU-1 does not bind at all. Binding of gelatin is not a rate-limiting step in its degradation, as DMSO inhibits gelatin binding, but not its degradation. However, when the fibronectin-like domain is deleted, the resultant enzyme loses its gelatinolytic activity (Pourmotabbed, 1994). The fibronectin-like domain is also important in the elastolytic activity of gelatinase B. The FN type II repeats can competitively inhibit gelatinase B binding to elastin, and a recombinant enzyme lacking this domain can neither bind to, nor degrade, elastin (Shipley *et al.*, 1996).

The type V collagen-like domain does not appear to be important for enzyme activity or substrate specificity. A mutant enzyme with this domain deleted has similar activity against gelatins and type V collagens compared to wild type enzymes (Pourmotabbed, 1994).

Interaction of gelatinase B with TIMP-1 is thought to be mediated by the COOH-terminal hemopexin domain. Deletion of as few as eight amino acids from the COOH terminus of gelatinase B inhibits complex formation with TIMP-1, and marked reduction in the rate constant for TIMP-1 inhibition of gelatinase B activity. The C-terminal sequence contains a cysteine, $Cys^{704,}$ the mutation of which also abolishes the interaction with TIMP-1 (O'Connell *et al.*, 1994). Cys^{704} and Cys^{516} are conserved in all hemopexin domains of the MMP family. Truncation of the COOH-terminal sequence, up to amino acid 426, has no effect on the activation or activity of gelatinase B. In fact, full activation of the enzyme is associated with truncation of the COOH domain.

IV.   SUBSTRATES AND INHIBITORS

A.   *Collagen Substrates*

Gelatinase B is widely thought of as a type IV collagenase, owing to numerous reports of its ability to cleave native type IV collagen molecules into discrete fragments. However, there is disagreement as to whether gelatinase B can truly cleave native full-length type IV collagen and function as a type IV collagenase *in vivo*. Most reports on the type IV collagenolytic activity of gelatinase B use type IV collagens that had been solubilized from tissues by pepsin as substrate. Even though these resist degradation by trypsin at 30°C, and therefore are considered nondenatured collagens, they may not represent true native type IV collagens. With full-length EHS type IV collagens, no cleavage by gelatinase B is seen at 25–32°C, whereas pepsin-extracted type IV collagens are readily cleaved at 25°C (Mackay *et al.*, 1990; Moll *et al.*, 1990). On the other hand, EHS type IV collagens are cleaved by gelatinase B at 32 and 35°C (Murphy *et al.*, 1989; Okada *et al.*, 1992; Wilhelm *et al.*, 1989). In addition, a gelatinase from neutrophils is able to solubilize a glomerular basement membrane preparation at 37°C (Vissers and Winterbourn, 1988).

Pepsin-solubilized type V collagens are also cleaved by gelatinase B at 25–37°C (Hibbs *et al.*, 1987; Niyibizi *et al.*, 1994; Pourmotabbed *et al.*, 1994). Cleavage of type XI collagen by gelatinase B has also been observed at 36°C (Pourmotabbed *et al.*, 1994). Unlike gelatinase A, gelatinase B does not cleave type I collagen.

B. Other Extracellular Matrix Proteins

Gelatinase B cleaves aggrecan, a cartilage proteoglycan, at a single cleavage site in the interglobular domain. However, its activity against aggrecan is lower than that of matrilysin, gelatinase A, and stromelysin-1 (Fosang et al., 1992). Link protein, a glycoprotein that stabilizes the interaction between aggrecans and hyaluronate in proteoglycan aggregates, is also a substrate for gelatinase B, but again, matrilysin, collagenase-1, gelatinase A, and stromelysin-2 are more active against link protein than gelatinase B (Nguyen et al., 1993). Gelatinase B also has significant elastinolytic activity, approximately 30% that of leukocyte elastase (Senior et al., 1991).

C. Other Substrates

A variety of non-ECM macromolecules are cleaved by gelatinase B. Myelin basic protein is cleaved at several sites, resulting in several peptides, one of which has been shown to be immunogenic and may act as an encephalitogen (Gijbels et al., 1993; Proost et al., 1993b). Gelatinase B also cleaves members of the galactoside-binding proteins, including CBP30, a protein implicated in cell–substratum interactions, and CBP35 (Mehul et al., 1994; Ochieng et al., 1994). Proteolytic inactivation by gelatinase B has been observed for α_1-antitrypsin, α_1-antichymotrypsin, and α_1-proteinase inhibitor (Desrochers et al., 1992; Sires et al., 1994). Gelatinase B may also regulate the activity of cytokines and hormones because it cleaves IL-1β and substance P (Backstrom and Tokes, 1995; Ito et al., 1996). It has also been found to cleave amyloid β peptide (Backstrom et al., 1996).

D. Tissue Inhibitors of Metalloproteinases

The tissue inhibitors of metalloproteinases (TIMPs) are a family of polypeptides that form noncovalent complexes with either active or latent MMPs and inhibit their activity and/or activation. Four members of the TIMP family have been cloned and sequenced. The TIMPs have only about 25% peptide sequence identity, but they all have 12 conserved cysteines that may be important in forming the secondary structures of the molecules. All the TIMPs appear to have activities against all members of the MMPs family, with few individual differences. Gelatinase A is secreted as a complex with TIMP-2, whereas gelatinase B is secreted with TIMP-1 (Ward et al., 1991). As mentioned earlier, complexing with TIMP-1 influences gelatinase B processing during activation and affects its activity as well. All the TIMPs can inhibit gelatinase B activity, but TIMP-2 appears to have a higher specific activity compared to TIMP-1, whereas that of TIMP-3 is about the same (Apte *et al.*, 1995; Howard *et al.*, 1991; Ward *et al.*, 1991).

E. Synthetic Inhibitors

Chelating agents such as 1,10 phenanthroline and EDTA interact with the zinc molecule at the active site, blocking MMP activity. Peptide substrate analogues that contain chelating moieties such as thiol, hydroxamate, or phosphoramidate can also act as potent and specific inhibitors of MMPs. Several of these have been synthesized and utilized effectively to inhibit MMPs (Delaisse *et al.*, 1985; Gray *et al.*, 1986; Grobelny *et al.*, 1992). Of interest, the thiol *N*-acetylcysteine has also been shown to inhibit MMP activity (Albini *et al.*, 1995).

Tetracycline and its derivatives have also been observed to inhibit MMPs, including interstitial collagenase and gelatinase B. The inhibitory activity does not correlate with the antimicrobial activity of these compounds (Golub *et al.*, 1995; Paemen *et al.*, 1996).

Neutralizing antibodies have also been used successfully as selective inhibitors of gelatinase B (Behrendtsen *et al.*, 1992; Librach *et al.*, 1991; Ramos-DeSimone and French, 1994).

F. α-Macroglobulins

α2-Macroglobulins are widely considered inhibitors of proteinases, but they function more as a molecular trap than true enzyme inhibitors (Sottrup-Jensen, 1989). The α-macroglobulins are large glycoproteins found in the serum. They bind proteinases through an exposed peptide stretch, known as the "bait" region. Binding of a proteinase to the bait sequence leads to its cleavage and large conformational changes in the macroglobulin molecule. This results in entrapment of the proteinase, and exposes binding sites for high affinity cell surface receptors that target the whole complex for degradation. Thus α-macroglobulins can serve to sequester and clear gelatinase B.

V. Regulation of Expression

Unlike gelatinase A, which is constitutively expressed at low level by many cells, gelatinase B is normally expressed only by trophoblasts, osteoclasts, neutrophils, and macrophages. However, its expression can be highly induced in a variety of cells by many agents, including growth factors, cytokines, ECM molecules, cell–cell and cell–ECM adhesion molecules, and agents that alter cell shape.

A. Growth Factors and Cytokines

Many growth factors and cytokines have been found to affect gelatinase B expression in a variety of cell types. These are summarized in Table I. The granulocyte chemotactic protein GCP-2 of the IL-8 family is both chemotactic for and induces gelatinase B expression in neutrophils, and may therefore be involved in the recruitment of neutrophils into tissues (Proost et al., 1993a). The expression of gelatinase B by activated T cells may be mediated in an autocrine fashion by the secretion of IL-2 (Montgomery et al., 1993). The induction of gelatinase B in rabbit articular chondrocytes in culture by IL-1 is augmented by cyclooxygenase inhibitors, and the addition of exogenous PGE_1 and PGE_2 decreases gelatinase B induction (Ito et al., 1995). The downregulation by IL-10 of gelatinase B induction by concanavalin A (conA) in monocytes appears to be mediated through a prostaglandin pathway (Mertz et al., 1994). ConA up-regulates the synthesis of inducible prostaglandin H synthase 2 (iPGHS-2), and this effect is attenuated by IL-10. The addition of exogeneous PGE_2 or dibutyryl cAMP restores the induction of iPGHS-2 and gelatinase B by conA in the presence of IL-10. Similarly, IL-4 suppresses conA induction of gelatinase B by monocytes through the same pathway, because exogenous PGE_2 and dibutyryl cAMP can overcome this suppression (Corcoran et al., 1992). Further evidence in support of the prostaglandin pathways was shown by the suppression of conA induction of gelatinase B by indomethacin. G proteins may be involved in this pathway, because cholera toxin enhances the production of iPGHS-2 and gelatinase B in conA-treated monocytes, and can reverse indomethacin suppression of the conA induction of these enzymes, while pertussin toxin treatment suppresses their induction (Corcoran et al., 1994). Lipopolysaccharide (LPS) induction of gelatinase B expression is probably through induction of cytokines that act in an autocrine fashion. Tumor promoter also acts synergistically with other agents such as TGF-β, TNF-α, and IL-1α.

Other soluble factors that induce gelatinase B expression have been tentatively identified. In coculture of an oral cavity squamous cell carcinoma cell line, UMSCC1, and oral cavity fibroblasts, a trypsin-sensitive soluble factor of 3–10 kDa in the fibroblast-conditioned medium increases gelatinase B expression by the carcinoma cells (Lengyel et al., 1995). Induction of gelatinase B in a breast cancer cell line by coculture with fibroblasts or by fibroblast-conditioned medium results from a trypsin-sensitive factor migrating at 30–100 kDa (Himelstein and Muschel, 1996). Gelatinase B is also induced in cocultures of fibroblasts and a human osteosarcoma cell line (Kurogi et al., 1996). Fibroblast-conditioned medium increases gelatinase B expression by the osteosar-

TABLE I

REGULATION OF GELATINASE B EXPRESSION BY GROWTH FACTORS AND CYTOKINES

Agent	Effect on *GelB* expression	Cell types
TGF-β	Increases	Peripheral blood monocytes,[28] skin fibroblasts,[22] colon carcinoma cells,[27] melanoma cells[16]
LIF	Increases	Blastocyst outgrowths[7]
EGF	Increases	Blastocyst outgrowths,[7] skin fibroblasts,[22] colon carcinoma cells,[27] breast carcinoma cells[13]
bFGF	Increases	Monocytes[29]
GCP-2	Increases	Neutrophils[26]
IL1β	Increases	Monocytes,[19] skin fibroblasts,[22] kidney mesangial cells,[18] colon carcinoma cells,[27] melanoma cells,[16] astrocytes[5]
IL1α	Increases	Aortic smooth muscle cells,[1] articular cartilage explants,[23] chondrocytes,[12] osteoblasts,[15] astrocytes[5]
IL-2	Increases	Activated T cells[24]
IL-10 and IL-4	Decreases the induction by ConA	Monocytes[3,21]
TNF-α	Increases	Skin fibroblasts,[22] osteosarcoma cells,[25] colon carcinoma cells,[27] monocytic leukemic cells,[19] osteoblasts,[15] melanoma cells,[14] fibrosarcoma cells,[25] astrocytes[5]
IFN-α and IFN-γ	Increases or decreases	Melanoma cells, increases with short-term treatment, decreases with long-term treatment[11]
LPS	Increases	Macrophages,[10] monocytic leukemic cells,[19] microglial cells,[4] astrocytes[5]
PTH	Increases	Osteoclasts and monocytic cells in bone explants,[31] osteoblasts[20]
TPA and PMA	Increases	Capillary endothelial cells,[8] HUVEC,[6] aortic smooth muscle cells,[1] corneal epithelial cells,[2] Kupffer cells,[30] mesothelial cells,[17] kidney mesangial cells,[18] melanoma cells[9,14,16]

[1] Fabunmi *et al.*, 1996; [2] Fini *et al.*, 1995; [3] Corcoran *et al.*, 1992; [4] Gottschall *et al.*, 1995a; [5] Gottschall *et al.*, 1995b; [6] Hanemaijer *et al.*, 1993; [7] Harvey *et al.*, 1995; [8] Herron *et al.*, 1986; [9] Houde *et al.*, 1993; [10] Houde *et al.*, 1996; [11] Hujanen *et al.*, 1994; [12] Ito *et al.*, 1995; [13] Kondapake *et al.*, 1997; [14] Lauricella-Lefebvre *et al.*, 1993; [15] Lorenzo *et al.*, 1992; [16] MacDougall *et al.*, 1995; [17] Marshall *et al.*, 1993; [18] Martin *et al.*, 1994; [19] McMillan *et al.*, 1996b; [20] Meikle *et al.*, 1992; [21] Mertz *et al.*, 1994; [22] Miyagi *et al.*, 1995; [23] Mohtai *et al.*, 1993; [24] Montgomery *et al.*, 1993; [25] Okada *et al.*, 1990; [26] Proost *et al.*, 1993a; [27] Shimizu *et al.*, 1996; [28] Wahl *et al.*, 1993; [29] Weston and Weeks, 1996; [30] Winwood *et al.*, 1995; [31] Witty *et al.*, 1996.

coma cells, but osteosarcoma-conditioned medium does not increase gelatinase B expression by fibroblasts. The activity in the fibroblast-conditioned medium may be mediated by bFGF, as it is inhibited by anti-FGF antibody.

B. Cell–Cell and Cell–Extracellular Matrix Interactions

Expression of gelatinase B is also regulated by cell adhesion molecules, ECM, and agents that change cell shape. Transfection of NCAM-B, a transmembrane NCAM, decreases gelatinase B secretion in a glioma cell line (Edvardsen et al., 1993). Following transfection, the cells become more rounded and spread less efficiently. Interestingly, transfection of the nontransmembrane NCAM-C, which is expressed at the cell surface through GPI linkage, has no effect on gelatinase B secretion. SPARC, a molecule secreted by rapidly renewing cell populations, also induces gelatinase B expression (Tremble et al., 1993). SPARC destabilizes actin in focal contacts. A laminin peptide, SIKVIV, induces gelatinase B in human monocytes; interestingly, this induction depends on culture conditions (Corcoran et al., 1995). If cells are allowed to adhere before the peptide is added, gelatinase B induction does not occur. However, the peptide retains its ability to act synergistically with conA. Intact laminin does not induce this response. SIKVIV has been shown to promote cell attachment, migration, and angiogenesis. Similarly, a fibronectin fragment that contains the central RGD cell-binding region also induces gelatinase B expression, while whole fibronectin does not (Huhtala et al., 1995). On the other hand, cells plated on a mixture of tenascin and fibronectin up-regulate gelatinase B (Tremble et al., 1994). Thus the composition of ECM and the state of degradation of the individual components act cooperatively in the regulation of expression of MMPs. Some of the interactions with ECM or cell adhesion molecules may be mediated by reorganization of the actin cytoskeleton, as treatment with cytochalasin D decreases gelatinase B expression in a human melanoma cell line, while colchicine has no effect (MacDougall and Kerbel, 1995). Cytochalasin D can also abolish the induction of gelatinase B in HL-60 cell lines treated with TPA.

Cell–cell contacts have also been shown to result in gelatinase B up-regulation. Fibroblasts cocultured with colon carcinoma cells show induction of gelatinase B (Segain et al., 1996). This effect requires direct cell–cell contacts and is inhibited by antibodies against $\beta1$ integrins, cytochalasin D, a protein kinase C inhibitor, and dexamethasone. Himelstein et al. (1994b) cocultured rat embryo fibroblasts with a transformed rat embryo cell line. This cell line does not make gelatinase B, but it gives rise to a metastatic tumor that contains gelatinase B in the stroma. The coculture results in the induction of gelatinase B in the fibroblasts. Cell contact is required, because induction is not observed with conditioned medium. However, the induction does not require viability of the transformed cells, since even methanol-fixed cells can mediate the effect.

C.   Viral Proteins

Several viral proteins have been shown to regulate the expression of gelatinase B. When a temperature-sensitive mutant of the SV40 T antigen is used to transform human placental trophoblast-like cells, the cells down-regulate expression of gelatinase B at the permissive temperature and change into a noninvasive phenotype (Logan *et al.*, 1996). Adenovirus E1A protein expression in Ha-ras and v-myc transformed rat embryo fibroblasts also results in decreased gelatinase B expression (Bernhard *et al.*, 1994). HIV-infected human monocytes have increased gelatinase B expression compared to noninfected cells (Chapel *et al.*, 1994; Weeks *et al.*, 1993). MMLV, a retrovirus that causes T-cell lymphoma and leukemia, can transactivate cellular genes through a transcript expressed from the viral LTR. When this LTR is stably or transiently introduced into either Balb/c-3T3 fibroblasts or Hela cells, it results in increased expression of gelatinase B (Faller *et al.*, 1997). This effect may be mediated through binding of AP1 transcription factors, because the AP1 site in the gelatinase B promoter is required for the induction.

VI.   Function

A.   Implantation

During implantation, the embryonic trophoblasts invade the uterine epithelium. This process requires extensive ECM degradation and remodeling. There are several lines of evidence suggesting that gelatinase B may be the important MMP in this process. Gelatinase B is the major MMP secreted by mouse and human trophoblast cells in culture, and a function-perturbing antigelatinase B antibody blocks ECM degradation by trophoblast outgrowths (Behrendtsen *et al.*, 1992; Librach *et al.*, 1991). In the implanting mouse embryo, gelatinase B mRNA is strongly expressed only by invading trophoblast cells (Alexander *et al.*, 1996; Reponen *et al.*, 1995). Interestingly, TIMP-3 is expressed by the maternal cells surrounding the invading embryonic cells (Fig. 4). It is therefore quite surprising that the gelatinase B-deficient mouse develops to term. These mice are also fertile and give birth to near normal size litters. There are morphologic abnormalities in the implantation site, but these are not lethal (unpublished observations).

Gelatinase B is also expressed in the parietal endoderm, where it may play a role in expansion of Reichert's membrane (Behrendtsen and Werb, 1997).

B.   Bone Development

The long bones develop from mesenchymal condensations from which cartilage cells differentiate and form a cartilage model. The cartilage

GelB   TIMP-3

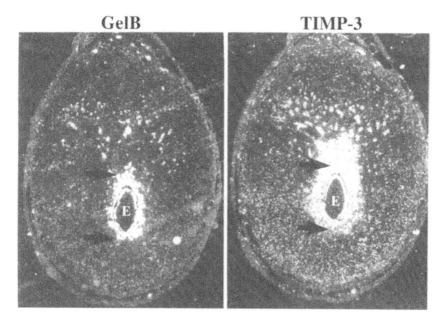

Fig. 4. *In situ* hybridizations with *GelB* and TIMP-3 antisense probes of mouse day 7.5 implantation site. The arrows point to the extent of trophoblast invasion into the maternal decidua. E, embryo.

model is then converted into bone by the process of endochondral ossification. This process requires invasion by blood vessels, degradation of the cartilage matrix, and the deposition and subsequent remodeling of bone matrix. The degradation of cartilage and the resorption of bone are thought to be carried out by specialized cells derived from hematopoietic stem cells, the osteoclasts. These cells express high levels of gelatinase B, both at sites of bone remodeling and during the initial migration from the perichondrium into the cartilagenous bone model prior to the initiation of the ossification process (Blavier and Delaisse, 1995; Okada *et al.*, 1995; Reponen *et al.*, 1994). Therefore, gelatinase B was expected to function in bone and cartilage resorption. Surprisingly, the gelatinase B-deficient mouse shows no evidence of osteopetrosis, indicating that gelatinase B is not the major bone resorbing enzyme. However, these mice have a very interesting phenotype in their skeletal growth plates (T.H. Vu *et al.*, manuscript in preparation). The vascular invasion into the cartilage ECM is delayed, resulting in delayed ossification and an excessively wide zone of hypertrophic cartilage (Fig. 5). Cartilage has always been considered to be resistant to vascular invasion. Gelatinase B appears to have an important role in this process, either by degrading the cartilage ECM to allow vessel ingrowth, or alternatively,

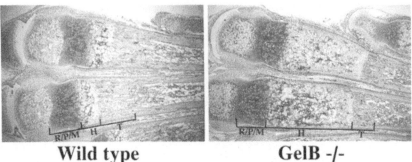

Wild type   GelB -/-

FIG. 5. Histological sections of the metatarsals of 2-week-old wild type and *GelB*-deficient mice. R/P/M, zones of resting, proliferative, and maturing chondrocytes; H, zone of hypertrophic cartilage; T, zone of trabecular bone.

by generating an angiogenic signal or removing an angiogenic inhibitor. The former seems less likely, since osteoclasts also express other ECM-degrading enzymes, and it would be surprising if the lack of gelatinase B alone were to cause a defect in the degradation of the cartilage ECM. It is a more attractive hypothesis that the function of gelatinase B is to generate a specific signal that allows vascular invasion to proceed.

C. Tissue Injury, Inflammation, and Wound Healing

Gelatinase B is expressed at high levels by several types of inflammatory cells: neutrophils, lymphocytes, eosinophils, mast cells, and macrophages. It may locally degrade the subendothelial basement membrane during the process of inflammatory cell extravasation from the vascular compartment into tissues. Migration of resting T cells across basement membrane *in vitro* is inhibited by an MMP inhibitor (Leppert *et al.*, 1995). This is likely due to inhibition of gelatinase B, because it is the major type IV collagenase expressed by resting T cells. IL-2 both induces gelatinase B expression by T cells and enhances their migration across basement membrane *in vitro* (Leppert *et al.*, 1996). Pretreatment of T cells with IFN-β attenuated both the IL-2 induction of gelatinase B and the enhancement of T-cell migration, most likely by down-regulation of surface IL-2 receptors. TPA also induces gelatinase B expression in T cells and causes T-cell adhesion to type IV collagen; these effects may be part of the mechanism of T-cell extravasation (Weeks *et al.*, 1993). Neutrophil migration across reconstituted basement membrane probably also involves gelatinase B. This process is inhibited both by TIMP-1 and by an inhibitor of leukocyte elastase, a possible activator of gelatinase B. The two effects are not additive, suggesting that they probably act through the same pathway (Delclaux *et al.*, 1996). When monocytic cells differentiate into macrophages, they express gelatinase B and acquire the ability to invade basement membrane (Pluznik *et al.*, 1992). Taken together, these studies suggest a role for gelatinase B in the process of cell recruitment to sites of inflammation. However, studies in the gelatinase B-deficient mice show that this enzyme does not appear to be essential for recruitment of neutrophils into tissues (Betsuyaku, T., *et al.*, manuscript in preparation). There is no difference in the number of neutrophils in bronchoalveolar lavage fluid in the gelatinase B-deficient animals compared to wild type 24 h after intratracheal administration of LPS or of the chemoattractant fMLP. There are also no differences between neutrophils from wild type and deficient animals in chemotaxis assays across Matrigel.

MMPs, among many products of inflammatory cells, may also participate in the induction of tissue injury. In human small intestine explants, severe mucosal injury with shedding of epithelium and loss of villi occurs when lamina propria T cells are activated by PMNs, resulting in increased secretion of gelatinase B, interstitial collagenase, and stromelysin-1 into the medium (Pender *et al.*, 1997). The addition of MMP inhibitors prevents this mucosal injury.

Increased gelatinase B expression has also been demonstrated in several other forms of tissue injury. Increased levels of gelatinase B are found in infarcted heart tissue compared to normal (Tyagi *et al.*, 1996). Patients with acute respiratory distress syndrome (ARDS) have higher levels of gelatinase B in their bronchoalveolar lavage fluid (Ricou *et al.*, 1996). In patients with partial to full thickness burns, gelatinase B activity is detected in burn fluid as early as 4–8 h after injury, and markedly increases over the subsequent 2 days, whereas gelatinase A and stromelysin-1 appear much later, suggesting that gelatinase B is involved in early injury, whereas gelatinase A and stromelysin-1 may function in wound repair (Young and Grinnell, 1994). Exposure of rats to 100% oxygen causes lethal lung injury by 72 h, increased gelatinase B levels are detected in bronchoalveolar lavage fluid, and *in situ* hybridizations show increased expression of gelatinase B, gelatinase A, and interstitial collagenase mRNA in alveolar macrophages, epithelial, and interstitial cells (Pardo *et al.*, 1996). Injury to a peripheral nerve is followed by degeneration of the nerve segment distal to the injury site, and is associated with an increase in gelatinase B activity in the distal segment. The majority of enzyme originates from infiltrating macrophages (La Fleur *et al.*, 1996).

Gelatinase B has also been implicated in the destruction of the endothelial cell barrier seen in many types of tissue injury. Bovine microvascular endothelial cells grown as a monolayer on filters form a barrier to the diffusion of macromolecules. Incubation of these cells with TNF-α for 24 h increases permeability (Partridge *et al.*, 1993). This is associated with the secretion of a 96-kDa gelatinase that may be gelatinase B. Coculture of endothelial cells with HIV-infected monocytes, which secrete high levels of gelatinase B, also results in increased permeability (Dhawan *et al.*, 1995). This effect is blocked by TIMP-1 and TIMP-2. Cerebral ischemic injury is associated with edema caused by capillary leakage at 1–2 days post injury. Occlusion of the middle cerebral artery in rats results in ischemic injury and secondary cerebral edema that is associated with an increase in gelatinase B in the infarcted hemisphere (Rosenberg *et al.*, 1996b). Infusion of TNF-α intracerebrally in rats also results in increased capillary permeability, associated with an incre. ; in gelatinase B at 24 h. The increased permeability is inhibited by treatment with an MMP inhibitor (Rosenberg *et al.*, 1995).

Gelatinase B may also be involved in normal wound healing. In several models of wound healing, it is up-regulated in the epithelial cells that are migrating to close a wound (Agren, 1994; Buisson *et al.*, 1996; Oikarinen *et al.*, 1993; Salo *et al.*, 1994). Preliminary studies in the gelatinase B-deficient mice shows that cutaneous wounds in these mice may have delayed reepithelization (K. Bullard, unpublished observations).

D. *Arthritis*

High levels of gelatinase B have been detected in serum and synovial fluid of patients with inflammatory arthritis such as rheumatoid arthritis compared to healthy patients or patients with osteoarthritis (Ahrens *et al.*, 1996; Gruber *et al.*, 1996). There is a correlation between the arthritic activity score of the joints and the amount of gelatinase B in the aspirated synovial fluid (Koolwijk *et al.*, 1995). Immunostaining for gelatinase B in synovium from rheumatoid joints shows staining of fibroblasts, neutrophils, and macrophages (Ahrens *et al.*, 1996; Gruber *et al.*, 1996). Of interest, synovial fibroblasts from normal joints do not normally make gelatinase B, but can be induced to do so by PMA and IL-1, whereas fibroblasts isolated from rheumatoid joints produce gelatinase B without exogenous stimulation (Tetlow *et al.*, 1993). In rat adjuvant arthritis, an animal model of chronic inflammatory arthritis, neutrophils and macrophages infiltrate the articular cartilage following adjuvant administration (Stein-Picarella *et al.*, 1994). These cells express increasing levels of gelatinase B, as do synovial fibroblasts. Therefore, gelatinase B may be produced in the inflammatory joints by inflammatory cell infiltrates or induced in the synovial fibroblasts by inflammatory cytokines, and once present, may participate in the joint destruction.

E. Inflammatory Diseases of the Central Nervous System

Multiple sclerosis is an inflammatory demyelinating disease of the central nervous system. Prominent expression of gelatinase B is seen in reactive astrocytes and macrophages in the demyelinating lesions, compared to normal brain tissue (Cuzner et al., 1996). High levels of gelatinase B are detected in the cerebrospinal fluid (CSF) of patients with multiple sclerosis (Rosenberg et al., 1996a). The levels decrease after treatment with corticosteroids. In patients with amyotrophic lateral sclerosis (ALS), gelatinase B expression is found in the pyramidal neurons in the motor cortex and in the motor neurons in the spinal cord (Lim et al., 1996). Its release at the synapses may cause matrix degradation, as part of the pathogenesis of the disease. Gelatinase B is found in the CSF in a mouse model of experimental allergic encephalomyelitis (EAE), and treatment of the mice with D-penicillamine decreases the morbidity and mortality (Norga et al., 1995). D-penicillamine inhibits gelatinase B activity in vitro in a dose-dependent fashion. Gelatinase B may participate in the pathogenesis of these inflammatory diseases either in tissue destruction or generation of an inflammatory signal. As noted earlier, gelatinase B cleaves myelin basic protein into fragments, some of which may act as encephalitogens.

F. Other Inflammatory Diseases

A high level of gelatinase B activity is found in the wall of aortic aneurysms (Freestone et al., 1995; Newman et al., 1994; Sakalihasan et al., 1996). The aneurysmic aortic wall is infiltrated by a large number of inflammatory cells, mainly macrophages and lymphocytes; these may be the source of gelatinase B production, which may participate in the process of connective tissue remodeling leading to the aneurysm. Patients with giant cell arteritis have increased level of gelatinase B and gelatinase B mRNA is found in smooth muscle cells and fibroblasts in the regions of fragmented elastic tissue in the lamina media of inflammed vessels (Sorbi et al., 1996). Increased levels of gelatinase B are also found in sputum of patients with cystic fibrosis and in bronchoalveolar lavage fluids of those with bronchiectasis (Delacourt et al., 1995; Sepper et al., 1994). Blister fluids from the skin lesions of bullous pemphigoid patients contain high levels of gelatinase B (Stahle-Backdahl et al., 1994), and interestingly, the gelatinase B-deficient mice do not develop blisters in an animal model of the disease (Z. Liu et al., manuscript in preparation).

G. Cancer

Because of the ability to cleave components of the extracellular matrix, and remodel the cellular microenvironment, MMPs are thought to play a role in the development and progression of tumors. The type IV collagenases in particular are implicated in tumor invasion and metastasis due to their ability to degrade basement membrane collagens (Himelstein et al., 1994a). Gelatinase B, while not commonly expressed in normal cells, has been found to be expressed in tumors from diverse sites, including skin, lungs, breast, colorectum, liver, prostate, brain, bone marrow, and bone. In some tumors, for example, squamous cell carcinoma of the skin and lungs, adenocarcinoma of the breast, hepatocellular carcinoma, gliomas, and giant cell tumor of bones, the tumor cells themselves express the enzyme (Ashida et al., 1996; Canete-Soler et al., 1994; Iwata et al., 1996; Nakagawa and Yagihashi, 1994; Pyke et al., 1992; Soini et al., 1994). Of interest, in many cases only the tumor cells at the margins adjacent to the tumor–stroma interface express gelatinase B, and loss of collagen type IV staining is sometimes observed in the surrounding ECM (Ueda et al., 1996). In other cases, stromal cells surrounding the tumors express gelatinase B. In the skin, it is found in macrophages in squamous cell carcinoma, and in esosinophils in basal cell carcinoma (Pyke et al., 1992; Stahle-Backdahl et al., 1994). Macrophages and neutrophils surrounding colorectal adenocarcinoma and prostate carcinoma express gelatinase B (Gallegos et al., 1995; Jeziorska et al., 1994; Nagle et al., 1994; Pyke et al., 1993; Zeng and Guillem, 1995). In squamous cell carcinoma of the lungs and adenocarcinoma of the breast, gelatinase B mRNA is present in stromal fibroblasts and endothelial cells (Heppner et al., 1996; Nakagawa and Yagihashi, 1994; Soini et al., 1994). In many tumors, high expression of gelatinase B correlates with invasiveness and metastatic potential. In gliomas, higher expression of gelatinase B is seen in the high-grade tumors such as malignant astrocytomas and glioblastoma multiforme, compared to low-grade tumors (Rao et al., 1996). In breast cancer, a high level of expression is seen in lymph node metastases (Iwata et al., 1996). Some patients with leukemia develop extramedullary tumors; a cell line derived from such a granulocytic sarcoma shows high levels of constitutive expression of gelatinase B and had high invasiveness potential in vitro (Kobayashi et al., 1995). Cell lines established from early and late stages of melanoma show different levels of gelatinase B expression. Most advanced-stage melanoma cells express gelatinase

B, whereas none of the early-stage cells do. Expression of gelatinase B appears to be specifically suppressed in the early stages, as somatic cell hybrids between cells from early and late stages show no expression (MacDougall *et al.,* 1995).

If gelatinase B expression in tumors is important for growth and progression, it is not clear whether expression by tumor cells, stromal cells, or both, are important. Several studies have shown induction of gelatinase B expression in tumor cells by stromal cells, and vice versa.

Despite the expression data, direct evidence implicating gelatinase B in tumor growth and invasion is lacking. However, there are several suggestive studies. Two cell lines are derived from a murine colon adenocarcinoma cell line: the LuM1 line, derived from a metastatic lung nodule, shows high incidence of spontaneous metastasis and se-cretes high levels of gelatinase B; whereas the NM11 line, derived from the primary tumor, rarely metastasizes and does not secrete gelatinase B (Sakata *et al.,* 1996). In a model of multistage carcinogenesis of squamous cell carcinoma of the skin, high levels of gelatinase B are detected in the dysplastic stage (L. Coussens *et al.,* unpublished obser-vations). When mice are treated with either an oral MMP inhibitor, a cytotoxic agent, or both, prior to injection with murine Lewis lung carcinoma cells, the combination treatment is more effective than either single agent alone in delaying local tumor growth and reducing the number and size of the lung metastases (Anderson *et al.,* 1996). The human osteosarcoma cell line HT1080 secretes gelatinase B and can invade basement membranes *in vitro.* When DNA from these cells was transfected into human fibroblast cell lines that are not invasive and do not make gelatinase B, several invasive transfectants were isolated, and all these had acquired the expression of gelatinase B (Kubota *et al.,* 1991). The gene(s) responsible for this induction is not known, but appears not to be the activated N-ras gene present in the HT1080 cells. Treatment of HT1080 cells with ursolic acid results in down-regulation of MMP-9, MMP-2, and MT1-MMP. This also results in decreased invasion *in vitro* through reconstituted basement membrane (Cha *et al.,* 1996). The thiol agent *N*-acetylcysteine (NAC) down-regulates gela-tinase B in melanoma and in Lewis lung carcinoma cells. When these cells are treated with NAC prior to injection into nude mice, or when mice are given NAC in drinking water before injection with tumor cells, local weight of the tumors and metastases are both reduced (Albini *et al.,* 1995). Human osteosarcoma cell line (OST) up-regulates gelatinase B in response to TNF-α and becomes more invasive through reconsti-tuted basement membrane *in vitro.* Treatment of these cells with TNF-α prior to injection into nude mice results in an increased number of lung metastases in a dose-dependent manner (Kawashima *et al.,*

1994). Injection of rat intramammary fat pads with mammary adenocarcinoma cell lines shows a correlation between the extent of metastases in the lungs and lymph nodes with gelatinase B levels in the serum
(Nakajima et al., 1993). Rat embryo fibroblasts, when transformed with
Ha-ras or Ha-ras and v-myc become tumorigenic and metastatic when
injected into nude mice (Bernhard et al., 1994); these cells make high
levels of gelatinase B. On the other hand, rat embryo fibroblasts transformed with Ha-ras and adenovirus E1A gene are tumorigenic but
nonmetastatic; these cells do not secrete gelatinase B. However, if these
cells are transfected with an expression vector for gelatinase B, they
acquire metastatic ability. Multiple passages of these cells in vivo result
in two populations; those that lost the gelatinase B expression vector
also lost metastatic potential, but those that retained gelatinase B
expression also retained metastatic potential. Inhibition of gelatinase
B expression in the rat embryo fibroblast cell line transformed with
Ha-ras and v-myc by introduction of a gelatinase B specific ribozyme
results in no detectable gelatinase B mRNA or protein secretion (Hua
and Muschel, 1996). The cells lose metastatic potential, but retain their
tumorigenicity when injected into animals.

H. Other Diseases

Gelatinase B has also been implicated in the pathogenesis of polycystic kidney disease (Murray et al., 1996), membranous nephropathy
(McMillan et al., 1996), and Alzheimer's (Lim et al., 1997).

VII. Conclusion and Future Directions

Much has been learned about gelatinase B since its identification.
However, the question about its in vivo substrates and functions remains. The gelatinase B null mice have begun to give us significant
insights about the potential role of this enzyme during development
and in tissue response to injury. Future studies utilizing this mouse
in specific disease models, especially in models of tumor invasion and
metastasis, may provide important information as to the role of gelatinase B in their pathogenesis. Generating animals deficient in multiple
MMPs will give further insights into the interplay among these ECM
modifying enzymes in particular developmental and disease processes.

ACKNOWLEDGMENTS

We thank all the authors who communicated their results prior to publication. We
thank Ole Behrendtsen and Mark Sternlicht for assistance with graphics. This work

was supported by grants from the National Institutes of Health (DE10306 and HD26732) and by an institutional National Research Service Award (HL07185).

References

Agren, M. S. (1994). Gelatinase activity during wound healing. *Br. J. Dermatol.* **131,** 5, 634–640.

Ahrens, D., Koch, A. E., Pope, R. M., Stein-Picarella, M., and Niedbala, M. J. (1996). Expression of matrix metalloproteinase 9 (96-kd gelatinase B) in human rheumatoid arthritis. *Arthritis Rheum.* **39,** 9, 1576–1587.

Albini, A., D'Agostini, F., Giunciuglio, D., Paglieri, I., Balansky, R., and De Flora, S. (1995). Inhibition of invasion, gelatinase activity, tumor take and metastasis of malignant cells by N-acetylcysteine. *Int. J. Cancer* **61,** 1, 121–129.

Alexander, C. M., Hansell, E. J., Behrendtsen, O., Flannery, M. L., Kishnani, N. S., Hawkes, S. P., and Werb, Z. (1996). Expression and function of matrix metalloproteinases and their inhibitors at the maternal-embryonic boundary during mouse embryo implantation. *Development* **122,** 6, 1723–1736.

Anderson, I. C., Shipp, M. A., Docherty, A. J., and Teicher, B. A. (1996). Combination therapy including a gelatinase inhibitor and cytotoxic agent reduces local invasion and metastasis of murine Lewis lung carcinoma. *Cancer Res.* **56,** 4, 715–718.

Apte, S. S., Olsen, B. R., and Murphy, G. (1995). The gene structure of tissue inhibitor of metalloproteinases (TIMP)-3 and its inhibitory activities define the distinct TIMP gene family. *J. Biol. Chem.* **270,** 24, 14313–14318.

Ashida, K., Nakatsukasa, H., Higashi, T., Ohguchi, S., Hino, N., Nouso, K., Urabe, Y., Yoshida, K., Kinugasa, N., and Tsuji, T. (1996). Cellular distribution of 92-kd type IV collagenase/gelatinase B in human hepatocellular carcinoma. *Am. J. Pathol.* **149,** 6, 1803–1811.

Backstrom, J. R., and Tokes, Z. A. (1995). The 84-kDa form of human matrix metalloproteinase-9 degrades substance P and gelatin. *J. Neurochem.* **64,** 3, 1312–1318.

Backstrom, J. R., Lim, G. P., Cullen, M. J., and Tokes, Z. A. (1996). Matrix metalloproteinase-9 (MMP-9) is synthesized in neurons of the human hippocampus and is capable of degrading the amyloid-beta peptide (1-40). *J. Neurosci.* **16,** 24, 7910–7919.

Behrendtsen, O., and Werb, Z. (1997). Metalloproteinases regulate parietal endoderm differentiating and migrating in cultured mouse embryos. *Dev. Dynamics* **208,** 255–265.

Behrendtsen, O., Alexander, C. M., and Werb, Z. (1992). Metalloproteinases mediate extracellular matrix degradation by cells from mouse blastocyst outgrowths. *Development* **114,** 2, 447–456.

Bernhard, E. J., Gruber, S. B., and Muschel, R. J. (1994). Direct evidence linking expression of matrix metalloproteinase 9 (92–kDa gelatinase/collagenase) to the metastatic phenotype in transformed rat embryo cells. *Proc. Natl. Acad. Sci. USA* **91,** 10, 4293–4297.

Blavier, L., and Delaisse, J. M. (1995). Matrix metalloproteinases are obligatory for the migration of preosteoclasts to the developing marrow cavity of primitive long bones. *J. Cell Sci.* **108,** 3649–3659.

Bu, C. H., and Pourmotabbed, T. (1995). Mechanism of activation of human neutrophil gelatinase B. Discriminating between the role of Ca^{2+} in activation and catalysis. *J. Biol. Chem.* **270,** 31, 18563–18569.

Bu, C. H., and Pourmotabbed, T. (1996). Mechanism of Ca^{2+}-dependent activity of human neutrophil gelatinase B. *J. Biol. Chem.* **271,** 24, 14308–14315.

Buisson, A. C., Zahm, J. M., Polette, M., Pierrot, D., Bellon, G., Puchelle, E., Birembaut, P., and Tournier, J. M. (1996). Gelatinase B is involved in the *in vitro* wound repair of human respiratory epithelium. *J. Cell Physiol.* **166**, 2, 413–426.

Canete-Soler, R., Litzky, L., Lubensky, I., and Muschel, R. J. (1994). Localization of the 92 kd gelatinase mRNA in squamous cell and adenocarcinomas of the lung using *in situ* hybridization. *Am. J. Pathol.* **144**, 3, 518–527.

Cha, H. J., Bae, S. K., Lee, H. Y., Lee, O. H., Sato, H., Seiki, M., Park, B. C., and Kim, K. W. (1996). Anti-invasive activity of ursolic acid correlates with the reduced expression of matrix metalloproteinase-9 (MMP-9) in HT1080 human fibrosarcoma cells. *Cancer Res.* **56**, 10, 2281–2284.

Chapel, C., Camara, V., Clayette, P., Salvat, S., Mabondzo, A., Leblond, V., Marce, D., Lafuma, C., and Dormont, D. (1994). Modulations of 92kDa gelatinase B and its inhibitors are associated with HIV-1 infection in human macrophage cultures. *Biochem. Biophys. Res. Commun.* **204**, 3, 1272–1278.

Collier, I. E., Wilhelm, S. M., Eisen, A. Z., Marmer, B. L., Grant, G. A., Seltzer, J. L., Kronberger, A., He, C. S., Bauer, E. A., and Goldberg, G. I. (1988). H-ras oncogene-transformed human bronchial epithelial cells (TBE-1) secrete a single metalloprotease capable of degrading basement membrane collagen. *J. Biol. Chem.* **263**, 14, 6579–6587.

Collier, I. E., Krasnov, P. A., Strongin, A. Y., Birkedal-Hansen, H., and Goldberg, G. I. (1992). Alanine scanning mutagenesis and functional analysis of the fibronectin-like collagen-binding domain from human 92-kDa type IV collagenase. *J. Biol. Chem.* **267**, 10, 6776–6781.

Corcoran, M. L., Stetler-Stevenson, W. G., Brown, P. D., and Wahl, L. M. (1992). Interleukin 4 inhibition of prostaglandin E2 synthesis blocks interstitial collagenase and 92-kDa type IV collagenase/gelatinase production by human monocytes. *J. Biol. Chem.* **267**, 1, 515–519.

Corcoran, M. L., Stetler-Stevenson, W. G., DeWitt, D. L., and Wahl, L. M. (1994). Effect of cholera toxin and pertussis toxin on prostaglandin H synthase-2, prostaglandin E2, and matrix metalloproteinase production by human monocytes. *Arch. Biochem. Biophys.* **310**, 2, 481–488.

Corcoran, M. L., Kibbey, M. C., Kleinman, H. K., and Wahl, L. M. (1995). Laminin SIKVAV peptide induction of monocyte/macrophage prostaglandin E2 and matrix metalloproteinases. *J. Biol. Chem.* **270**, 18, 10365–10368.

Cuzner, M. L., Gveric, D., Strand, C., Loughlin, A. J., Paemen, L., Opdenakker, G., and Newcombe, J. (1996). The expression of tissue-type plasminogen activator, matrix metalloproteases and endogenous inhibitors in the central nervous system in multiple sclerosis: Comparison of stages in lesion evolution. *J. Neuropathol. Exp. Neurol.* **55**, 12, 1194–1204.

Delacourt, C., Le Bourgeois, M., D'Ortho, M. P., Doit, C., Scheinmann, P., Navarro, J., Harf, A., Hartmann, D. J., and Lafuma, C. (1995). Imbalance between 95 kDa type IV collagenase and tissue inhibitor of metalloproteinases in sputum of patients with cystic fibrosis. *Am. J. Respir. Crit. Care Med.* **152**, 2, 765–774.

Delaisse, J. M., Eeckhout, Y., Sear, C., Galloway, A., McCullagh, K., and Vaes, G. (1985). A new synthetic inhibitor of mammalian tissue collagenase inhibits bone resorption in culture. *Biochem. Biophys. Res. Commun.* **133**, 2, 483–490.

Delclaux, C., Delacourt, C., D'Ortho, M. P., Boyer, V., Lafuma, C., and Harf, A. (1996). Role of gelatinase B and elastase in human polymorphonuclear neutrophil migration across basement membrane. *Am. J. Respir. Cell Mol Biol.* **14**, 3, 288–295.

Desrochers, P. E., Mookhtiar, K., Van Wart, H. E., Hasty, K. A., and Weiss, S. J. (1992). Proteolytic inactivation of alpha 1-proteinase inhibitor and alpha 1-antichymotrypsin by oxidatively activated human neutrophil metalloproteinases. *J. Biol. Chem.* **267**, 7, 5005–5012.

Dhawan, S., Weeks, B. S., Soderland, C., Schnaper, H. W., Toro, L. A., Asthana, S. P., Hewlett, I. K., Stetler-Stevenson, W. G., Yamada, S. S., Yamada, K. M., *et al.* (1995). HIV-1 infection alters monocyte interactions with human microvascular endothelial cells. *J. Immunol.* **154**, 1, 422–432.

Edvardsen, K., Chen, W., Rucklidge, G., Walsh, F. S., Obrink, B., and Bock, E. (1993). Transmembrane neural cell-adhesion molecule (NCAM), but not glycosyl-phosphatidylinositol-anchored NCAM, down-regulates secretion of matrix metalloproteinases. *Proc. Natl. Acad. Sci. USA* **90**, 24, 11463–11467.

Fabunmi, R. P., Baker, A. H., Murray, E. J., Booth, R. F., and Newby, A. C. (1996). Divergent regulation by growth factors and cytokines of 95 kDa and 72 kDa gelatinases and tissue inhibitors or metalloproteinases-1, -2, and -3 in rabbit aortic smooth muscle cells. *Biochem. J.* **315**, Pt 1), 335–342.

Faller, D. V., Weng, H., and Choi, S. Y. (1997). Activation of collagenase IV gene expression and enzymatic activity by the Moloney murine leukemia virus long terminal repeat. *Virology* **227**, 2, 331–342.

Fang, K. C., Raymond, W. W., Lazarus, S. C., and Caughey, G. H. (1996). Dog mastocytoma cells secrete a 92-kD gelatinase activated extracellularly by mast cell chymase. *J. Clin. Invest.* **97**, 7, 1589–1596.

Ferry, G., Lonchampt, M., Pennel, L., de Nanteuil, G., Canet, E., and Tucker, G. C. (1997). Activation of MMP-9 by neutrophil elastase in an in vivo model of acute lung injury. *FEBS Lett.* **402**, 2–3, 111–115.

Fini, M. E., Girard, M. T., Matsubara, M., and Bartlett, J. D. (1995). Unique regulation of the matrix metalloproteinase, gelatinase B. *Invest. Ophthalmol. Vis. Sci.* **36**, 3, 622–633.

Fosang, A. J., Neame, P. J., Last, K., Hardingham, T. E., Murphy, G., and Hamilton, J. A. (1992). The interglobular domain of cartilage aggrecan is cleaved by PUMP, gelatinases, and cathepsin B. *J. Biol. Chem.* **267**, 27, 19470–19474.

Freestone, T., Turner, R. J., Coady, A., Higman, D. J., Greenhalgh, R. M., and Powell, J. T. (1995). Inflammation and matrix metalloproteinases in the enlarging abdominal aortic aneurysm. *Arterioscler. Thromb. Vasc. Biol.* **15**, 8, 1145–1151.

Fridman, R., Toth, M., Pena, D., and Mobashery, S. (1995). Activation of progelatinase B (MMP-9) by gelatinase A (MMP-2). *Cancer Res.* **55**, 12, 2548–2555.

Gallegos, N. C., Smales, C., Savage, F. J., Hembry, R. M., and Boulos, P. B. (1995). The distribution of matrix metalloproteinases and tissue inhibitor of metalloproteinases in colorectal cancer. *Surg. Oncol.* **4**, 1, 21–29.

Gijbels, K., Proost, P., Masure, S., Carton, H., Billiau, A., and Opdenakker, G. (1993). Gelatinase B is present in the cerebrospinal fluid during experimental autoimmune encephalomyelitis and cleaves myelin basic protein. *J. Neurosci. Res.* **36**, 4, 432–440.

Goldberg, G. I., Strongin, A., Collier, I. E., Genrich, L. T., and Marmer, B. L. (1992). Interaction of 92-kDa type IV collagenase with the tissue inhibitor of metalloproteinases prevents dimerization, complex formation with interstitial collagenase, and activation of the proenzyme with stromelysin. *J. Biol. Chem.* **267**, 7, 4583–4591.

Golub, L. M., Sorsa, T., Lee, H. M., Ciancio, S., Sorbi, D., Ramamurthy, N. S., Gruber, B., Salo, T., and Konttinen, Y. T. (1995). Doxycycline inhibits neutrophil (PMN)-type matrix metalloproteinases in human adult periodontitis gingiva. *J. Clin. Periodont.* **22**, 2, 100–109.

Gottschall, P. E., Yu, X., and Bing, B. (1995a). Increased production of gelatinase B (matrix metalloproteinase-9) and interleukin-6 by activated rat microglia in culture. *J. Neurosci. Res.* **42**, 335–342.

Gottschall, P. E., and Yu, X. (1995b). Cytokines regulate gelatinase A and B (matrix metalloproteinase 2 and 9) activity in cultured rat astrocytes. *J. Neurochem.* **64**, 1513–1520.

Gray, R. D., Miller, R. B., and Spatola, A. F. (1986). Inhibition of mammalian collagenases by thiol-containing peptides. *J. Cell. Biochem.* **32**, 71–77.

Grobelny, D., Poncz, L., and Galardy, R. E. (1992). Inhibition of human skin fibroblast collagenase, thermolysin, and *Pseudomonas aeruginosa* elastase by peptide hydroxamic acids. *Biochemistry* **31**, 31, 7152–7154.

Gross, J., and Lapiere, C. M. (1962). Collagenolytic activity in amphibian tissues: A tissue culture assay. *PNAS* **48**, 1014–1022.

Gruber, B. L., Sorbi, D., French, D. L., Marchese, M. J., Nuovo, G. J., Kew, R. R., and Arbeit, L. A. (1996). Markedly elevated serum MMP-9 (gelatinase B) levels in rheumatoid arthritis: a potentially useful laboratory marker. *Clin. Immunol. Immunopathol.* **78**, 2, 161–171.

Gum, R., Lengyel, E., Juarez, J., Chen, J. H., Sato, H., Seiki, M., and Boyd, D. (1996). Stimulation of 92–kDa gelatinase B promoter activity by ras is mitogen-activated protein kinase 1-independent and requires multiple transcription factor binding sites including closely spaced PEA3/ets and AP-1 sequences. *J. Biol. Chem.* **271**, 18, 10672–10680.

Hanemaaijer, R., Koolwijk, P., le Clercq, L., de Vree, W. J., and van Hinsbergh, V. W. (1993). Regulation of matrix metalloproteinase expression in human vein and microvascular endothelial cells. Effects of tumour necrosis factor alpha, interleukin 1 and phorbol ester. *Biochem. J.* **296**, Pt 3, 803–809.

Harvey, M. B., Leco, K. J., Arcellana-Panlilio, M. Y., Zhang, X., Edwards, D. R., and Schultz, G. A. (1995). Proteinase expression in early mouse embryos is regulated by leukaemia inhibitory factor and epidermal growth factor. *Development* **121**, 4, 1005–1014.

He, C. (1996). Molecular mechanism of transcriptional activation of human gelatinase B by proximal promoter. *Cancer Lett.* **106**, 2, 185–191.

Heppner, K. J., Matrisian, L. M., Jensen, R. A., and Rodgers, W. H. (1996). Expression of most matrix metalloproteinase family members in breast cancer represents a tumor-induced host response. *Am. J. Pathol.* **149**, 1, 273–282.

Herron, G. S., Banda, M. J., Clark, E. J., Gavrilovic, J., and Werb, Z. (1986). Secretion of metalloproteinases by stimulated capillary endothelial cells. II. Expression of collagenase and stromelysin activities is regulated by endogenous inhibitors. *J. Biol. Chem.* **261**, 6, 2814–2818.

Hibbs, M. S., Hoidal, J. R., and Kang, A. H. (1987). Expression of a metalloproteinase that degrades native type V collagen and denatured collagens by cultured human alveolar macrophages. *J. Clin. Invest.* **80**, 6, 1644–1650.

Himelstein, B. P., and Muschel, R. J. (1996). Induction of matrix metalloproteinase 9 expression in breast carcinoma cells by a soluble factor from fibroblasts. *Clin. Exp. Metastasis* **14**, 3, 197–208.

Himelstein, B. P., Canete-Soler, R., Bernhard, E. J., Dilks, D. W., and Muschel, R. J. (1994a). Metalloproteinases in tumor progression: the contribution of MMP-9. *Invasion Metastasis* **14**, 1–6, 246–258.

Himelstein, B. P., Canete-Soler, R., Bernhard, E. J., and Muschel, R. J. (1994b). Induction of fibroblast 92 kDa gelatinase/type IV collagenase expression by direct contact with metastatic tumor cells. *J. Cell Sci.* **107**, Pt 2, 477–486.

Houde, M., de Bruyne, G., Bracke, M., Ingelman-Sundberg, M., Skoglund, G., Masure, S., van Damme, J., and Opdenakker, G. (1993). Differential regulation of gelatinase B and tissue-type plasminogen activator expression in human Bowes melanoma cells. *Int. J. Cancer* **53**, 3, 395–400.

Houde, M., Tremblay, P., Masure, S., Opdenakker, G., Oth, D., and Mandeville, R. (1996). Synergistic and selective stimulation of gelatinase B production in macrophages by lipopolysaccharide, trans-retinoic acid and CGP 41251, a protein kinase C regulator. *Biochim. Biophys. Acta* **1310**, 2, 193–200.

Howard, E. W., Bullen, E. C., and Banda, M. J. (1991). Preferential inhibition of 72- and 92-kDa gelatinases by tissue inhibitor of metalloproteinases-2. *J. Biol. Chem.* **266**, 20, 13070–13075.

Hua, J., and Muschel, R. J. (1996). Inhibition of matrix metalloproteinase 9 expression by a ribozyme blocks metastasis in a rat sarcoma model system. *Cancer Res.* **56**, 22, 5279–5284.

Huhtala, P., Tuuttila, A., Chow, L. T., Lohi, J., Keski-Oja, J., and Tryggvason, K. (1991). Complete structure of the human gene for 92-kDa type IV collagenase. Divergent regulation of expression for the 92- and 72-kilodalton enzyme genes in HT-1080 cells. *J. Biol. Chem.* **266**, 25, 16485–16490.

Huhtala, P., Humphries, M. J., McCarthy, J. B., Tremble, P. M., Werb, Z., and Damsky, C. H. (1995). Cooperative signaling by alpha 5 beta 1 and alpha 4 beta 1 integrins regulates metalloproteinase gene expression in fibroblasts adhering to fibronectin. *J. Cell Biol.* **129**, 3, 867–879.

Hujanen, E. S., Vaisanen, A., Zheng, A., Tryggvason, K., and Turpeenniemi-Hujanen, T. (1994). Modulation of *M*(r) 72,000 and *M*(r) 92,000 type-IV collagenase (gelatinase A and B) gene expression by interferons alpha and gamma in human melanoma. *Int. J. Cancer* **58**, 4, 582–586.

Ito, A., Nose, T., Takahashi, S., and Mori, Y. (1995). Cyclooxygenase inhibitors augment the production of pro-matrix metalloproteinase 9 (progelatinase B) in rabbit articular chondrocytes. *FEBS Lett.* **360**, 1, 75–79.

Ito, A., Mukaiyama, A., Itoh, Y., Nagase, H., Thogersen, I. B., Enghild, J. J., Sasaguri, Y., and Mori, Y. (1996). Degradation of interleukin 1beta by matrix metalloproteinases. *J. Biol. Chem.* **271**, 25, 14657–14660.

Iwata, H., Kobayashi, S., Iwase, H., Masaoka, A., Fujimoto, N., and Okada, Y. (1996). Production of matrix metalloproteinases and tissue inhibitors of metalloproteinases in human breast carcinomas. *Jpn. J. Cancer Res.* **87**, 6, 602–611.

Jeziorska, M., Haboubi, N. Y., Schofield, P. F., Ogata, Y., Nagase, H., and Woolley, D. E. (1994). Distribution of gelatinase B (MMP-9) and type IV collagen in colorectal carcinoma. *Int. J. Colorectal Dis.* **9**, 3, 141–148.

Kawashima, A., Nakanishi, I., Tsuchiya, H., Roessner, A., Obata, K., and Okada, Y. (1994). Expression of matrix metalloproteinase 9 (92-kDa gelatinase/type IV collagenase) induced by tumour necrosis factor alpha correlates with metastatic ability in a human osteosarcoma cell line. *Virchows Arch.* **424**, 5, 547–552.

Kjeldsen, L., Johnsen, A. H., Sengelov, H., and Borregaard, N. (1993). Isolation and primary structure of NGAL, a novel protein associated with human neutrophil gelatinase. *J. Biol. Chem.* **268**, 14, 10425–10432.

Kobayashi, M., Hamada, J., Li, Y. Q., Shinobu, N., Imamura, M., Okada, F., Takeichi, N., and Hosokawa, M. (1995). A possible role of 92 kDa type IV collagenase in the extramedullary tumor formation in leukemia. *Jpn. J. Cancer Res.* **86**, 3, 298–303.

Kolkenbrock, H., Orgel, D., Hecker-Kia, A., Zimmermann, J., and Ulbrich, N. (1995). Generation and activity of the ternary gelatinase B/TIMP-1/LMW-stromelysin-1 complex. *Biol Chem Hoppe Seyler* **376**, 8, 495–500.

Kondapaka, S. B., Fridman, R., and Reddy, K. B. (1997). Epidermal growth factor and amphiregulin up-regulate matrix metalloproteinase-9 (MMP-9) in human breast cancer cells. *Int. Nat. J. Cancer* **70**, 6, 722–726.

Koolwijk, P., Miltenburg, A. M., van Erck, M. G., Oudshoorn, M., Niedbala, M. J., Breedveld, F. C., and van Hinsbergh, V. W. (1995). Activated gelatinase-B (MMP-9) and urokinase-type plasminogen activator in synovial fluids of patients with arthritis. Correlation with clinical and experimental variables of inflammation. *J. Rheumatol.* **22**, 3, 385–393.

Kubota, S., Mitsudomi, T., and Yamada, Y. (1991). Invasive human fibrosarcoma DNA mediated induction of a 92 kDa gelatinase/type IV collagenase leads to an invasive phenotype. *Biochem. Biophys. Res. Commun.* **181**, 3, 1539–1547.

Kurogi, T., Nabeshima, K., Kataoka, H., Okada, Y., and Koono, M. (1996). Stimulation of gelatinase B and tissue inhibitors of metalloproteinase (TIMP) production in co-culture of human osteosarcoma cells and human fibroblasts: Gelatinase B production was stimulated via up-regulation of fibroblast growth factor (FGF) receptor. *Int. J. Cancer* **66**, 1, 82–90.

La Fleur, M., Underwood, J. L., Rappolee, D. A., and Werb, Z. (1996). Basement membrane and repair of injury to peripheral nerve: defining a potential role for macrophages, matrix metalloproteinases, and tissue inhibitor of metalloproteinases-1. *J. Exp. Med.* **184**, 6, 2311–2326.

Lauricella-Lefebvre, M. A., Castronovo, V., Sato, H., Seiki, M., French, D. L., and Merville, M. P. (1993). Stimulation of the 92-kD type IV collagenase promoter and enzyme expression in human melanoma cells. *Invasion Metastasis* **13**, 6, 289–300.

Lengyel, E., Gum, R., Juarez, J., Clayman, G., Seiki, M., Sato, H., and Boyd, D. (1995). Induction of M(r) 92,000 type IV collagenase expression in a squamous cell carcinoma cell line by fibroblasts. *Cancer Res.* **55**, 4, 963–967.

Leppert, D., Waubant, E., Burk, M. R., Oksenberg, J. R., and Hauser, S. L. (1996). Interferon beta-1b inhibits gelatinase secretion and *in vitro* migration of human T cells: a possible mechanism for treatment efficacy in multiple sclerosis. *Ann. Neurol.* **40**, 6, 846–852.

Leppert, D., Waubant, E., Galardy, R., Bunnett, N. W., and Hauser, S. L. (1995). T cell gelatinases mediate basement membrane transmigration in vitro. *J. Immunol.* **154**, 9, 4379–4389.

Librach, C. L., Werb, Z., Fitzgerald, M. L., Chiu, K., Corwin, N. M., Esteves, R. A., Grobelny, D., Galardy, R., Damsky, C. H., and Fisher, S. J. (1991). 92-kD type IV collagenase mediates invasion of human cytotrophoblasts. *J. Cell Biol.* **113**, 2, 437–449.

Lim, G. P., Backstrom, J. R., Cullen, M. J., Miller, C. A., Atkinson, R. D., and Tokes, Z. A. (1996). Matrix metalloproteinases in the neocortex and spinal cord of amyotrophic lateral sclerosis patients. *J. Neurochem.* **67**, 1, 251–259.

Lim, G. P., Russell, M. J., Cullen, M. J., and Tokes, Z. A. (1997). Matrix metalloproteinases in dog brains exhibiting Alzheimer-like characteristics. *J. Neurochem.* **68**, 4, 1606–1611.

Logan, S. K., Hansell, E. J., Damsky, C. H., and Werb, Z. (1996). T-antigen inhibits metalloproteinase expression and invasion in human placental cells transformed with temperature-sensitive simian virus 40. *Matrix Biol.* **15**, 2, 81–89.

Lorenzo, J. A., Pilbeam, C. C., Kalinowski, J. F., and Hibbs, M. S. (1992). Production of both 92- and 72-kDa gelatinases by bone cells. *Matrix* **12**, 282–290.

MacDougall, J. R., and Kerbel, R. S. (1995). Constitutive production of 92-kDa gelatinase B can be suppressed by alterations in cell shape. *Exp. Cell Res.* **218**, 2, 508–515.

MacDougall, J. R., Bani, M. R., Lin, Y., Rak, J., and Kerbel, R. S. (1995). The 92-kDa gelatinase B is expressed by advanced stage melanoma cells: suppression by somatic cell hybridization with early stage melanoma cells. *Cancer Res.* **55**, 18, 4174–4181.

Mackay, A. R., Hartzler, J. L., Pelina, M. D., and Thorgeirsson, U. P. (1990). Studies on the ability of 65-kDa and 92-kDa tumor cell gelatinases to degrade type IV collagen. *J. Biol. Chem.* **265**, 35, 21929–21934.

Mainardi, C. L., Hibbs, M. S., Hasty, K. A., and Seyer, J. M. (1984). Purification of a type V collagen degrading metalloproteinase from rabbit alveolar macrophages. *Coll. Relat. Res.* **4**, 6, 479–492.

Marshall, B. C., Santana, A., Xu, Q. P., Petersen, M. J., Campbell, E. J., Hoidal, J. R., and Welgus, H. G. (1993). Metalloproteinases and tissue inhibitor of metalloproteinases in mesothelial cells. Cellular differentiation influences expression. *J. Clin. Invest.* **91**, 4, 1792–1799.

Martin, J., Knowlden, J., Davies, M., and Williams, J. D. (1994). Identification and independent regulation of human mesangial cell metalloproteinases. *Kidney Int.* **46**, 3, 877–885.

Masure, S., Nys, G., Fiten, P., Van Damme, J., and Opdenakker, G. (1993). Mouse gelatinase B. cDNA cloning, regulation of expression and glycosylation in WEHI-3 macrophages and gene organisation. *Eur. J. Biochem.* **218**, 1, 129–141.

McMillan, J. I., Riordan, J. W., Couser, W. G., Pollock, A. S., and Lovett, D. H. (1996a). Characterization of a glomerular epithelial cell metalloproteinase as matrix metalloproteinase-9 with enhanced expression in a model of membranous nephropathy. *J. Clin. Invest.* **97**, 4, 1094–1101.

McMillan, J. I., Weeks, R., West, J. W., Bursten, S., Rice, G. C., and Lovett, D. H. (1996b). Pharmacological inhibition of gelatinase B induction and tumor cell invasion. *Int. Natl. J. Cancer* **67**, 4, 523–531.

Mehul, B., Bawumia, S., Martin, S. R., and Hughes, R. C. (1994). Structure of baby hamster kidney carbohydrate-binding protein CBP30, an S-type animal lectin. *J. Biol. Chem.* **269**, 27, 18250–18258.

Meikle, M. C., Bord, S., Hembry, R. M., Compston, J., Croucher, P. I., and Reynolds, J. J. (1992). Human osteoblasts in culture synthesize collagenase and other matrix metalloproteinases in response to osteotropic hormones and cytokines. *J. Cell Sci.* **103**, 1093–1099.

Menashi, S., Fridman, R., Desrivieres, S., Lu, H., Legrand, Y., and Soria, C. (1994). Regulation of 92-kDa gelatinase B activity in the extracellular matrix by tissue kallikrein. *Ann. NY Acad. Sci.* **732**, 466–468.

Mertz, P. M., DeWitt, D. L., Stetler-Stevenson, W. G., and Wahl, L. M. (1994). Interleukin 10 suppression of monocyte prostaglandin H synthase-2. Mechanism of inhibition of prostaglandin-dependent matrix metalloproteinase production. *J. Biol. Chem.* **269**, 33, 21322–21329.

Miyagi, E., Yasumitsu, H., Hirahara, F., Nagashima, Y., Minaguchi, H., Miyazaki, K., and Umeda, M. (1995). Marked induction of gelatinases, especially type B, in host fibroblasts by human ovarian cancer cells in athymic mice. *Clin. Exp. Metastasis* **13**, 89–96.

Mohtai, M., Smith, R. L., Schurman, D. J., Tsuji, Y., Torti, F. M., Hutchinson, N. I., Stetler-Stevenson, W. G., and Goldberg, G. I. (1993). Expression of 92-kD type IV collagenase/gelatinase (gelatinase B) in osteoarthritic cartilage and its induction in normal human articular cartilage by interleukin 1. *J. Clin. Invest.* **92**, 1, 179–185.

Moll, U. M., Youngleib, G. L., Rosinski, K. B., and Quigley, J. P. (1990). Tumor promoter-stimulated Mr 92,000 gelatinase secreted by normal and malignant human cells: isolation and characterization of the enzyme from HT1080 tumor cells. *Cancer Res.* **50**, 19, 6162–6170.

Montgomery, A. M., Sabzevari, H., and Reisfeld, R. A. (1993). Production and regulation of gelatinase B by human T-cells. *Biochim. Biophys. Acta* **1176**, 3, 265–268.

Munaut, C., Reponen, P., Huhtala, P., Kontusaari, S., Foidart, J. M., and Tryggvason, K. (1994). Structure of the mouse 92-kDa type IV collagenase gene. In vitro and in vivo expression in transient transfection studies and transgenic mice. *Ann. NY Acad. Sci.* **732**, 369–371.

Murphy, G., Cawston, T. E., Galloway, W. A., Barnes, M. J., Bunning, R. A., Mercer, E., Reynolds, J. J., and Burgeson, R. E. (1981). Metalloproteinases from rabbit bone culture medium degrade types IV and V collagens, laminin and fibronectin. *Biochem. J.* **199**, 3, 807–811.

Murphy, G., Reynolds, J. J., Bretz, U., and Baggiolini, M. (1982). Partial purification of collagenase and gelatinase from human polymorphonuclear leucocytes. Analysis of their actions on soluble and insoluble collagens. *Biochem. J.* **203**, 1, 209–210.

Murphy, G., Ward, R., Hembry, R. M., Reynolds, J. J., Kuhn, K., and Tryggvason, K. (1989). Characterization of gelatinase from pig polymorphonuclear leucocytes. A metalloproteinase resembling tumour type IV collagenase. *Biochem. J.* **258**, 2, 463–472.

Murray, S. L., Grubman, S. A., Perrone, R. D., Rojkind, M., Moy, E., Lee, D. W., and Jefferson, D. M. (1996). Matrix metalloproteinase activity in human intrahepatic biliary epithelial cell lines from patients with autosomal dominant polycystic kidney disease. *Connect. Tissue Res.* **33**, 4, 249–256.

Nagase, H., Barrett, A. J., and Woessner, J. F., Jr. (1992). Nomenclature and glossary of the matrix metalloproteinases. *Matrix Suppl.* **1**, 421–424.

Nagle, R. B., Knox, J. D., Wolf, C., Bowden, G. T., and Cress, A. E. (1994). Adhesion molecules, extracellular matrix, and proteases in prostate carcinoma. *J. Cell Biochem. Suppl.* **19**, 232–237.

Nakagawa, H., and Yagihashi, S. (1994). Expression of type IV collagen and its degrading enzymes in squamous cell carcinoma of lung. *Jpn. J. Cancer Res.* **85**, 9, 934–938.

Nakajima, M., Welch, D. R., Wynn, D. M., Tsuruo, T., and Nicolson, G. L. (1993). Serum and plasma $M(r)$ 92,000 progelatinase levels correlate with spontaneous metastasis of rat 13762NF mammary adenocarcinoma. *Cancer Res.* **53**, 23, 5802–5807.

Newman, K. M., Malon, A. M., Shin, R. D., Scholes, J. V., Ramey, W. G., and Tilson, M. D. (1994). Matrix metalloproteinases in abdominal aortic aneurysm: characterization, purification, and their possible sources. *Connect. Tissue Res.* **30**, 4, 265–276.

Nguyen, Q., Murphy, G., Hughes, C. E., Mort, J. S., and Roughley, P. J. (1993). Matrix metalloproteinases cleave at two distinct sites on human cartilage link protein. *Biochem. J.* **295**, Pt 2, 595–598.

Niyibizi, C., Chan, R., Wu, J. J., and Eyre, D. (1994). A 92 kDa gelatinase (MMP-9) cleavage site in native type V collagen. *Biochem. Biophys. Res. Commun.* **202**, 1, 328–333.

Norga, K., Paemen, L., Masure, S., Dillen, C., Heremans, H., Billiau, A., Carton, H., Cuzner, L., Olsson, T., Van Damme, J., *et al.* (1995). Prevention of acute autoimmune encephalomyelitis and abrogation of relapses in murine models of multiple sclerosis by the protease inhibitor D-penicillamine. *Inflamm. Res.* **44**, 12, 529–534.

Ochieng, J., Fridman, R., Nangia-Makker, P., Kleiner, D. E., Liotta, L. A., Stetler-Stevenson, W. G., and Raz, A. (1994). Galectin-3 is a novel substrate for human matrix metalloproteinases-2 and -9. *Biochemistry* **33**, 47, 14109–14114.

O'Connell, J. P., Willenbrock, F., Docherty, A. J., Eaton, D., and Murphy, G. (1994). Analysis of the role of the COOH-terminal domain in the activation, proteolytic activity, and tissue inhibitor of metalloproteinase interactions of gelatinase B. *J. Biol. Chem.* **269**, 21, 14967–14973.

Ogata, Y., Enghild, J. J., and Nagase, H. (1992). Matrix metalloproteinase 3 (stromelysin) activates the precursor for the human matrix metalloproteinase 9. *J. Biol. Chem.* **267**, 6, 3581–3584.

Ogata, Y., Itoh, Y., and Nagase, H. (1995). Steps involved in activation of the pro-matrix metalloproteinase 9 (progelatinase B)-tissue inhibitor of metalloproteinases-1 complex by 4-aminophenylmercuric acetate and proteinases. *J. Biol. Chem.* **270**, 31, 18506–18511.

Oikarinen, A., Kylmaniemi, M., Autio-Harmainen, H., Autio, P., and Salo, T. (1993). Demonstration of 72–kDa and 92–kDa forms of type IV collagenase in human skin: Variable expression in various blistering diseases, induction during re-epithelialization, and decrease by topical glucocorticoids. *J. Invest. Dermatol.* **101**, 2, 205–210.

Okada, Y., Tsuchiya, H., Shimizu, H., Tomita, K., Nakanishi, I., Sato, H., Seiki, M., Yamashita, K., and Hayakawa, T. (1990). Induction and stimulation of 92-kDa gelatinase/type IV collagenase production in osteosarcoma and fibrosarcoma cell lines by tumor necrosis factor alpha. *Biochem. Biophys. Res. Commun.* **171**, 2, 610–617.

Okada, Y., Gonoji, Y., Naka, K., Tomita, K., Nakanishi, I., Iwata, K., Yamashita, K., and Hayakawa, T. (1992). Matrix metalloproteinase 9 (92-kDa gelatinase/type IV collagenase) from HT 1080 human fibrosarcoma cells. Purification and activation of the precursor and enzymic properties. *J. Biol. Chem.* **267**, 30, 21712–21719.

Okada, Y., Naka, K., Kawamura, K., Matsumoto, T., Nakanishi, I., Fujimoto, N., Sato, H., and Seiki, M. (1995). Localizalton of matrix metalloproteinase 9 (92-kilodalton gelatinase/type IV collagenase = gelatinase B) in osteoclasts: Implications for bone resorption. *Lab. Invest.* **72**, 3, 311–322.

Paemen, L., Martens, E., Norga, K., Masure, S., Roets, E., Hoogmartens, J., and Opdenakker, G. (1996). The gelatinase inhibitory activity of tetracyclines and chemically modified tetracycline analogues as measured by a novel microtiter assay for inhibitors. *Biochem. Pharmacol.* **52**, 1, 105–111.

Pardo, A., Selman, M., Ridge, K., Barrios, R., and Sznajder, J. I. (1996). Increased expression of gelatinases and collagenase in rat lungs exposed to 100% oxygen. *Am. J. Respir. Crit. Care Med.* **154**, 4 Pt 1, 1067–1075.

Partridge, C. A., Jeffrey, J. J., and Malik, A. B. (1993). A 96-kDa gelatinase induced by TNF-alpha contributes to increased microvascular endothelial permeability. *Am. J. Physiol.* **265**, 5 Pt 1, L438–L447.

Pender, S. L., Tickle, S. P., Docherty, A. J., Howie, D., Wathen, N. C., and MacDonald, T. T. (1997). A major role for matrix metalloproteinases in T cell injury in the gut. *J. Immunol.* **158**, 4, 1582–1590.

Pluznik, D. H., Fridman, R., and Reich, R. (1992). Correlation in the expression of type IV collagenase and the invasive and chemotactic abilities of myelomonocytic cells during differentiation into macrophages. *Exp. Hematol.* **20**, 1, 57–63.

Pourmotabbed, T. (1994). Relation between substrate specificity and domain structure of 92-kDa type IV collagenase. *Ann. NY Acad. Sci.* **732**, 372–374.

Pourmotabbed, T., Solomon, T. L., Hasty, K. A., and Mainardi, C. L. (1994). Characteristics of 92 kDa type IV collagenase/gelatinase produced by granulocytic leukemia cells: structure, expression of cDNA in E. coli and enzymic properties. *Biochim. Biophys. Acta* **1204**, 1, 97–107.

Pourmotabbed, T., Aelion, J. A., Tyrrell, D., Hasty, K. A., Bu, C. H., and Mainardi, C. L. (1995). Role of the conserved histidine and aspartic acid residues in activity and stabilization of human gelatinase B: an example of matrix metalloproteinases. *J. Protein Chem.* **14**, 7, 527–535.

Proost, P., De Wolf-Peeters, C., Conings, R., Opdenakker, G., Billiau, A., and Van Damme, J. (1993a). Identification of a novel granulocyte chemotactic protein (GCP-2) from human tumor cells. *In vitro* and *in vivo* comparison with natural forms of GRO, IP-10, and IL-8. *J. Immunol.* **150**, 3, 1000–1010.

Proost, P., Van Damme, J., and Opdenakker, G. (1993b). Leukocyte gelatinase B cleavage releases encephalitogens from human myelin basic protein. *Biochem. Biophys. Res. Commun.* **192**, 3, 1175–1181.

Pyke, C., Ralfkiaer, E., Huhtala, P., Hurskainen, T., Dano, K., and Tryggvason, K. (1992). Localization of messenger RNA for Mr 72,000 and 92,000 type IV collagenases in human skin cancers by in situ hybridization. *Cancer Res.* **52**, 5, 1336–1341.

Pyke, C., Ralfkiaer, E., Tryggvason, K., and Dano, K. (1993). Messenger RNA for two type IV collagenases is located in stromal cells in human colon cancer. *Am. J. Pathol.* **142**, 2, 359–365.

Ramos-DeSimone, N., and French, D. L. (1994). Monoclonal antibodies to human MMP-9. *Ann. NY Acad. Sci.* **732**, 469–471.

Rao, J. S., Yamamoto, N., Mohaman, S., Gokaslan, Z. L., Fuller, Z. N., Stetler-Stevenson, W. G., Rao, V. H., Liotta, L. A., Nicolson, G. L., and Sawaya, R. E. (1996). Expression and localization of 92 kDa type IV collagenase/gelatinase B (MMP-9) in human gliomas. *Clin. Exp. Metastasis* **14**, 1, 12–18.

Reponen, P., Sahlberg, C., Munaut, C., Thesleff, I., and Tryggvason, K. (1994). High expression of 92-kDa type IV collagenase (gelatinase) in the osteoclast lineage during mouse development. *Ann. NY Acad. Sci.* **732**, 472–475.

Reponen, P., Leivo, I., Sahlberg, C., Apte, S. S., Olsen, B. R., Thesleff, I., and Tryggvason, K. (1995). 92-kDa type IV collagenase and TIMP-3, but not 72-kDa type IV collagenase or TIMP-1 or TIMP-2, are highly expressed during mouse embryo implantation. *Dev. Dynamics* **202**, 4, 388–396.

Ricou, B., Nicod, L., Lacraz, S., Welgus, H. G., Suter, P. M., and Dayer, J. M. (1996). Matrix metalloproteinases and TIMP in acute respiratory distress syndrome. *Am. J. Respir. Crit. Care Med.* **154**, 2 Pt 1, 346–352.

Rosenberg, G. A., Estrada, E. Y., Dencoff, J. E., and Stetler-Stevenson, W. G. (1995). Tumor necrosis factor-alpha-induced gelatinase B causes delayed opening of the blood-brain barrier: an expanded therapeutic window. *Brain Res.* **703**, 1–2, 151–155.

Rosenberg, G. A., Dencoff, J. E., Correa, N., Jr., Reiners, M., and Ford, C. C. (1996a). Effect of steroids on CSF matrix metalloproteinases in multiple sclerosis: Relation to blood–brain barrier injury. *Neurology* **46**, 6, 1626–1632.

Rosenberg, G. A., Navratil, M., Barone, F., and Feuerstein, G. (1996b). Proteolytic cascade enzymes increase focal cerebral ischemia in rat. *J. Cereb. Blood Flow Metab.* **16**, 3, 360–366.

Sakalihasan, N., Delvenne, P., Nusgens, B. V., Limet, R., and Lapiere, C. M. (1996). Activated forms of MMP2 and MMP9 in abdominal aortic aneurysms. *J. Vasc. Surg.* **24**, 1, 127–133.

Sakata, K., Kozaki, K., Iida, K., Tanaka, R., Yamagata, S., Utsumi, K. R., Saga, S., Shimizu, S., and Matsuyama, M. (1996). Establishment and characterization of high-and low-lung-metastatic cell lines derived from murine colon adenocarcinoma 26 tumor line. *Jpn. J. Cancer Res.* **87**, 1, 78–85.

Salo, T., Makela, M., Kylmaniemi, M., Autio-Harmainen, H., and Larjava, H. (1994). Expression of matrix metalloproteinase-2 and -9 during early human wound healing. *Lab. Invest* **70**, 2, 176–182.

Sang, Q. X., Birkedal-Hansen, H., and Van Wart, H. E. (1995). Proteolytic and non-proteolytic activation of human neutrophil progelatinase B. *Biochim. Biophys. Acta* **1251**, 2, 99–108.

Sato, H., and Seiki, M. (1993). Regulatory mechanism of 92 kDa type IV collagenase gene expression which is associated with invasiveness of tumor cells. *Oncogene* **8**, 2, 395–405.

Sato, H., Kita, M., and Seiki, M. (1993). v-Src activates the expression of 92-kDa type IV collagenase gene through the AP-1 site and the GT box homologous to retinoblastoma control elements. A mechanism regulating gene expression independent of that by inflammatory cytokines. *J. Biol. Chem.* **268**, 31, 23460–23468.

Segain, J. P., Harb, J., Gregoire, M., Meflah, K., and Menanteau, J. (1996). Induction of fibroblast gelatinase B expression by direct contact with cell lines derived from primary tumor but not from metastases. *Cancer Res.* **56**, 23, 5506–5512.

Senior, R. M., Griffin, G. L., Fliszar, C. J., Shapiro, S. D., Goldberg, G. I., and Welgus, H. G. (1991). Human 92– and 72–kilodalton type IV collagenases are elastases. *J. Biol. Chem.* **266**, 12, 7870–7875.

Sepper, R., Konttinen, Y. T., Sorsa, T., and Koski, H. (1994). Gelatinolytic and type IV collagenolytic activity in bronchiectasis. *Chest* **106**, 4, 1129–1133.

Shapiro, S. D., Fliszar, C. J., Broekelmann, T. J., Mecham, R. P., Senior, R. M., and Welgus, H. G. (1995). Activation of the 92-kDa gelatinase by stromelysin and 4-aminophenylmercuric acetate. Differential processing and stabilization of the carboxyl-terminal domain by tissue inhibitor of metalloproteinases (TIMP). *J. Biol. Chem.* **270**, 11, 6351–6356.

Shimizu, S., Nishikawa, Y., Kuroda, K., S, T., K, K., Hyuga, S., Saga, S., and Matsuyama, M. (1996). Involvement of transforming growth factor beta 1 in autocrine enhancement of gelatinase B secretion by murine metastatic colon carcinoma cells. *Cancer Res.* **56**, 14, 3366–3370.

Shipley, J. M., Doyle, G. A. R., Fliszar, C. J., Ye, Q.-Z., Johnson, L. L., Shapiro, S. D., Welgus, H. G., and Senior, R. M. (1996). The structural basis for the elastolytic activity of the 92-kDa and 72-kDa gelatinases. *J. Biol. Chem.* **271**, 8, 4335–4341.

Sires, U. I., Murphy, G., Baragi, V. M., Fliszar, C. J., Welgus, H. G., and Senior, R. M. (1994). Matrilysin is much more efficient than other matrix metalloproteinases in the proteolytic inactivation of alpha 1-antitrypsin. *Biochem. Biophys. Res. Commun.* **204**, 2, 613–620.

Soini, Y., Hurskainen, T., Hoyhtya, M., Oikarinen, A., and Autio-Harmainen, H. (1994). 72 KD and 92 KD type IV collagenase, type IV collagen, and laminin mRNAs in breast cancer: A study by *in situ* hybridization. *J. Histochem. Cytochem.* **42**, 7, 945–951.

Sopata, I., and Dancewicz, A. M. (1974). Presence of a gelatin-specific proteinase and its latent form in human leucocytes. *Biochim. Biophys. Acta* **370**, 510–523.

Sorbi, D., French, D. L., Nuovo, G. J., Kew, R. R., Arbeit, L. A., and Gruber, B. L. (1996). Elevated levels of 92-kd type IV collagenase (matrix metalloproteinase 9) in giant cell arteritis. *Arthritis Rheum.* **39**, 10, 1747–1753.

Sottrup-Jensen, L. (1989). Alpha-macroglobulins: structure, shape, and mechanism of proteinase complex formation. *J. Biol. Chem.* **264**, 20, 11539–11542.

Stahle-Backdahl, M., Inoue, M., Guidice, G. J., and Parks, W. C. (1994). 92-kD gelatinase is produced by eosinophils at the site of blister formation in bullous pemphigoid and cleaves the extracellular domain of recombinant 180-kD bullous pemphigoid autoantigen. *J. Clin. Invest.* **93**, 5, 2022–2030.

Stein-Picarella, M., Ahrens, D., Mase, C., Golden, H., and Niedbala, M. J. (1994). Localization and characterization of matrix metalloproteinase-9 in experimental rat adjuvant arthritis. *Ann. NY Acad. Sci.* **732**, 484–485.

Strongin, A. Y., Collier, I. E., Krasnov, P. A., Genrich, L. T., Marmer, B. L., and Goldberg, G. I. (1993). Human 92 kDa type IV collagenase: functional analysis of fibronectin and carboxyl-end domains. *Kidney Int.* **43**, 1, 158–162.

Tetlow, L. C., Less, M., Ogata, Y., Nagase, H., and Wooley, D. E. (1993). Differential expression of gelatinase B (MMP-9) and stromelysin (MMP-3) by rheumatoid synovial cells *in vitro* and *in vivo*. *Rheum Int* **13**, 2, 53–59.

Tremble, P. M., Lane, T. F., Sage, E. H., and Werb, Z. (1993). SPARC, a secreted protein associated with morphogenesis and tissue remodeling, induces expression of metalloproteinases in fibroblasts through a novel extracellular matrix-dependent pathway. *J. Cell Biol.* **121**, 6, 1433–1444.

Tremble, P., Chiquet-Ehrismann, R., and Werb, Z. (1994). The extracellular matrix ligands fibronectin and tenascin collaborate in regulating collagenase gene expression in fibroblasts. *Mol. Biol. Cell* **5**, 4, 439–453.

Triebel, S., Blaser, J., Reinke, H., Knauper, V., and Tschesche, H. (1992). Mercurial activation of human PMN leucocyte type IV procollagenase (gelatinase). *FEBS Lett.* **298**, 2–3, 280–284.

Tyagi, S. C., Kumar, S. G., Haas, S. J., Reddy, H. K., Voelker, D. J., Hayden, M. R., Demmy, T. L., Schmaltz, R. A., and Curtis, J. J. (1996). Post-transcriptional regulation of extracellular matrix metalloproteinase in human heart end-stage failure secondary to ischemic cardiomyopathy. *J. Mol. Cell Cardiol.* **28**, 7, 1415–1428.

Ueda, Y., Imai, K., Tsuchiya, H., Fujimoto, N., Nakanishi, I., Katsuda, S., Seiki, M., and Okada, Y. (1996). Matrix metalloproteinase 9 (gelatinase B) is expressed in multinucleated giant cells of human giant cell tumor of bone and is associated with vascular invasion. *Am. J. Pathol.* **148**, 2, 611–622.

Van Wart, H. E., and Birkedal-Hansen, H. (1990). The cysteine switch: A principle of regulation of metalloproteinase activity with potential applicability to the entire matrix metalloproteinase gene family. *Proc. Natl. Acad. Sci. USA* **87**, 14, 5578–5582.

148 GELATINASE B

Vissers, M. C., and Winterbourn, C. C. (1988). Gelatinase contributes to the degradation of glomerular basement membrane collagen by human neutrophils. *Coll. Relat. Res.* **8**, 2, 113–122.

Wahl, S. M., Allen, J. B., Weeks, B. S., Wong, H. L., and Klotman, P. E. (1993). Transforming growth factor beta enhances integrin expression and type IV collagenase secretion in human monocytes. *Proc. Natl. Acad. Sci. USA* **90**, 10, 4577–4581.

Ward, R. V., Hembry, R. M., Reynolds, J. J., and Murphy, G. (1991). The purification of tissue inhibitor of metalloproteinases-2 from its 72 kDa progelatinase complex. Demonstration of the biochemical similarities of tissue inhibitor of metalloproteinases-2 and tissue inhibitor of metalloproteinases-1. *Biochem. J.* **278**, Pt 1, 179–187.

Weeks, B. S., Schnaper, H. W., Handy, M., Holloway, E., and Kleinman, H. K. (1993). Human T lymphocytes synthesize the 92 kDa type IV collagenase (gelatinase B). *J. Cell Physiol.* **157**, 3, 644–649.

Weston, C. A., and Weeks, B. S. (1996). bFGF stimulates U937 cell adhesion to fibronectin and secretion of gelatinase B. *Biochem. Biophys. Res. Commun.* **228**, 2, 318–323.

Wilhelm, S. M., Collier, I. E., Marmer, B. L., Eisen, A. Z., Grant, G. A., and Goldberg, G. I. (1989). SV40-transformed human lung fibroblasts secrete a 92-kDa type IV collagenase which is identical to that secreted by normal human macrophages [published erratum appears in *J. Biol. Chem.* 1990 Dec 25;265(36):22570]. *J. Biol. Chem.* **264**, 29, 17213–17221.

Winwood, P. J., Schuppan, D., Iredale, J. P., Kawser, C. A., Docherty, A. J. and Arthur, M. J. (1995). Kupffer cell-derived 95-kd type IV collagenase/gelatinase B: Characterization and expression in cultured cells. *Hepatology* **22**, 1, 304–315.

Witty, J. P., Foster, S. A., Stricklin, G. P., Matrisian, L. M., and Stern, P. H. (1996). Parathyroid hormone-induced resorption in fetal rat limb bones is associated with production of the metalloproteinases collagenase and gelatinase B. *J. Bone Miner. Res.* **11**, 72–78.

Yamagata, S., Tanaka, R., Ito, Y., and Shimizu, S. (1989). Gelatinases of murine metastatic tumor cells. *Biochem. Biophys. Res. Commun.* **158**, 228–234.

Young, P. K., and Grinnell, F. (1994). Metalloproteinase activation cascade after burn injury: A longitudinal analysis of the human wound environment. *J. Invest. Dermatol.* **103**, 5, 660–664.

Zeng, Z. S., and Guillem, J. G. (1995). Distinct pattern of matrix metalloproteinase 9 and tissue inhibitor of metalloproteinase 1 mRNA expression in human colorectal cancer and liver metastases. *Br. J. Cancer* **72**, 3, 575–582.

Matrilysin

Carole L. Wilson* and Lynn M. Matrisian†

*Dermatology Division, Washington University School of Medicine, St. Louis, Missouri 63110; and
†Department of Cell Biology, Vanderbilt University School of Medicine, Nashville, Tennessee 37232

I. Introduction

In the last decade, we have witnessed an explosion of knowledge about the matrix metalloproteinase (MMP) matrilysin, formerly known as pump-1 (for putative metalloproteinase-1). Although we recently provided a summary of the history and current knowledge about matrilysin (Wilson and Matrisian, 1996), many significant developments since then have led to new ideas about and reassessments of this MMP, as well as the entire family in general. Our aim in this chapter is to discuss the properties and functions of matrilysin in light of more recent experimental data and to provide our view of the picture that is emerging of matrilysin's possible roles. An in-depth description of biochemical assays for matrilysin purification and activity can be found in a review by Woessner (1995).

II. Amino Acid Sequence and Gene Organization

Complementary DNA encoding human matrilysin or pump-1 was first isolated by Muller et al. (1988) by screening a mixed tumor library in an effort to clone stromelysin-related genes. The cDNA was determined to be 49% homologous to stromelysin-1, but, at 1078 bp, it was considerably shorter than the cDNAs for stromelysin-1 and -2

149

and collagenase-1 (see Chapters 2 and 3). This sequence similarity, and the ability of recombinant protein to be activated with 4-aminophenylmercuric acetate (APMA) and to cleave typical metalloproteinase substrates, such as casein, fibronectin, type I, III, IV, and V gelatins, and procollagenase-1 (Quantin *et al.*, 1989), together led to the supposition that this was another, albeit unusual, member of the MMP family. Once the connection between this molecule and a rat small uterine metalloproteinase (ump) described by Woessner and coworkers was made (Sellers and Woessner, 1980; Woessner and Taplin, 1988), the stage was set for further studies to characterize this MMP. Of particular interest was the observation that the level of mRNA and enzyme activity correlated well with the time course of uterine involution and rate of proteoglycan degradation (Sellers and Woessner, 1980; Quantin *et al.*, 1989; Woessner and Taplin, 1988), suggesting that pump-1 has an active role in tissue remodeling events in the reproductive tract. This finding was exciting in that it identified another MMP family member as a potential player in complementing the degradative activity of collagenase, which is responsible for the initial cleavage of triple helical collagens. In fact, recent evidence indicates that almost all members of the MMP family are expressed during this process in rodents (Wolf *et al.*, 1996; Rudolph-Owen *et al.*, 1997a), and these enzymes appear to be important in the menstrual cycle of primates as well (Rodgers *et al.*, 1994; Brenner *et al.*, 1996; reviewed in Hulboy *et al.*, 1997).

In 1991, the Enzyme Commission named pump-1/ump matrilysin (E.C. 3.4.24.23) to describe its ability to degrade general matrix components, although other names had also been used (punctuated MMP, matrin). Matrilysin is also known as MMP-7 in the numeric designation scheme for the MMP family members. These two designations appear to be the ones used most frequently in the current literature, and "MMP-7" denotes the genetic locus for *matrilysin* in both humans (*MMP7*) and mice (*Mmp7*). Although the term matrilysin was suggested as another name for gelatinase A (Alexander and Werb, 1991), it should be noted that this name is the accepted term for MMP-7.

In addition to the cDNA clone first isolated by Muller *et al.* (1988), another clone with an additional 3' sequence was derived from a human mesangial cell cDNA library (Marti *et al.*, 1992). Both cDNA sequences aligned well with the genomic sequence (Gaire *et al.*, 1994). In addition, rat and mouse cDNA sequences have been determined (Abramson *et al.*, 1995; Wilson *et al.*, 1995). As might be expected, the mouse and rat deduced amino acid sequences are more homologous to each other (87%) than either sequence is to the human (70%). However, as shown in Fig. 1, motifs involved in enzyme latency and activation (the PRCGVDV and HEXGHXXGXXH sequences) are conserved among the three species. The alignment between the rodent and human cDNA sequences does diverge at the 3′ end of the coding sequence, in that there are two adjacent termination codons in the rodent cDNAs, whereas the human cDNA encodes three additional amino acids (Arg–Lys–Lys) at the corresponding position (Fig. 1). However, it was reported that recombinant human matrilysin produced in Chinese hamster ovary cells and its corresponding activated form lack these 3 carboxyl-terminal amino acids (Barnett *et al.*, 1994), which suggests that these residues do not contribute to the proteolytic activity of the enzyme.

The structures of the human and mouse genes show the same arrangement of exons (Gaire *et al.*, 1994; Wilson *et al.*, 1995). The gene is composed of six exons, the first five of which are organized in a

```
                          ↓
Mouse  (MAA)MQLTLFCFVCLLPGHLALPLSQEAGDVSAHQWEQAQNYLRKFYPHDSKTK
Rat    (···)R·····RI·····C···········E·T·L··········L·····
Human   --- MR··VL·A······S·····P····GM·EL·····D··KR··LY··E··

                                              ↓
Mouse  KVNSLVDNLKEMQKFFGLPMTGKLSPYIMEIMQKPRCGVPDVAEYSLMPNSPKW
Rat    ·AT·A··K·R···········E····RV··············F········
Human  NA···EAK·········I··M·NSRVI··············F·······

Mouse  HSRIVTYRIVSYTSDLPRIVVDQIVKKALRMWSMQIPLNFKRVSWGTADIIIGF
Rat    ···T············T····FL·····R·················
Human  T·KV··········R···H·T··RL·S···N··GKE···H·RK·V·······M··

Mouse  ARRDHGDSFPFDGPGNTLGHAFAPGPGLGGDAHFDKDEYWTDGEDAGVNFLFAA
Rat    ··G····N·····································S·····V·
Human  ··GA····Y···········A·······T········E··R····SSL·I···Y··

Mouse  THEFGHSLGLSHSSVPGTVMYPTYQRDYSEDFSLTKDDIAGIQKLYGKRNTL---
Rat    ···L·····G·····SS······G·H·················K·---
Human  ···L·····MG···D·NA······GNGDPQN·K·SQ···K········SNSRKK
```

FIG. 1. Alignment of deduced amino acid sequences of mouse, rat, and human matrilysin. The predicted amino acid sequence of mouse matrilysin is shown compared to the rat and human sequences. The symbol (·) denotes rat and human residues identical to those in the mouse sequence. The shaded areas represent residues that are identical among all three species. In the rodent sequences, initiation at the first ATG of two that are present in the 5′ region of the cDNA would result in the three additional residues shown in parentheses at the amino terminus (MAA). The putative boundaries of the pre- and pro-domains are indicated by the regular and bold arrows, respectively. The conserved PRCGVPDV and HEXGHXXGXXH motifs are underlined. The human sequence contains three additional residues at the carboxyl terminus.

manner homologous to other MMP family members for which the genomic structure is known (see Chapter 1). Although the sequence contained within the first five exons shows similarity to other MMPs,
exon 6 does not have a corresponding equivalent in other MMP genes.
Although exons 1 and 6 differ in size between mouse and human, the
primary difference lies in the length of the introns. Southern blotting
has confirmed that human matrilysin is encoded by a single gene (Gaire
et al., 1994), and the same is true for the mouse homologue (Wilson et
al., 1997). Attempts to isolate matrilysin-related genes in the mouse
by low stringency Southern analysis have proven fruitless, indicating
that matrilysin may exist as a unique member of the MMP family
(C.L.W. and L.M.M., unpublished observations). This finding is interesting since most other MMPs, with the exception of metalloelastase,
have closely related counterparts in the family. Other significant differences exist between matrilysin and the rest of the MMPs, which we
will elaborate on further in this chapter.

The genes encoding human and mouse matrilysin were mapped to
chromosomes 11 and 9, respectively (Knox et al., 1996a; Wilson et al.,
1997). The q region of human Chromosome 11 contains a cluster of MMP
loci, including the genes for collagenase-1, stromelysin-1, stromelysin-2
(Formstone et al., 1993), and metalloelastase (Belaaouaj et al., 1995),
in addition to matrilysin. The region of mouse chromosome 9 where
the matrilysin locus (Mmp7) resides is synteneic with human chromosome 11q. The mouse MMP genes, like the human, may also be organized into a cluster: The matrilysin gene is closely linked to that encoding metalloelastase (Shapiro et al., 1992), as well as the gene for
stromelysin-1 (D.L. Hulboy and L.M.M., unpublished observations).
The mouse collagenase-3 gene has been localized to chromosome 9 as
well (Schorpp et al., 1995), but to our knowledge has not been genetically
mapped. However, the expectation is that it and at least some other
MMP genes will be found at the proximal end of chromosome 9 along
with matrilysin, metalloelastase, and stromelysin-1.

III. REGULATION OF EXPRESSION

A limited number of studies have been done to determine the structural elements of the matrilysin gene promoter and the effector molecules important in controlling the expression of the gene. As in most
other MMP genes, the human matrilysin promoter contains a TATA
box and an activator protein 1 (AP-1) site (TGAGTCA) at equivalent
positions (Gaire et al., 1994; reviewed in Crawford and Matrisian,
1996). In addition, there are two adjacent inverted polyomavirus enhancer A-binding protein 3 (PEA-3) elements [consensus sequence

(C/G)AGGAAG(T/C)] to which members of the c-*ets* family can bind. Although this motif occurs in several other promoters, the number, position, and orientation of the element varies considerably among MMP genes (Crawford and Matrisian, 1996; see Chapter 13). This AP-1/PEA-3 combinatorial motif appears to be the primary element in conferring responsiveness to growth factors, oncogenes, and phorbol esters (Chapter 13). The necessity of this element for expression of the matrilysin gene has been examined by functional dissection of the promoter region. Gaire *et al.* (1994) found that both the PEA-3 and AP-1 sequences are required for induction of matrilysin/CAT reporter constructs in response to the tumor promoter 12-*O*-tetradecanoylphorbol-13-acetate (TPA) and epidermal growth factor (EGF). Both of these molecules have been shown to up-regulate matrilysin expression in human colon adenocarcinoma cell lines (Gaire *et al.*, 1994). In addition, Sundareshan and co-workers (1997) demonstrated that TPA mediates both an increase in matrilysin expression and apoptosis in the prostate cancer cell line LNCaP; because a synthetic MMP inhibitor did not prevent TPA-induced apoptosis of these cells, the authors concluded that there is no direct relationship between matrilysin and regulated cell death under these conditions. It has also been shown that introduction of an activated Ki-*ras* oncogene into SW1417 colon carcinoma cells induces matrilysin transcription and enzymatic activity (Yamamoto *et al.*, 1995a). Using CAT assays, these workers found that an increase in AP-1 activity and high levels of AP-1 binding protein were associated with expression of the mutant Ki-*ras*. These findings led to the suggestion that induction of matrilysin expression in cells containing an activation in *ras* is mediated through an AP-1-dependent pathway. These results are also significant in light of the association of *ras* mutations with colon tumor progression; as will be discussed later, a significant majority of colorectal tumors express matrilysin.

It has been demonstrated that matrilysin is regulated by transforming growth factor β (TGF-β), potentially by elements that show similarity to the TGF-β inhibitory element (TIE) originally identified in the rat stromelysin-1 promoter (Kerr *et al.*, 1990). In general, TGF-β down-regulates expression of MMPs as it stimulates the elaboration of extracellular matrix (ECM) molecules such as fibronectin, although these effects can vary depending on factors such as cell type and state of transformation (reviewed in Massagué, 1990). Marti and co-workers (1994) demonstrated that in cultured human glomerular mesangial cells, TGF-β_1 suppressed both the steady-state levels of matrilysin and stromelysin mRNAs and secretion of the zymogens in a dose-dependent manner. In fact, TGF-β isoforms were identified as the principal factors mediating the progesterone-induced inhibition of matrilysin mRNA and protein in the human endometrium (Bruner *et al.*, 1995). However, TGF-β appears to have the opposite effect on transformed cells. For example, Nakano *et al.* (1995) observed that matrilysin mRNA was up-regulated in some human glioma cell lines by this growth factor, and that, in general, TGF-β stimulated the invasive behavior of these cells in a Matrigel invasion assay. Indeed, Borchers and colleagues (1994) found that a fibroblast-derived soluble factor, possibly TGF-β, induced expression of matrilysin mRNA and protein in the human squamous cell carcinoma cell line II-4.

Other suggested candidates for the fibroblast-derived stimulatory factor found by Borchers *et al.* (1994) include PDGF and interleukins (IL) 1, 3, and 6, all of which are potentially produced by dermal fibroblasts. The promoter region of the human matrilysin gene does contain at least two sites which are homologous to the NF-IL6 consensus binding sequence (T(T/G)NNGNAA(T/G)), indicating that expression is potentially responsive to IL-1 and IL-6. In cultured human mesangial cells, matrilysin mRNA levels increased following treatment of the cells simultaneously with tumor necrosis factor a (TNF-α) and IL-1β for 48 h (Marti *et al.*, 1992). In addition, there are other sites in the promoter region of the human matrilysin gene with similarity to consensus binding sequences for factors such as glucocorticoids, GATA-1 (required for erythroid differentiation), interferon γ (IFN-γ), and C/EBP. Indeed, Busiek *et al.* (1995) determined that matrilysin expression in cultured monocyte-derived macrophages was suppressed by glucocorticoids and IFN-γ among others. These workers also found that retinoids had an inhibitory effect, and this same effect was observed using *trans*-retinoic acid on BM314 colon cancer cells (Yamamoto *et al.*, 1995b). Unlike mesangial cells, matrilysin expression in macrophages was not influenced by IL-1, IL-6, or TNF-α. Instead, lipopolysaccharide (LPS) and opsonized zymosan, a particulate preparation of yeast cell wall material, were major stimulators of matrilysin production in these phagocytes, whereas IL-4 and IL-10 inhibited expression. Taken together, these findings emphasize that the effect of specific growth factors, cytokines, and pharmacologic agents on matrilysin expression varies depending on the cell type, its environment, and its characteristic physiologic response to inflammatory mediators.

There are several intriguing examples of matrilysin expression in isolated cells being dependent on culture conditions. As mentioned earlier, it was shown that mesangial cells synthesize a low level of matrilysin, with those levels being augmented by either TNF-α or IL-1 (Marti *et al.*, 1992). Abdel and Mason (1996) recently reported that matrilysin mRNA and protein were significantly decreased when these cells were grown in high concentrations of glucose. Another study suggests that matrilysin expression may be influenced by cell–cell contact. The squamous carcinoma cell line II-4, which, when cocultured with fibroblasts, produces matrilysin mRNA and protein, was shown to synthesize significantly higher levels in confluent versus log phase cultures (Borchers *et al.*, 1997a). Importantly, a neutralizing antibody against E-cadherin was able to inhibit expression, as was growth in medium disruptive for E-cadherin-mediated cell–cell interactions. Although these experiments do not show a direct relationship between E-cadherin and matrilysin, they do raise the possibility of a link between these proteins which may have implications for the development and growth of neoplastic cells.

From the examples discussed, one might surmise that matrilysin expression is limited only to mesangial cells, monocyte-derived macrophages, and carcinoma cells. As will become clear in the ensuing sections of this chapter, this is not the case; however, there is a cell-type specificity in the expression of this metalloenzyme in that it is rarely produced by fibroblasts or stromal cells. To our knowledge, only two studies have shown expression of matrilysin in fibroblastic cells: Overall and Sodek (1990) reported that the lectin concanavalin A (con A) induced transcription of matrilysin mRNA in human gingival fibroblasts, which normally do not express matrilysin. This effect appeared to be mediated by changes in cell shape induced by con A. In our laboratory, we detected matrilysin mRNA in fibroblastic cells adjacent to tumor epithelium in some cases of breast carcinoma *in situ* (Heppner *et al.*, 1996). Could expression in these cells also be due to alterations in cell morphology, perhaps mediated by changes in the remodeling extracellular environment? Or could it be the result of aberrant epithelial–mesenchymal interactions in confined tumor nests? Fibroblastic expression of matrilysin was not detected in cases of invasive cancer (Heppner *et al.*, 1996) and has not been reported in other analyses of breast cancer specimens (Wolf *et al.*, 1993). Although there are no clear answers as yet, the majority of data supports the observation that matrilysin is primarily expressed by parenchymal, rather than stromal, cells.

Finally, there is one report that indicates that matrilysin expression may also be regulated at the post-transcriptional level. Wallon and coworkers (1994) showed that treatment of the human colon adenocarcinoma cell line SW1116 with an inhibitor of polyamine synthesis (difluoromethylornithine, DMFO) for several days resulted in a reduction of secreted matrilysin levels without a concomitant decrease in the steady-state level of mRNA. This reduction was thought to be due to decreased translation and/or increased degradation of the protein rather than an alteration in the protein export process, since intracellular proenzyme did not accumulate. Overall, there is little information about the kinetics of promatrilysin synthesis and secretion. Although the zymogen lacks Asn-linked glycosylation (Crabbe *et al.*, 1992), presumably it follows an endoplasmic reticulum-mediated secretory pathway, because it contains a consensus leader sequence. However, recent evidence suggests that, at least in glandular epithelial cells and tumor cells of epithelial origin, secretion of matrilysin is directionally controlled. For example, in our laboratory, we detected immunoreactive protein on the apical side of uterine epithelial cells (Fig. 2a) and in the lumen of

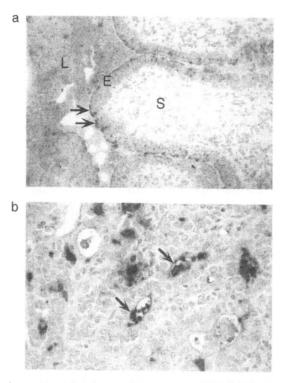

FIG. 2. Apical secretion of matrilysin protein from epithelial cells in mouse tissue. (a) A transverse section of postpartum uterus was immunostained for matrilysin using a polyclonal antibody against the carboxyl terminus. Protein can be seen on the apical side of epithelial cells (E) lining the lumen (L) of the uterus. The arrows indicate areas of intense punctate staining, which can be seen throughout the section. Lumenal contents are also stained with the antibody. Protein was not detected in the adjacent stroma (S). (b) The same antibody was used to localize matrilysin in a section of a small intestinal tumor from a *Min* mouse. A significant degree of immunoreactivity is evident within the lumen (arrows) of dysplastic glands in the tumor.

dysplastic intestinal glands (Fig. 2b; Wilson *et al.*, 1997) in mice; *in situ* hybridization studies have confirmed that the mRNA is produced by the epithelial cells in these samples (Wilson *et al.*, 1995, 1997). Using a Transwell® system, Parks and co-workers (1997) showed that matrilysin is secreted apically from rat type II pneumocytes *in vitro*. However, treatment of the cells with phorbol ester resulted in nearly equivalent levels of both basal and apical secretion (Parks *et al.*, 1997). These observations indicate that the regulation of matrilysin also involves control of vectorial secretion of the proteinase and suggest that the delivery of this enzyme to different tissue compartments is associated with distinct functions.

IV. PROTEIN STRUCTURE AND PROTEOLYTIC ACTIVITIES

The ability to generate high levels of recombinant matrilysin protein has been critical for studies of the protein biochemistry and *in vitro* proteolytic activities, as well as analysis of the protein structure. The structure of the catalytic domain of matrilysin complexed with substrate-analogue inhibitors was determined (Browner *et al.*, 1995), as has been done for collagenase-1, collagenase-2, and stromelysin-1 (see Chapter 9). All show a remarkable similarity in secondary and tertiary structure. In agreement with Soler *et al.* (1994), Browner and colleagues (1995) determined that there are four metal ions bound to the matrilysin molecule, a catalytic zinc ion, a structural zinc ion, and two calcium ions. As expected, the catalytic zinc is complexed by a tetracoordinate bond among the three His residues in the HEXGHXXGXXH region. The structural zinc is ligated in a similar fashion, and binding of the calcium ions stabilizes the secondary structure. Like collagenase-1, the substrate-binding site (S_1') of matrilysin consists of a shallow hydrophobic pocket in which the amino acid residue at position 214 forms the bottom of the pocket. In contrast, stromelysin-1 and collagenase-2 are examples of MMPs containing a deep hydrophobic pocket. These differences in S_1' structure have been proposed to play a role in substrate specificity, and suggest that matrilysin and collagenase-1 may share some preferences for particular residues in the P_1' position (Welch *et al.*, 1996).

Like other MMP zymogens, promatrilysin contains an amino-terminal domain of approximately 9 kD, which is involved in maintaining its inactive state (Chapter 1). Destabilization of the complex between the cysteine in the pro-domain and the active site $Zn2+$, primarily by stepwise cleavage of about the first 30 amino acids of the pro segment, results in a conformational change that leads to autoproteolysis and removal of the entire pro-domain. Crabbe *et al.* (1992)

determined that activation of matrilysin occurs by a similar stepwise cleavage mechanism. Like other proMMPs, the matrilysin precursor is activated by APMA, trypsin, and high temperature (53°C). The intermediates obtained appear to depend on the reagent used for activation, and include products from cleavage at Glu^{12}–Leu^{13} (heat activation), Glu^{63}–Ile^{64} (APMA), and Lys^{33}–Asn^{34} and Arg^{69}–Cys^{70} (trypsin). Regardless of the method used, Crabbe *et al.* (1992) found that the final autocatalytic cleavage occurs at Glu^{77}–Tyr^{78} to yield the mature form of the protein.

Endoproteinases are believed to be the physiological activators of promatrilysin and other MMP zymogens, and plasmin is the most probable candidate, because it cleaves at sites also recognized by trypsin. Plasmin was shown to activate promatrilysin *in vitro* to about 50% of its full activity (Imai *et al.*, 1995a). In addition, there is circumstantial evidence linking serine and metal-dependent proteinases *in vivo*. For example, the promoter region of the urokinase plasminogen activator (uPA) gene contains AP-1 and PEA-3 elements, suggesting that it may be coordinately regulated with MMPs, whose genes also contain these motifs. In fact, these gene products are frequently coexpressed in remodeling and diseased tissue (e.g., see Wolf *et al.*, 1993; Airola *et al.*, 1995). A recent study by Borchers *et al.* (1997b) showed that both matrilysin and uPA were induced in the squamous cell carcinoma line II-4 when cocultured with normal melanocytes or melanoma cells. Although there is experimental evidence for processing of uPA forms by matrilysin (see later discussion), data for the reverse interaction have not been forthcoming. Therefore, other proteinases and factors yet to be identified must be involved in producing mature matrilysin. Imai *et al.* (1995a) found that stromelysin-1 and leukocyte elastase activated promatrilysin to varying degrees *in vitro,* but it is not known if these cleavage reactions occur *in vivo.*

More information is available about potential substrates for matrilysin, rather than activators of the protease. As with other MMPs, the proteolytic activities of matrilysin have been defined primarily by *in vitro* experiments using activated recombinant promatrilysin and purified substrates. Investigators have been particularly interested in determining the level of matrilysin activity against these substrates in comparison to other MMPs. Matrilysin has been found to cleave a variety of protein substrates, which can be grouped into three major categories: ECM components, proMMPs, and nonmatrix proteins. Each group is considered in more detail in the following subsections, along with some of the potential biological effects of matrilysin-mediated degradation.

A. ECM Components

Activated matrilysin has been shown to degrade the ECM proteins fibronectin (Woessner and Taplin, 1988; Quantin *et al.,* 1989), gelatins of types I, III, IV, and V (Woessner and Taplin, 1988; Quantin *et al.,* 1989), collagen type IV (Miyazaki *et al.,* 1991; Murphy *et al.,* 1991), laminin (Miyazaki *et al.,* 1991), entactin/nidogen (Mayer *et al.,* 1993; Sires *et al.,* 1993), fibulin-1 and -2 (Sasaki *et al.,* 1996), elastin (human matrilysin only; Murphy *et al.,* 1991), vitronectin (Imai *et al.,* 1995b), the cartilage proteoglycan aggrecan (Sellers and Woessner, 1980; Fosang *et al.,* 1992), cartilage link protein (Nguyen *et al.,* 1993), the chondroitin sulfate proteoglycan versican (Halpert *et al.,* 1996), and tenascin-C isoforms (Siri *et al.,* 1995). Matrilysin digests fibronectin more efficiently than stromelysin-1 (Woessner and Taplin, 1988; Quantin *et al.,* 1989), but does not hydrolyze collagen IV at the same rate (Murphy *et al.,* 1991). Furthermore, it was determined that matrilysin has little, if any, activity against native fibrillar collagens, whereas stromelysin-1 is at least able to degrade collagen III to a limited extent (Quantin *et al.,* 1989). Sires *et al.*(1993) showed that matrilysin cleaves entactin/nidogen, a glycoprotein that links laminin and collagen type IV, from 100- to 600-fold more rapidly than either collagenase-1 or gelatinase B. Matrilysin attacks the link region between two of the major globular domains, as well as several sites within these domains (Mayer *et al.,* 1993; Sires *et al.,* 1993). Although stromelysin-1 also cleaves entactin/nidogen, the majority of the sites do not appear to overlap with those found for matrilysin (Mayer *et al.,* 1993). Fibulin-1 and -2, which are found in basement membranes and in fibronectin-rich fibrils, are also sensitive to matrilysin, according to a recent report by Sasaki *et al.* (1996). While fibulin-2 is cleaved by a variety of proteases and MMP family members, fibulin-1 shows a more limited susceptibility, in that, among the proteases tested, only leukocyte elastase and matrilysin were active against the protein. Imai *et al.* (1995b) recently showed that matrilysin digests vitronectin from 8- to 44-fold more efficiently than either gelatinase A, gelatinase B, or stromelysin-1. Interestingly, it appears that human matrilysin, but not the rat homologue, degrades insoluble elastin (Woessner and Taplin, 1988; Murphy *et al.,* 1991; Imai *et al.,* 1995a), which is one of the few indications that there may be subtle differences between the rodent and human enzymes.

Several other ECM molecules are cleaved by matrilysin, including the proteoglycan aggrecan and its link protein, versican, and tenascin. In fact, bovine nasal cartilage proteoglycan (now termed aggrecan) was the first substrate that matrilysin was shown to degrade (Sellers and

Woessner, 1980). In comparison to other MMPs, matrilysin most extensively degrades the components of cartilage proteoglycan aggregates, which include the link protein that stabilizes the interaction between aggrecan and hyaluronic acid in these aggregates (Nguyen *et al.*, 1993). Like entactin/nidogen, aggrecan is composed of globular domains separated by a protease-sensitive region. Stromelysin-1, the gelatinases, collagenase-1 and -2, and matrilysin all cleave at the same bond in this interglobular region; however, matrilysin and the collagenases recognize an additional site in proximity to the second globular domain (Fosang *et al.*, 1992, 1993). Matrilysin also degrades the proteoglycan versican more efficiently than equimolar concentrations of collagenase-1, gelatinase B, or stromelysin-1 (Halpert *et al.*, 1996), making it one of the most potent proteoglycanases among the MMPs. Finally, the 200-kD isoform of tenascin-C, which is generated by alternative splicing of seven fibronectin-like type III repeats, is susceptible to degradation by matrilysin, but not other MMPs; in contrast, the large form (300 kD), which includes those repeats, is digested by matrilysin, gelatinase A, and collagenase-1 within the type III repeats (Siri *et al.*, 1995). From all these studies, we can infer that the carboxyl-terminal hemopexin-like domain absent in matrilysin does not play a role in determining bond recognition in all substrates, because matrilysin and other MMPs cleave some of the same sites. However, it is clear that differential cleavage of molecules by MMPs, even those grouped by similar substrate specificity, is possible, and may be mediated by residues adjacent to the cleavage site and/or secondary structure at that site in the substrate.

From the preceding discussion, it appears that matrilysin targets components of basement membranes and has the potential to degrade proteins associated with remodeling or provisional matrices. As discussed later, since expression of matrilysin is observed in a variety of epithelial tissue and epithelial-derived tumor cells, it is not too surprising that this metalloenzyme has such an affinity for basement membrane proteins. These findings have led us and others to speculate that matrilysin may have a role in epithelial cell homeostasis, and, when aberrantly expressed in neoplastic lesions, may facilitate tissue breakdown associated with the growth and spread of tumor cells. Do the cleavage activities defined by *in vitro* biochemical assays occur *in vivo*, and are they biologically relevant? Although the answers to these questions are actively being pursued, a few studies have attempted to address some of these issues. For example, von Bredow and co-workers (1995) showed that recombinant matrilysin added to fibroblast cultures degrades fibronectin fibrils assembled by the cells. Several fragments were generated, two of which were derived from the cell-binding region.

Similarly, Fukai *et al.* (1995) found that matrilysin liberated cell-binding activity from fibronectin, whereas stromelysin-1, collagenase-1, and gelatinase A tended to produce amino- and carboxyl-terminal fragments. In addition, cleavage of entactin/nidogen by matrilysin generates biologically active peptides that promote neutrophil chemotaxis and phagocytosis (Gresham *et al.*, 1996). These experiments further support the idea that matrilysin-mediated degradation of ECM components has physiological significance *in vivo*.

B. ProMMPs

In addition to its ability to degrade ECM components, matrilysin may have an indirect role in matrix remodeling by activating the latent forms of other MMPs. Quantin *et al.*(1989) found that treatment of procollagenase-1 with activated matrilysin and APMA increased the level of collagenase-1 activity five-fold over that observed using APMA alone. In contrast, stromelysin-1 elicited a higher level of activity at a lower concentration of enzyme, indicating that it is a more efficient collagenase-1 activator than matrilysin, although both enzymes generate the same amino terminus (Sang *et al.*, 1996). Neither human nor rat matrilysin appear to cleave procollagenase-3 (Abramson *et al.*, 1995), which is the only identified interstitial collagenase present in rodents (see Chapter 2). Matrilysin excises the propeptide directly from the latent form of gelatinase A in the absence of any autoproteolysis (Crabbe *et al.*, 1994); cleavage of progelatinase A by both matrilysin and collagenase-1 has been confirmed by Sang *et al.* (1996). Another report indicates that, in a manner similar to that of stromelysin-1 and trypsin, both matrilysin and collagenase-1 generate an active 65-kDa form of neutrophil gelatinase B (Sang *et al.*, 1995). Thus, in situations where these metalloenzymes are coexpressed, matrilysin may act as a physiological activator of other MMP family members.

C. Nonmatrix Proteins

In vitro, matrilysin cleaves several nonmatrix proteins such as casein, insulin, transferrin (Woessner and Taplin, 1988), α_1-antitrypsin/α_1-proteinase inhibitor (Sires *et al.*, 1994; Zhang *et al.*, 1994), pro-uPA and uPA (Marcotte *et al.*, 1992), proTNF-α (Gearing *et al.*, 1994), and myelin basic protein (Chandler *et al.*, 1995). Cleavage of some of these proteins by matrilysin may be involved in modulating cellular behavior. For example, matrilysin was identified as the proteinase responsible for converting the high-molecular-weight form of both pro-uPA and active uPA into two fragments (Marcotte *et al.*, 1992), one of which makes up the receptor-binding domain and the other the serine proteinase region. The receptor-binding fragment has been shown to have growth factor-like activities for certain cell types (e.g., see Anichini *et al.*, 1994). Similarly, Gearing *et al.* (1994) showed that purified matrilysin, along with stromelysin-1 and collagenase-1, is particularly adept at processing a recombinant, truncated form of the TNF-α precursor to its mature form. Normally, proteolytic processing of the precursor results in release of the membrane-bound precursor as the active cytokine, and this processing is prevented by synthetic MMP inhibitors (Gearing *et al.*, 1994). Although a cell-surface member of the ADAM (a disintegrin and metalloproteinase) or adamalysin family of metalloproteinases has been identified as the primary TNF-α convertase (Black *et al.*, 1996, 1997; Moss *et al.*, 1997), matrix proteinases may still be involved at some stage in the processing in specific tissue environments. Several investigators have shown that activated matrilysin may regulate the activity of the serine proteinase leukocyte elastase through its ability to cleave the major physiological inhibitor of this enzyme, α_1-antitrypsin/α_1-proteinase inhibitor (Sires *et al.*, 1994; Zhang *et al.*, 1994). Matrilysin was found to be from 30- to 180-fold more effective at digesting α_1-antitrypsin than other MMPs tested (Sires *et al.*, 1994). Therefore, as an elastase itself and as an inactivator of elastase inhibitors, matrilysin may act to enhance elastin degradation. Matrilysin has been proposed to be a general serpinase, or inactivator of serine proteinase inhibitors, that could play a role in inflammation and tissue damage (Zhang *et al.*, 1994).

D. Cleavage Specificity

The composition of the peptide bonds recognized by matrilysin in the substrates described earlier follows the general trend first noted by Woessner and Taplin (1988) for the B chain of insulin, which has cleavage sites at Ala[14]–Leu[15] and Tyr[16]–Leu[17]. Matrilysin shows a preference for a Leu residue in the P_1' position and generally a hydrophobic residue in the P_1' position. Accordingly, in pro-uPA and α_1-antitrypsin, matrilysin cleaves at a single site containing Leu in the P_1' position. It also recognizes Leu-containing bonds in aggrecan and cartilage link protein. To dissect further the bond specificity of matrilysin, several groups have analyzed its cleavage efficiency against synthetic peptide substrates. Using a series of octapeptides, Netzel-Arnett *et al.* (1993) found that matrilysin exhibited a particular affinity for sites with Leu at P_1' and Met at P_3, and, unlike the gelatinases, was able to accommodate residues with bulkier side groups than that of Gly or Ala in the P_1 position. Of the peptides tested, the ones showing the highest activity as compared to the reference peptide Gly–Pro–Gln–Gly ↓ Ile–Ala–

Gly–Gln (from P_4 to P_4') were Gly–Pro–Gln–**Ala** ↓ Ile–Ala–Gly–Gln > Gly–Pro–Gln–Gly ↓ Ile–Ala–**Met**–Gln > Gly–Pro–**Leu**–Gly ↓ Ile–Ala–Gly–Gln > Gly–Pro–**Met**–Gly ↓ Ile–Ala–Gly–Gln (where the altered residues are shown in bold). To identify potentially new cleavage sites, as well as confirm the site preferences observed previously, Smith and co-workers (1995) constructed a phage display library to test with activated recombinant matrilysin. They obtained the highest k_{cat}/K_M ratio with the hexapeptide Pro–Leu–Glu ↓ Leu–Arg–Ala (from P_3 to P_3'). From this assay, they could clearly distinguish between the preference of stromelysin-1 for Phe and Met at positions P_2 and P_1', respectively, and the preference of matrilysin for Leu at both of these positions, despite the similarity in substrate recognition normally shared by these enzymes. Because matrilysin and stromelysin-1 differ at residues 214 and 215 in the substrate-binding region (S_1'), Welch *et al.* (1996) introduced mutations into this region to attempt to determine the basis for the P_1' preferences. When either residue 214 or both residues 214 and 215 in matrilysin were replaced by amino acids present in stromelysin-1, the mutants showed P_1' preferences characteristic of stromelysin-1. Precise molecular studies such as those described here could lead to the design of specific synthetic inhibitors potentially useful in therapeutic intervention in cancer and other diseases.

E. Inhibition of Activity

Matrilysin and other MMPs are inhibited by metal-chelating agents such as EDTA and 1, 10-phenanthroline, which are most useful for *in vitro* biochemical studies. A number of synthetic MMP inhibitors that act as substrate analogues have also been developed and are active against matrilysin (see Chapter 11). Serum inhibitors of matrilysin include the a-macroglobulins and α_1-inhibitor$_3$ (Zhu and Woessner, 1991). In addition, there are tissue inhibitors of MMPs (TIMPs) of which TIMP-1 and TIMP-2 are known to form noncovalent 1 : 1 complexes with the active form of matrilysin at the catalytic site. However, the mechnism by which binding of TIMPs inhibits MMP activity is still not entirely clear. Using high-resolution gel filtration chromatography, Baragi *et al.*(1994) compared the ability of matrilysin, truncated stromelysin-1, and full-length stromelysin-1 to bind TIMP-1. While each of these molecules was able to form a complex with TIMP-1 that was resistant to separation by gel filtration, both matrilysin and truncated stromelysin-1 showed a reduced ability to bind TIMP-1 when competing with full-length stromelysin-1. These data indicate that the carboxyl-terminal domain absent in matrilysin contributes to the affinity of MMPs for their physiological inhibitors and lead to the hypothesis that matrilysin may be more resistant to inhibition than other MMPs. Note that other MMPs, metalloelastase in particular (see Chapter 7), can undergo carboxyl-terminal processing to remove the hemopexin-like domain at the hinge region.

V.   EXPRESSION IN CANCER AND OTHER DISEASES

Human matrilysin was first cloned from a human cDNA library prepared from a pool of primary and metastatic tumor RNAs. In fact, until a few years ago, matrilysin expression in human tissue had been associated *primarily* with neoplastic lesions in a variety of organs, suggesting that this metalloenzyme plays a role in tumor invasion and metastasis. Matrilysin has been detected in lesions of the breast (Basset *et al.*, 1990; Wolf *et al.*, 1993; Heppner *et al.*, 1996), stomach (McDonnell *et al.*, 1991; Honda *et al.*, 1996), colon (McDonnell *et al.*, 1991; Yoshimoto *et al.*, 1993; Newell *et al.*, 1994; Yamamoto *et al.*, 1994; Mori *et al.*, 1995; Ishikawa *et al.*, 1996), prostate (Pajouh *et al.*, 1991; Knox *et al.*, 1996b), upper aerodigestive tract (oral cavity, oropharynx, hypophar-ynx, and endolarynx) and lung (Muller *et al.*, 1991; Bolon *et al.*, 1996), and skin (Karelina *et al.*, 1994). *In situ* hybridization and immunohisto-chemistry localized expression primarily to the tumor cells in both benign and malignant neoplasms; other MMP family members usually originate in inflammatory cells or are induced in the surrounding stroma, although epithelial expression of stromal MMPs in advanced lesions has been observed (e.g., see Wright *et al.*, 1994; reviewed in MacDougall and Matrisian, 1995; Powell and Matrisian, 1996). In addi-tion, Wang *et al.* (1995) detected matrilysin mRNA in neoplastic cells of several osteosarcoma subtypes, and protein was found in cholangio-cellular carcinoma of the liver (Lichtinghagen *et al.*, 1995). Matrilysin is also expressed in glioma specimens and several cell lines established from these lesions, which are characteristically highly destructive to tissues (Nakano *et al.*, 1993, 1995). Varying levels of the mRNA have been detected in some cell lines derived from colon and breast adenocar-cinomas (Gaire *et al.*, 1994; and L. M. M., unpublished observations), and the protein was isolated from a rectal carcinoma-derived cell line (Miyazaki *et al.*, 1991). Another group used casein zymography to corre-late the levels of enzyme with tumor stage of colorectal lesions, and found that while matrilysin was readily detectable in adenomatous and cancerous tissue, it was not present in hyperplastic polyps or nor-mal colon tissue (Itoh *et al.*, 1996). Together, these observations clearly show that matrilysin expression is a characteristic feature of many cancerous lesions, particularly those of epithelial and glandular epithelial origin. In addition, its appearance in adenomas suggests that it may function early in tumor development.

Several groups have used gene transfer approaches to study the effect of matrilysin overexpression and underexpression on tumor cell behavior. One of the first definitive experiments was carried out by Powell *et al.* (1993), who showed that when the nonmetastatic prostate tumor cell line DU-145 was transfected with human matrilysin cDNA, the cells were significantly more invasive in an *in vivo* model of cell invasion. This finding suggested that matrilysin may be involved in basement membrane degradation as an initial step in tumor cell extravasation. In support of this hypothesis, Imai and co-workers found that introduction of sense or antisense matrilysin constructs into BM314 colon cancer cells resulted in an increase or a decrease, respectively, in the ability of the cells to invade an artificial membrane *in vitro* (Yamamoto *et al.*, 1995b; Itoh *et al.*, 1996). In contrast, overexpression of a similar cDNA construct in the SW480 colon tumor cell line (which, like the DU-145 cells, does not endogenously express matrilysin) did not reproducibly stimulate invasion when the cells were injected into the cecum of nude mice (Witty *et al.*, 1994); instead, an increase in the tumorigenicity of the cells was observed. This finding was confirmed by the observation that orthotopic injection of antisense SW620 clones led to a decrease both in tumorigenicity and incidence of metastasis (Witty *et al.*, 1994). Unlike the Imai studies, we and others saw no statistically significant difference between matrilysin-transfected and control cells in their ability to traverse an artificial basement membrane. Even when a cDNA containing an activating mutation in sequence encoding the pro-domain was introduced into the SW480 cells, only one clone of five showed an increase in invasion (Witty *et al.*, 1994), indicating that matrilysin alone is not always sufficient for cell migration through a matrix barrier. It is possible that the differences in cellular responses observed in these studies may reflect differences in cell type and cell line, in the systems used for measuring cell invasion, and/or in the other MMPs that may be elaborated by these cells. Multiple members of the MMP family probably cooperate to degrade basement membrane components to enhance the invasive process. Degradation of the matrix may, in turn, increase expression of specific enzymes by a positive feedback mechanism. For example, Yamamoto *et al.* (1994) showed that the level of matrilysin mRNA was up-regulated in WiDr colon carcinoma cells plated on immobilized FN fragments containing the cell-binding domain. As mentioned earlier, a fibroblast-derived factor stimulates expression of matrilysin in a human squamous cell line (Borchers *et al.*, 1994), suggesting that its production in epithelial cells is not completely independent of the stromal component in all cases.

The differences noted earlier, along with the observation that matrilysin is expressed early in tumorigenesis, also suggest that matrilysin has dual roles in tumor progression, depending on the temporal and spatial pattern of its appearance. In addition to the effects on tumorigenicity observed *in vivo,* expression of matrilysin in early adenomatous lesions is indicative of its involvement in processes such as cell growth and proliferation. We envision that the induction of matrilysin, perhaps as a result of oncogenic mutations and/or alterations in the matrix environment, may favor the growth and expansion of premalignant cells under the proper conditions by an as yet unknown mechanism. Then, as development of the tumor progresses, sustained production of matrilysin, probably in conjunction with stromal and leukocyte-derived MMPs, could lead to tissue degradation, cell invasion, and metastasis to distant sites via classical matrix-degrading activities. To begin to test these ideas, and to attempt to define the mechanism by which matrilysin could promote tumor formation, our laboratory has generated several mouse models in which levels of matrilysin have been genetically manipulated.

Mice lacking matrilysin were created using standard methods of gene ablation in embryonic stem cells (Wilson *et al.,* 1997). Pups homozygous for the targeted allele ($Mmp7^{m1Vu}$) were produced in the expected Mendelian ratio from heterozygote matings, and, under conditions of conventional barrier housing, these mice have shown no obvious defects throughout their life span. Because matrilysin is constitutively expressed in the glandular epithelium of the reproductive tract of both males and females (see later discussion), we expected there might be deleterious effects on fertility, but none were noted (Rudolph-Owen *et al.,* 1997a). To assess the role of matrilysin in the development of adenomatous lesions, we elected to focus on intestinal tumors that spontaneously form in mice with the Apc^{Min} mutation. We demonstrated by *in situ* hybridization that the majority of these tumors (90%) express high levels of matrilysin (Wilson *et al.,* 1997). Interestingly, the protein was predominantly localized to the apical face of these cells and to the lumen of the dysplastic glands, rather than to the basement membrane (Fig. 2b). Matrilysin-deficient mice were bred to these mice and examined for tumors at an age (4 months) at which lesions are clearly evident in wild-type ($Mmp7+/+$) *Min* animals. On the matrilysin-null background, we found that the number of tumors was significantly reduced (by 58%), and that the tumors were on average smaller in diameter than those in control *Min* mice (Wilson *et al.,* 1997). Matrilysin knock-out/*Min* animals analyzed at 6 months, an age to which most wild-type *Min* animals do not survive, exhibited many tumors, indicating that a deficiency in matrilysin results in a *delay* in tumor development. Although the mechanism by which matrilysin promotes tumor formation in the gastrointestinal tract has yet to be elucidated, we speculate that it may affect the proliferation of transformed epithelial cells, perhaps by modulating growth factor accessibility and/or production of stimulatory degradative fragments of ECM molecules.

Complementary experiments in which matrilysin has been overexpressed *in vivo* have solidified the concept of this metalloenzyme as a critical regulator of the tumorigenic phenotype. Transgenic mice in which a human matrilysin mini-gene was targeted to the mammary gland epithelium via the mouse mammary tumor virus promoter were generated recently (Rudolph-Owen *et al.*, 1997b); a significant number of multiparous transgenic females showed focal areas of epithelial hyperplasia, or hyperplastic alveolar nodules (HANs), which are believed to be precursors of mammary carcinomas. Formation of these HANs may be related to the precocious epithelial cell differentiation that was observed in these animals, because virgin mammary glands were found to produce milk proteins prematurely (Rudolph-Owen *et al.*, 1997b). Furthermore, when these transgenic mice were bred to animals expressing the oncogene *neu* under the same promoter, the resulting progeny developed mammary tumors 13 weeks earlier, and at a higher frequency (100% versus 80%), than the *neu* controls. However, no obvious differences were observed between these two groups in the growth rate of *neu*-induced tumors or in the number of lung metastases (Rudolph-Owen and L.M.M., manuscript in preparation). Therefore, these results point to an effect of matrilysin on *development* of the mammary tumors, somewhat analogous to its effect on the formation of intestinal tumors. *Neu*-induced tumors alone do not express matrilysin; indeed, Goto *et al.* (1995) detected matrilysin in pregnancy-dependent mammary tumors of mice, but not in hormone-independent lesions. In these studies, matrilysin expression was found to be highest during the initiation and regression stages of these tumors.

Recent studies have shown that the expression of matrilysin is also a feature of diseases other than cancer, in particular disorders characterized by active tissue remodeling. Saarialho-Kere and co-workers (1996) examined samples of gastric ulcers, Crohn's disease, and ulcerative colitis for matrilysin expression by *in situ* hybridization and immunohistochemistry. While another group was unable to detect matrilysin expression in *mild* cases of ulcerative colitis (Yoshimoto *et al.*, 1993; Yamamoto *et al.*, 1994; Itoh *et al.*, 1996), Saarialho-Kere *et al.*(1996) found matrilysin localized to the mucosal epithelium bordering the ulcer in the majority of lesions, and signal for this metalloenzyme was not evident in the stroma or inflammatory cells. In addition, immunostaining for laminin and fibronectin in the basement membrane beneath the matrilysin-positive cells was weak compared to distal sites showing no matrilysin expression, suggesting that the enzyme may have degraded these glycoproteins. Furthermore, absence of a signal for the proliferation marker Ki-67 led these investigators to hypothesize that matrilysin expression is associated with cell migration, rather than proliferation, in this type of lesion. Support for this hypothesis was obtained in a recent study by Parks *et al.* (1997), who showed expression of matrilysin in migrating airway epithelial cells in an *ex vivo* model of tracheal wound repair. They also found high levels of matrilysin mRNA and protein in bronchiolar cells from patients with cystic fibrosis (Fig. 3), suggesting that this MMP in particular is up-regulated in response to epithelial damage and that it plays a role in reparative processes in the lung. In general, it appears that matrilysin expression is limited to glandular or lumenal epithelium in tissue repair, and is not expressed in the mature epidermis or in skin wounds or lesions (Saarialho-Kere *et al.*, 1993; Vaalamo *et al.*, 1996). However, some cutaneous and lung inflammatory diseases, such as cystic fibrosis, do show matrilysin production in occasional blood vessel-associated monocytes and freshly extravasated tissue macrophages (Busiek *et al.*, 1995). In atherosclerotic plaques, matrilysin, along with metalloelastase, is expressed in lipid-laden macrophages only at junctions between the lipid core and fibrous cap (Halpert *et al.*, 1996). This precise and confined localization implicates matrilysin activity in the structural instability of these lesions. The finding that versican, a blood vessel chondroitin sulfate proteoglycan, is abundant in these plaques and is degraded more readily by matrilysin than other MMPs tested, suggests one potential mechanism for plaque rupture (Halpert *et al.*, 1996). Interestingly, other MMPs, namely, collagenase-1, stromelysin-1, and gelatinase B, are also expressed by lipid-laden macrophages in atherosclerotic plaques, but *within* the cellular areas of the fibrous cap and shoulder (Henney *et al.*, 1991; Galis *et al.*, 1994; Brown *et al.*, 1995; Nikkari *et al.*, 1995). Thus, subpopulations of macrophages can be defined by their location and profile of MMP expression, suggesting distinct regulatory mechanisms for and functions of the family members. Matrilysin expression is associated with several other pathologies: The protein was detected in cells of the synovial lining in a single case of atypical rheumatoid arthritis characterized by heavy infiltration of inflammatory cells (Hembry *et al.*, 1995). Also, Vettraino and co-workers (1996) found that matrilysin, which is normally expressed in cytotrophoblastic cells during pregnancy, is produced at increased levels and by several cell types in the placenta in preeclampsia.

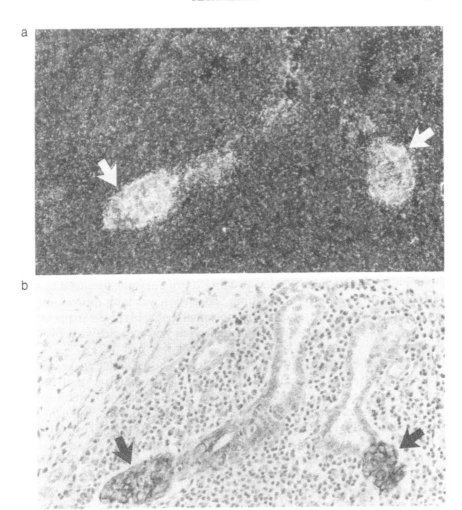

Fig. 3. Colocalization of matrilysin mRNA and protein within damaged epithelium. Serial sections of airway tissue from a patient with cystic fibrosis were processed for (a) *in situ* hybridization and (b) immunohistochemistry using probes specific for human matrilysin as described in Saarialho-Kere *et al.*, 1995. The arrows point to glandular structures positive for matrilysin by both assays. (Photomicrographs courtesy of W. C. Parks.)

VI.   Role in Normal Tissue Remodeling and Homeostasis

Early analysis of matrilysin expression in human neoplasms revealed that the mRNA appears in the surrounding normal tissue in a few limited cases. For example, while matrilysin was not present in normal gastric or colonic tissue (McDonnell *et al.*, 1991), the message was detected at low levels in the bronchial mucosa (Muller *et al.*, 1991) and in 3 of 11 normal prostate samples (Pajouh *et al.*, 1991) by Northern hybridization. However, this type of analysis precluded identification of the cell type responsible for matrilysin synthesis. Because it had been shown earlier that matrilysin is expressed in the early involuting uterus of the rat (Woessner and Taplin, 1988; Quantin *et al.*, 1989), Rodgers *et al.* (1993) used a combination of Northern blotting, *in situ* hybridization, and immunohistochemistry to examine the expression pattern of matrilysin in the human endometrium, a tissue that undergoes rapid remodeling during the normal menstrual cycle. These investigators found that matrilysin mRNA was expressed in the proliferative, late secretory, and menstrual phases of the cycle. Furthermore, both the mRNA and protein localized to the epithelium of endometrial glands. A subsequent investigation (Rodgers *et al.*, 1994) showed that other MMPs are expressed in a cycle-specific pattern in the endometrium as well, but the mRNAs encoding these enzymes localized to the stroma rather than the epithelium. This work probably provided the first indication that the cell-type specificity of matrilysin expression that was observed in cancerous lesions would be recapitulated in normal tissue.

Several other studies have since confirmed and extended these observations to other organs and tissues in humans. For example, matrilysin protein was localized to epidermal cells in the developing fetal skin, where it was prominently expressed in cells invading the mesenchyme in budding hair follicles and sweat glands (Karelina *et al.*, 1994). In adult skin, the protein was detected in the outer root sheath of the hair follicles and in the secretory cells of the sweat glands (Karelina *et al.*, 1994). In fact, Saarialho-Kere *et al.* (1995) showed that matrilysin is produced by most exocrine glands in the body, including the mammary and parotid glands, the pancreas, liver, prostate, and peribronchial glands of the lung. Wolf and co-workers (1993) demonstrated by *in situ* hybridization that matrilysin is expressed in non-neoplastic as well as neoplastic cells in human breast tissue. In addition, Honda *et al.* (1996) observed low levels of matrilysin protein in normal gastric epithelial cells. Both matrilysin mRNA and protein were localized to ductal and secretory epithelium in the organs mentioned and were frequently found to colocalize, as depicted in a representative section of a damaged airway in Fig. 3. However, in some cases, such as the eccrine sweat glands, the mRNA was observed in ductal epithelial cells while protein was localized to secretory areas of the glands (Saarialho-Kere *et al.*, 1995). These studies also indicated that the protein is secreted in the lumenal direction, since staining was not observed in the underlying stroma. This interpretation was strengthened recently by experiments showing that alveolar type II cells secrete matrilysin from their apical surface when cultured *in vitro* (Parks *et al.*, 1997). Taken as a whole, these results led these investigators to propose that matrilysin participates in the normal function of glands by maintaining the patentability of the glandular lumen (acting as an "enzymatic pipe cleaner"). The constitutive pattern of expression does indicate that matrilysin functions in a capacity that is not related merely to a matrix remodeling process. This idea is reinforced by the pattern of matrilysin expression observed in rodent tissue, as described later. Matrilysin is also expressed by a few other cell types in humans. For example, matrilysin is produced by cultured fetal kidney cells, cytokine-stimulated glomerular mesangial cells, and in diseased kidney (Marcotte *et al.*, 1992; Marti *et al.*, 1992). Matrilysin is synthesized by promonocytes extracted from bone marrow and by peripheral blood monocytes following brief culture *in vitro* (Busiek *et al.*, 1992). As mentioned previously, tissue macrophages positive for matrilysin are seen in atherosclerosis and inflamed lung (Busiek *et al.*, 1995; Halpert *et al.*, 1996). However, it is not produced by fully differentiated pulmonary alveolar macrophages (Busiek *et al.*, 1992). Lastly, a recent report indicates that matrilysin protein is produced in cytotrophoblasts and syncytioblasts during the first trimester of pregnancy, and appears in intermediate trophoblasts and decidual cells during all stages (Vettraino *et al.*, 1996). Overall, however, matrilysin expression is restricted to glandular epithelium, with the repertoire of cell types producing the enzyme expanding to a very limited degree in tissue remodeling disorders.

The first study to show that normal tissue is capable of producing high levels of matrilysin under nonpathological conditions was that of Sellers and Woessner (1980) in their work on uterine involution in the rat. Analyzing matrilysin expression and function in rodents has been of great interest in the last few years due to the experimental and genetic malleability of these organisms. Based on Woessner's work, we examined matrilysin expression in the mouse postpartum uterus, from which the cDNA was cloned, as well as in the pregnant and cycling uterus. We found that the mRNA is expressed at high levels late in gestation, with levels increasing immediately after birth and then decreasing until about day 4.5 postpartum (Wilson *et al.*, 1995), in agreement with Woessner's results and more recently those of Wolf *et al.* (1996) in the rat. Matrilysin is also produced during several stages of the estrous cycle in both mice (Rudolph-Owen *et al.*, 1997a) and rats (Wolf *et al.*, 1996; Woessner, 1996). In addition, Wolf *et al.* noted expression of matrilysin in the rat cervix in a temporal pattern similar to that of the uterus, suggesting it may have a role in cervical ripening (Woessner, 1996). In all uterine samples, we found the message localized to epithelial cells lining the lumen of the uterus and associated glandular structures. Using a polyclonal antibody against the carboxyl-terminal segment of matrilysin, we detected protein in epithelial cells of the involuting uterus, and found that it appeared to be secreted *apically* into the lumen (Fig. 2a). This result was in marked contrast to our expectations; we had anticipated that matrilysin, like other MMPs expressed in the postpartum uterus, would have a role in tissue degradation and remodeling during the process of involution. One role that we had postulated for matrilysin in this tissue is that it might serve to activate procollagenase-3, a stromal MMP critical for clearance of excess collagen from the uterus. However, because procollagenase-3 is able to cleave itself efficiently at the site that, in procollagenase-1, is recognized by matrilysin, it has been difficult to determine whether matrilysin is truly capable of catalyzing this reaction (Abramson *et al.,* 1995). In addition, matrilysin-null animals appear to undergo involution normally (Rudolph-Owen *et al.,* 1997a), although a rigorous time course on the kinetics of tissue resorption has not been done. Secretion of matrilysin protein from the apical side of these cells suggests a role for the enzyme either on the cell surface or in the lumen.

In the male reproductive tract, matrilysin transcripts were localized to epithelial cells of the efferent ducts, and, in mature animals, to cells of the initial segment and cauda of the epididymis (Wilson *et al.,* 1995); the mRNA has also been detected in the seminal vesicles (W.C. Powell and L.M.M, unpublished observations). This localization pattern led us to speculate that matrilysin plays a role in the the maturation and progression of sperm through the extratesticular ducts. While this possibility has not been entirely ruled out, the observation that matrilysin-null animals show no overt deficiencies in either fertility (ability to reproduce) or fecundity (number of offspring) indicates that matrilysin function in these ducts is either dispensable, is replaced by another protease, or has not been revealed by the experiments carried out so far. Attempts to demonstrate alterations in the processing of fertilins, members of the ADAMs which are differentially expressed in the testis and associated ducts, in these mice have proven futile (C.L.W., R. Yuan, D. Myles, and L.M.M., unpublished observations). We have also examined the repertoire of proteins produced in both the tissue and lumenal fluid of the epididymis, as well as the sperm, and have not discovered any significant differences in the profile of processed proteins (C.L.W. and L.M.M., unpublished observations). In addition to the constitutive level of matrilysin produced by some extratesticular ducts, Powell *et al.* (1996) showed that matrilysin is induced in the rat ventral prostate as it undergoes resorption in response to androgen withdrawal. Again, expression was restricted to the epithelial cells of the involuting gland. As uPA is also expressed during prostate involution, these workers propose that both metallo- and serine proteases play a role in remodeling of the prostate tissue architecture. Matrilysin knock-out mice will be extremely useful for addressing these questions and deciphering the potential interaction between members of the different protease families.

Another organ in which matrilysin is constitutively expressed at high levels in the mouse is the small intestine (Wilson et al., 1995). Using in situ hybridization, we observed that the mRNA was localized only to the Paneth cells at the base of crypts, and was not present in other epithelial cell lineages in the small intestine (Wilson et al., 1995). Although message was also detected in the stomach and colon by reverse transcription–polymerase chain reaction (RT-PCR), we were unable to localize expression by in situ hybridization and concluded that expression in these organs is very low (Wilson et al., 1995). Along with other Paneth cell markers, namely, cryptdin-4, -5, and -6, matrilysin mRNA was detected in the P1 (postnatal day 1) small intestine by RT-PCR (Darmoul et al., 1997), suggesting that the undifferentiated intervillus epithelium is capable of producing these proteins prior to crypt formation. The precise role of the Paneth cell has been somewhat enigmatic, but, given that proteins such as lysozyme and the cryptdin peptides (crypt defensins) are synthesized by these cells, they likely act as specialized defense cells in the gastrointestinal system. Does matrilysin expression contribute to the function(s) of the Paneth cell in vivo? And, more important, does matrilysin's activity in these cells relate to its role in other epithelial cells? Again, with mice lacking matrilysin available, we can begin to delve into these issues.

As alluded to previously, rodents and humans show striking similarities in the pattern of matrilysin expression in normal organs and tissues. For example, the mRNA has been detected in ducts of the breast, the bile duct, and peribronchial glands of the lung in humans (Saarialho-Kere et al., 1995); similarly, we have found that the mammary gland (cycling, lactating, and involuting) and lung in mice are positive for matrilysin by RT-PCR, and transcripts have been localized to an extrahepatic branch of the bile duct in these animals (Wilson et al., 1995). However, there are some differences between the species in the organs that express matrilysin. While matrilysin was found in ducts of the parotid glands and pancreas in humans, it was absent from these organs in the mouse, at least at the detection level of in situ hybridization. In contrast to the mouse, matrilysin does not appear to be produced by Paneth cells in humans (Saarialho-Kere et al., 1996;

and personal communication); however, it can be detected in the decidua (Vettraino *et al.*, 1996), whereas this tissue is negative in mice (Wilson *et al.*, 1995). Some of these differences may reflect the techniques used for visualizing expression, and some may represent true disparities between the species. There are certainly species-specific events associated with tissue remodeling; for example, in humans and other primates, uterine tissue is restructured and the endometrial lining ultimately shed during the menstrual cycle, whereas in the mouse, the uterus undergoes an extremely rapid estrous cycle characterized by alternating phases of active cell growth, degeneration, and quiescence. Tissue variabilities such as these are probably important in dictating the repertoire and activity of molecules expressed *in vivo*. Equally possible is that other molecule(s) may substitute for matrilysin in some instances, as the pathways controlling expression of this protease may have evolved differently between rodents and humans. The tissue microenvironment and architecture differs in some organs of these mammals, as well, as has been observed in the skin (Cohen and Mast, 1990), and these differences probably govern the specific molecules involved in remodeling of those tissues. Despite these caveats, the overwhelming similarities in organ structure and function between the species justify using genetically altered mice as suitable model systems for human physiology and disease.

VII. Summary and Concluding Remarks

Among the MMP family members characterized thus far, matrilysin is unique in that the gene encodes only the minimal domains required for activity. Although the hemopexin-like domain present in other MMPs can be removed by proteolysis *in vitro*, it is not clear if and when this processing occurs *in vivo*, suggesting that the structure of matrilysin is most likely functionally distinctive. Matrilysin has traditionally been grouped with the stromelysins based on its ability to cleave a wide range of ECM glycoproteins, although it shows a greater similarity to the collagenase-1 catalytic domain with regard to the structure of its substrate-binding pocket. Of the stromelysins, matrilysin has primarily been compared with stromelysin-1 as to substrate bond specificity and rate of catalysis. For several ECM substrates, activated matrilysin was found to be a more efficient enzyme than stromelysin-1, at least *in vitro*, and often cleaved at sites distinct from those recognized by stromelysin-1. This observation, along with the finding that in both normal and neoplastic tissue, matrilysin is characteristically of epithelial rather than stromal origin, suggests that it may function in a manner unlike that of other MMPs. Although experiments *in vitro* show that matrilysin is able to cleave matrix components, the finding that the protein is primarily localized to the apical face or lumen of glandular epithelium implicates this MMP in extracellular activities unrelated to ECM reorganization. Although it may degrade matrix structures when secreted basally, particularly in actively remodeling tissue, the lumenal secretion of matrilysin suggests that this molecule participates in homeostatic processes such as host defense, cell proliferation, and protein turnover (Fig. 4). Because mice deficient in this MMP are viable, matrilysin is not absolutely essential for these processes in normal embryonic and postnatal development. Furthermore, under conditions of conventional housing, matrilysin-null mice have exhibited no overt phenotypic alterations. However, growth of genetically induced adenomatous lesions in the intestinal tract does appear to be delayed in the knock-out mice as compared to wild type animals, indicating there is an association between matrilysin expression and growth control *in vivo*. This finding also attests to the need for the appropriate stimulus or experimental test to reveal the potentially subtle differences that may exist between matrilysin knockout and wild-type mice, especially if we consider that compensatory mechanisms or redundant pathways may be involved.

Apical (lumenal) Secretion

- Maintenance of patency
- Protein activation
- Host defense
- Turnover of cell-surface proteins

Basal Secretion

- Matrix remodeling
- Cell turnover
- Injury response
- Growth factor release

FIG. 4. This schematic depicts the functions that have been proposed for matrilysin based on its vectorial secretion in polarized epithelial cells. Constitutive expression of matrilysin in a variety of glandular epithelia and its delivery to the apical cell surface and lumen suggest that it acts as a sentinel molecule in maintaining tissue homeostasis and properly functioning glands. In contrast, induction and basal secretion of the protease appear to be associated with dynamic processes of cell activation and tissue remodeling.

Because matrilysin is constitutively expressed by a variety of glandular epithelia, we propose that it is acting as a sentinel molecule in these cells, maintaining the gland or duct in a state poised for rapid response when the critical signals are received. The challenge for us is to identify those cellular responses, the signals that initiate them, and the precise role that matrilysin plays in them. As part of this challenge, it will be vital to determine the substrates that are recognized and cleaved by matrilysin *in vivo,* particularly since the emerging data indicate that activated matrilysin may be involved in processing of some protein precursors. Does matrilysin nonspecifically cleave lumenal contents in the various glandular organs in which it is expressed, or is there some function common to most glandular epithelia that serves as the focus for matrilysin activity? These and other questions can be addressed using genetically defined mice as we have described, as well as other mouse models and the appropriate tissue culture systems.

As with other MMPs, the overexpression of matrilysin has been found to correlate with tumor progression, and it appears to contribute to the ability of tumor cells to extravasate and metastasize, presumably by degradation of the basement membrane. However, the demonstration that matrilysin expression increases the tumorigenic potential of neoplastic colon cells, and our recent finding that mice deficient in matrilysin develop fewer intestinal adenomas than their wild type counterparts, support the idea that this MMP affects the *formation* of tumors, at least in the gastrointestinal tract, in a way that is as yet undefined. Furthermore, the prominent expression of matrilysin in other diseases, such as atherosclerosis, cystic fibrosis, and severe ulcerative colitis, underscores the potential association of this metalloenzyme with growth control, cell migration, and proliferation, in addition to tissue destruction. Perhaps the designation "matrix-degrading enzyme" is too limiting a description for this member of the MMP family. The coming years should hold many more exciting discoveries as details of the way matrilysin functions unfold.

Acknowledgments

The authors wish to thank members of the Matrisian laboratory, as well as the many investigators in this field, for their contributions to the work reviewed in this chapter. We are grateful to Bill Parks for sharing work in press and for providing artistic input into the figures and helpful comments on the manuscript. Research in the Matrisian laboratory is supported by the National Institutes of Health (CA 46843 and CA 60867), the Department of Defense (DAMD 17-94-J-4226), and the Vanderbilt Cancer Center (Support Grant P30 CA68485). C.L.W. was a postdoctoral fellow of the American Cancer Society.

REFERENCES

Abdel, W. N., and Mason, R. M. (1996). Modulation of neutral protease expression in human mesangial cells by hyperglycaemic culture. *Biochem. J.* **320,** 777–783.

Abramson, S. R., Conner, G. E., Nagase, H., Neuhaus, I., and Woessner, J. F., Jr. (1995). Characterization of rat uterine matrilysin and its cDNA. Relationship to human pump-1 and activation of procollagenases. *J. Biol. Chem.* **270,** 16016–16022.

Airola, K. Vaalamo, M., Reunala, T. A, and Saarialho-Kere, U. K. (1995). Enhanced expression of interstitial collagenase, stromelysin-1, and urokinase plasminogen activator in lesions of dermatitis herpetiformis. *J. Invest. Dermatol.* **105,** 184–189.

Alexander, C. M., and Werb, Z. (1991). Extracellular matrix degradation. *In* "Cell Biology of the Extracellular Matrix" (E. D. Hay, ed.), 2nd ed., pp. 255–302, Plenum Press, New York.

Anichini, E., Fibbi, G., Pucci, M., Caldini, R., Chevanne, M., and Del Rosso, M. (1994). Production of second messengers following chemotactic and mitogenic urokinase-receptor interaction in human fibroblasts and mouse fibroblasts transfected with human urokinase receptor. *Exp. Cell Res.* **213,** 438–448.

Baragi, V. M., Fliszar, C. J., Conroy, M. C., Ye, Q.-Z., Shipley, J. M., and Welgus, H. G. (1994). Contribution of the C-terminal domain of metalloproteinases to binding by tissue inhibitor of metalloproteinases. C-terminal truncated stromelysin and matrilysin exhibit equally compromised binding affinities as compared to full-length stromelysin. *J. Biol. Chem.* **269,** 12692–12697.

Barnett, J., Straub, K., Nguyen, B., Chow, J., Suttman, R., Thompson, K., Tsing, S., Benton, P., Schatzman R., and Chen, M. (1994). Production and characterization of human matrilysin (PUMP) from recombinant Chinese hamster ovary cells. *Protein Exp. Purif.* **5,** 27–36.

Basset, P., Bellocq, J. P., Wolf, C., Stoll, I., Hutin, P., Limacher, J. M., Podhajcer, O. L., Chenard, M. P., Rio, M. C., and Chambon, P. (1990). A novel metalloproteinase gene specifically expressed in stromal cells of breast carcinomas. *Nature* **348,** 699–704.

Belaaouaj, A., Shipley, J. M., Kobayashi, D. K., Zimonjic, D. B., Popescu, N., Silverman, G. A., and Shapiro, S. D. (1995). Human macrophage metalloelastase. Genomic organization, chromosomal location, gene linkage, and tissue-specific expression. *J. Biol. Chem.* **270,** 14568–14575.

Black, R. A., Durie, F. H., Otten-Evans, C., Miller, R., Slack, J. L., Lynch, D. H., Castner, B., Mohler, K. M., Gerhart, M., Johnson, R. S., Itoh, Y., Okada, Y., and Nagase, H. (1996). Relaxed specificity of matrix metalloproteinases (MMPs) and TIMP insensitivity of tumor necrosis factor-α (TNF-α) production suggest the major TNF-α converting enzyme is not an MMP. *Biochem. Biophys. Res. Commun.* **225,** 400–405.

Black, R. A., Rauch, C. T., Kozlosky, C. J., Peschon, J. J., Slack, J. L., Wolfson, M. F., Castner, B. J., Stocking, K. L., Reddy, P., Srinivasan, S., Nelson, N., Boiani, N., Schooley, K. A., Gerhart, M., Davis, R., Fitzner, J. N., Johnson, R. S., Paxton, R. J., March, C. J., and Cerretti, D. P. (1997). A metalloproteinase disintegrin that releases tumour-necrosis factor-a from cells. *Nature* **385,** 729–733.

Bolon, I., Brambilla, E., Vandenbunder, B., Robert, C., Lantuejoul, S., and Brambilla, C. (1996). Changes in the expression of matrix proteases and of the transcription factor c-Ets-1 during progression of precancerous bronchial lesions. *Lab. Invest.* **75,** 1–13.

Borchers, A. H., Powell, M. B., Fusenig, N. E., and Bowden, G. T. (1994). Paracrine factor and cell–cell contact-mediated induction of protease and c-*ets* gene expression in malignant keratinocyte/dermal fibroblast cocultures. *Exp. Cell Res.* **213,** 143–147.

Borchers, A. H., Sanders, L. A., and Bowden, G. T. (1997a). Regulation of matrilysin expression in cells of squamous cell carcinoma by E-cadherin-mediated cell–cell contact. *J. Cancer Res. Clin. Oncol.* **123,** 13–20.

Borchers, A. H., Sanders, L. A., Powell, M. B., and Bowden, G. T. (1997b). Melanocyte mediated paracrine induction of extracellular matrix degrading proteases in squamous cell carcinoma cells. *Exp. Cell Res.* **231**, 61–65.

Brenner, R. M., Rudolph, L. A., Matrisian, L. M., and Slayden, O. D. (1996). Nonhuman primate models: Artificial menstrual cycles, endometrial matrix metalloproteinases and subcutaneous endometrial grafts. *Hum. Reprod.* **11**, 150–164.

Brown, D. L., Hibbs, M. S., Kearney, M., Loushin, C., and Isner, J. M. (1995). Identification of 92-kD gelatinase in human coronary atherosclerotic lesions. Association of active enzyme synthesis with unstable angina. *Circulation* **91**, 2125–2131.

Browner, M. F., Smith, W. W., and Castelhano, A. L. (1995). Matrilysin-inhibitor complexes: Common themes among metalloproteases. *Biochemistry* **34**, 6602–6610.

Bruner, K. L., Rodgers, W. H., Gold, L. I., Korc, M., Hargrove, J. T., Matrisian, L. M., and Osteen, K. G. (1995). Transforming growth factor-β mediates the progesterone supression of an epithelial metalloproteinase by adjacent stroma in the human endometrium. *Proc. Natl. Acad. Sci. USA* **92**, 7342–7366.

Busiek, D. F., Ross, F. P., McDonnell, S., Murphy, G., Matrisian, L. M., and Welgus, H. G. (1992). The matrix metalloproteinase matrilysin (PUMP) is expressed in developing human mononuclear phagocytes. *J. Biol. Chem.* **267**, 9087–9092.

Busiek, D. F., Baragi, V., Nehring, L. C., Parks, W. C., and Welgus, H. G. (1995). Matrilysin expression by human mononuclear phagocytes and its regulation by cytokines and hormones. *J. Immunol.* **154**, 6484–6491.

Chandler, S. Coates, R. Gearing, A., Lury, J. Wells, G, and Bone, E. (1995). Matrix metalloproteinases degrade myelin basic protein. *Neurosci. Lett.* **201**, 223–226.

Cohen, I. K., and Mast, B. A. (1990). Models of wound healing. *J. Trauma* **30**, S149–S155.

Crabbe, T., Willenbrock, F., Eaton, D., Hynds, P., Carne, A. F., Murphy, G., and Docherty, A. J. P. (1992). Biochemical characterization of matrilysin. Activation conforms to the stepwise mechanisms proposed for other matrix metalloproteinases. *Biochemistry* **31**, 8500–8507.

Crabbe, T., Smith, B., O'Connell, J., and Docherty, A. (1994). Human progelatinase A can be activated by matrilysin. *FEBS Lett.* **345**, 14–16.

Crawford, H. C., and Matrisian, L. M. (1996). Mechanisms controlling the transcription of matrix metalloproteinase genes in normal and neoplastic cells. *Enzyme Protein* **49**, 20–37.

Darmoul, D., Brown, D., Selsted, M. E., and Ouellette, A. J. (1997). Cryptdin gene expression in developing mouse small intestine. *Am. J. Physiol.* **272**, G197–G206.

Formstone, C. J., Byrd, P. J., Ambrose, H. J., Riley, J. H., Hernandez, D., McConville, C. M., and Taylor, A. M. R. (1993). The order and orientation of a cluster of metalloproteinase genes, stromelysin-2, collagenase, and stromelysin, together with D11S385, on chromosome 11q22→q23. *Genomics* **16**, 289–291.

Fosang, A. J., Neame, P. J., Last, K., Hardingham, T. E., Murphy, G., and Hamilton, J. A. (1992). The interglobular domain of cartilage aggrecan is cleaved by PUMP, gelatinases, and cathepsin B. *J. Biol. Chem.* **267**, 19470–19474.

Fosang, A. J., Last, K., Knauper, V., Neame, P. J., Murphy, G., Hardingham, T. E., Tschesche, H., and Hamilton, J. A. (1993). Fibroblast and neutrophil collagenases cleave at two sites in the cartilage aggrecan interglobular domain. *Biochem. J.* **295**, 273–276.

Fukai, F., Ohtaki, M., Fujii, N., Yajima, H., Ishii, T., Nishizawa, Y., Miyazaki, K., and Katayama, T. (1995). Release of biological activities from quiescent fibronectin by a conformational change and limited proteolysis by matrix metalloproteinases. *Biochemistry* **34**, 11453–11459.

Gaire, M., Magbanua, Z., McDonnell, S., McNeil, L., Lovett, D. H., and Matrisian, L. M. (1994). Structure and expression of the human gene for the matrix metalloproteinase matrilysin. *J. Biol. Chem.* **269**, 2032–2040.

Galis, Z. S., Sukhova, G. K., Lark, M. W., and Libby, P. (1994). Increased expression of matrix metalloproteinases and matrix degrading activity in vulnerable regions of human atherosclerotic plaques. *J. Clin. Invest.* **94,** 2493–2503.

Gearing, A. J. H., Beckett, P., Christodoulou, M., Churchill, M., Clements, J., Davidson, A. H., Drummond, A. H., Galloway, W. A., Gilbert, R., Gordon, J. L., Leber, T. M., Mangan, M., Miller, K., Nayee, P., Owen, K., Patel, S., Thomas, W., Wells, G., Wood, L. M., and Woolley, K. (1994). Processing of tumour necrosis factor-a precursor by metalloproteinases. *Nature* **370,** 555–557.

Goto, Y., Nagasawa, H., Sasaki, T., Enami, J., and Iguchi, T. (1995). Biochemical changes during growth and regression of pregnancy-dependent mammary tumors of GR/A mice. *Proc. Soc. Exp. Biol. Med.* **209,** 343–353.

Gresham, H. D., Graham, I. L., Griffin, G. L., Hsieh, J. C., Dong, L. J., Chung, A. E., and Senior, R. M. (1996). Domain-specific interactions between entactin and neutrophil integrins. G2 domain ligation of integrin $\alpha3\beta1$ and E domain ligation of the leukocyte response integrin signal for different responses. *J. Biol. Chem.* **271,** 30587–30594.

Halpert, I., Sires, U. I., Roby, J. D., Potter-Perigo, S., Wight, T. N., Shapiro, S. D., Welgus, H. G., Wickline, S. A., and Parks, W. C. (1996). Matrilysin is expressed by lipid-laden macrophages at sites of potential rupture in atherosclerotic lesions and localizes to areas of versican deposition, a proteoglycan substrate for the enzyme. *Proc. Natl. Acad. Sci. USA* **93,** 9748–9753.

Hembry, R. M., Bagga, M. R., Reynolds, J. J., and Hamblen, D. L. (1995). Immunolocalisation studies on six matrix metalloproteinases and their inhibitors, TIMP-1 and TIMP-2, in synovia from patients with osteo- and rheumatoid arthritis. *Ann. Rheum. Dis.* **54,** 25–32.

Henney, A. M., Wakely, P. R., Davies, M. J., Foster, K., Hembry, R., Murphy, G., and Humphries, S. (1991). Localization of stromelysin gene expression in atherosclerotic plaques by *in situ* hybridization. *Proc. Natl. Acad. Sci. USA* **88,** 8154–8158.

Heppner, K. J., Matrisian, L. M., Jensen, R. A., and Rodgers, W. H. (1996). Expression of most matrix metalloproteinase family members in breast cancer represents a tumor-induced host response. *Am. J. Pathol.* **149,** 273–282.

Honda, M., Mori, M., Ueo, H., Sugimachi, K., and Akiyoshi, T. (1996). Matrix metalloproteinase-7 expression in gastric carcinoma. *Gut* **39,** 444–448.

Hulboy, D. L., Rudolph, L. A., and Matrisian, L. M. (1997). Matrix metalloproteinases as mediators of reproductive function. *Mol. Hum. Reprod.* **3,** 27–45.

Imai, K., Yokohama, Y., Nakanishi, I., Ohuchi, E., Fujii, Y., Nakai, N., and Okada, Y. (1995a). Matrix metalloproteinase 7 (matrilysin) from human rectal carcinoma cells. Activation of the precursor, interaction with other matrix metalloproteinases and enzymic properties. *J. Biol. Chem.* **270,** 6691–6697.

Imai, K., Shikata, H., and Okada, Y. (1995b). Degradation of vitronectin by matrix metalloproteinases-1, -2, -3, -7 and -9. *FEBS Lett.* **369,** 249–251.

Ishikawa, T., Ichikawa, Y., Mitsuhashi, M., Momiyama, N., Chishima, T., Tanaka, K., Yamaoka, H., Miyazaki, K., Nagashima, Y., Akitaya, T., and Shimada, H. (1996). Matrilysin is associated with progression of colorectal tumor. *Cancer Lett.* **107,** 5–10.

Itoh, F., Yamamoto, H., Hinoda, Y., and Imai, K. (1996). Enhanced secretion and activation of matrilysin during malignant conversion of human colorectal epithelium and its relationship with invasive potential of colon cancer cells. *Cancer* **77,** 1717–1721.

Karelina, T. V., Goldberg, G. I., and Eisen, A. Z. (1994). Matrilysin (PUMP) correlates with dermal invasion during appendageal development and cutaneous neoplasia. *J. Invest. Dermatol.* **103,** 482–487.

Kerr, L. D., Miller, D. B., and Matrisian, L. M. (1990). TGF-β_1 inhibition of transin/stromelysin gene expression is mediated through a fos binding sequence. *Cell* **61,** 267–278.

180   CAROLE L. WILSON AND LYNN M. MATRISIAN

Knox, J. D., Boreham, D. R., Wlaker, J. A., Morrison, D. P., Matrisian, L. M., Nagle, R. B., and Bowden, G. T. (1996a). Mapping of the metalloproteinase gene matrilysin (MMP7) to human chromosome 11q21–q22. *Cytogenet. Cell Genet.* **72,** 170–182.

Knox, J. D., Wolf, C., McDaniel, K., Clark, V., Loriot, M., Bowden, G. T., and Nagle, R. B. (1996b). Matrilysin expression in human prostate carcinoma. *Mol. Carcinog.* **15,** 57–63.

Lichtinghagen, R., Helmbrecht, T., Arndt. B., and Boker, K. H. (1995). Expression pattern of matrix metalloproteinases in human liver. *Eur. J. Clin. Chem. Clin. Biochem.* **33,** 65–71.

MacDougall, J. R., and Matrisian, L. M. (1995). Contributions of tumor and stromal matrix metalloproteinases to tumor progression, invasion and metastasis. *Cancer Metastasis Rev.* **14,** 351–362.

Marcotte, P. A., Kozan, I. M., Dorwin, S. A., and Ryan, J. M. (1992). The matrix metalloproteinase pump-1 catalyzes formation of low molecular weight (pro)urokinase in cultures of normal human kidney cells. *J. Biol. Chem.* **267,** 13803–13806.

Marti, H.-P., McNeil, L., Thomas, G., Davies, M., and Lovett, D. H. (1992). Molecular characterization of a low-molecular-mass matrix metalloproteinase secreted by glomerular mesangial cells as PUMP-1. *Biochem. J.* **285,** 899–905.

Marti, H.-P., Lee, L., Kashgarian, M., and Lovett, D. H. (1994). Transforming growth factor-β_1 stimulates glomerular mesangial cell synthesis of the 72-kd type IV collagenase. *Am. J. Pathol.* **144,** 82–94.

Massagué, J. (1990). The transforming growth factor-β family. *Ann. Rev. Cell Biol.* **6,** 597–641.

Mayer, U., Mann, K., Timpl, R., and Murphy, G. (1993). Sites of nidogen cleavage by proteases involved in tissue homeostasis and remodelling. *Eur. J. Biochem.* **217,** 877–884.

McDonnell, S., Navre, M., Coffey, R. J., and Matrisian, L. M. (1991). Expression and localization of the matrix metalloproteinase pump-1 (MMP-7) in human gastric and colon carcinomas. *Mol. Carcinogen.* **4,** 527–533.

Miyazaki, K., Hattori, Y., Umenishi, F., Yasumitsu, H., and Umeda, M. (1991). Purification and characterization of extracellular matrix-degrading metalloproteinase, matrin (Pump-1), secreted from human rectal carcinoma cell line. *Cancer Res.* **50,** 7758–7764.

Mori, M., Barnard, G. F., Mimori, K., Ueo, H., Akiyoshi, T., and Sugimachi, K. (1995). Overexpression of matrix metalloproteinase-7 mRNA in human colon carcinomas. *Cancer* **75,** 1516–1519.

Moss, M. L., Jin, S.-L. C., Milla, M. E., Burkhart, W., Carter, H. L., Chen, W.-J., Clay, W. C., Didsbury, J. R., Hassler, D., Hoffman, C. R., Kost, T. A., Lambert, M. H., Leesnitzer, M. A., McCauley, P., McGeehan, G., Mitchell, J., Moyer, M., Pahel, G., Rocque, W., Overton, L. K., Schoenen, F., Seaton, T., Su, J.-L., Warner, J., Willard, D., and Becherer, J. D. (1997). Cloning of a disintegrin metalloproteinase that processes precursor tumour-necrosis factor-α. *Nature* **385,** 733–736.

Muller, D., Quantin, B., Gesnel, M. C., Millon-Collard, R., Abecassis, J., and Breathnach, R. (1988). The collagenase gene family in humans consists of at least four members. *Biochem. J.* **253,** 187–192.

Muller, D., Breathnach, R., Engelmann, A., Millon, R., Bronner, G., Flesch, H., Dumont, P., Eber, M., and Abecassis, J. (1991). Expression of collagenase-related metalloproteinase genes in human lung or head and neck tumours. *Int. J. Cancer* **48,** 550–556.

Murphy, G., Cockett, M. I., Ward, R. V., and Docherty, A. J. P. (1991). Matrix metalloproteinase degradation of elastin, type IV collagen, and proteoglycan: A quantitative comparison of the activities of 95 kDa and 72 kDa gelatinases, stromelysins-1 and -2 and punctuated metalloproteinase (PUMP). *Biochem. J.* **277,** 277–279.

Nakano, A., Tani, E., Miyazaki, K., Furuyama, J., and Matsumoto, T. (1993). Expressions of matrilysin and stromelysin in human glioma cells. *Biochem. Biophys. Res. Commun.* **192,** 999–1003.

Nakano, A., Tani, E., Miyazaki, K., Yamamoto, Y., and Furuyama, J. (1995). Matrix metalloproteinases and tissue inhibitors of metalloproteinases in human gliomas. *J. Neurosurg.* **83,** 298–307.

Netzel-Arnett, S., Sang, Q.-X., Moore, W. G. I., Navre, M., Birkedal-Hansen, H., and Van Wart, H. E. (1993). Comparative sequence specificities of human 72- and 92–kDa gelatinases (type IV collagenases) and PUMP (matrilysin). *Biochemistry* **32,** 6427–6432.

Newell, K., Witty, J., Rodgers, W. H., and Matrisian, L. M. (1994). Expression and localization of matrix-degrading metalloproteinases during colorectal tumorigenesis. *Mol. Carcinogen.* **10,** 199–206.

Nguyen, Q., Murphy, G., Hughes, C. E., Mort, J. S., and Roughley, P. J. (1993). Matrix metalloproteinases cleave at two distinct sites on human cartilage link protein. *Biochem. J.* **295,** 595–598.

Nikkari, S. T., O'Brien, K. D., Ferguson, M., Hatsukami, T., Welgus, H. G., Alpers, C. E., and Clowes, A. W. (1995). Interstitial collagenase (MMP-1) expression in human carotid atherosclerosis. *Circulation* **92,** 1393–1398.

Overall, C. M., and Sodek, J. (1990). Concanavalin A produces a matrix-degradative phenotype in human fibroblasts. Induction and endogenous activation of collagenase, 72–kDa gelatinase, and *pump*-1 is accompanied by the suppression of the tissue inhibitor of matrix metalloproteinases. *J. Biol. Chem.* **265,** 21141–21151.

Pajouh, M S., Nagle, R B., Breathnach, R., Finch, J. S., Brawer, M.K., and Bowden, GT. (1991). Expression of metalloproteinase genes in human prostate cancer. *J. Cancer Res. Clin. Oncol.* **117,** 144–150.

Parks, W. C., Dunsmore, S. E., Saarialho-Kere, U. K., Roby, J. D., and Welgus, H. G. (1997). Matrilysin is expressed by conducting airway epithelium and alveolar type II pneumocytes in cystic fibrotic lung: Potential role in lung repair. Submitted for publication.

Powell, W. C., and Matrisian, L. M. (1996). Complex roles of matrix metalloproteinases in tumor progression. *In* "Current Topics in Microbiology and Immunology: I. Attempts to Understand Metastasis Formation" (U. Günthert and W. Birchmeier, eds.), Vol. 213 (Pt. 1), pp. 1–21, Springer-Verlag, Heidelberg.

Powell, W. C., Knox, J. D., Navre, M., Grogan, T. M., Kittelson, J., Nagle, R. B., and Bowden, G. T. (1993). Expression of the metalloproteinase matrilysin in DU-145 cells increases their invasive potential in severe combined immunodeficient mice. *Cancer Res.* **53,** 417–422.

Powell, W. C., Domann, F. E., Jr., Mitchen, J. M., Matrisian, L. M., Nagle, R. B., and Bowden, G.T. (1996). Matrilysin expression in the involuting rat ventral prostate. *The Prostate* **29,** 159–168.

Quantin, B., Murphy, G., and Breathnach, R. (1989). Pump-1 cDNA codes for a protein with characteristics similar to those of classical collagenase family members. *Biochem. J.* **28,** 5327–5333.

Rodgers, W. H., Osteen, K. G., Matrisian, L. M., Navre, M., Giudice, L. C., and Gorstein, F. (1993). Expression and localization of matrilysin, a matrix metalloproteinase, in human endometrium during the reproductive cycle. *Am. J. Obstet. Gynecol.* **168,** 253–260.

Rodgers, W. H., Matrisian, L. M., Giudice, L. C., Dsupin, B., Cannon, P., Svitek, C., Gorstein, F., and Osteen, K. G. (1994). Patterns of matrix metalloproteinase expression in cycling endometrium imply differential functions and regulation by steroid hormones. *J. Clin. Invest.* **94,** 946–953.

Rudolph-Owen, L. A., Hulboy, D. L., Wilson, C. L., Mudgett, J., and Matrisian, L. M. (1997a). Coordinate expression of matrix metalloproteinase family members in the uterus of normal, matrilysin-deficient, and stromelysin-1 deficient mice. *Endocrinology* **138,** 4902–4911.

Rudolph-Owen, L. A., Cannon, P., and Matrisian, L. M. (1997b). Overexpression of the matrix metalloproteinase matrilysin results in premature mammary gland differentiation and male infertility. Submitted for publication.

Saarialho-Kere, U. K., Kovacs, S. O., Pentland, A. P., Olerud, J. E., Welgus, H. G., and Parks, W. C. (1993). Cell–matrix interactions modulate interstitial collagenase expression by human keratinocytes actively involved in wound healing. *J. Clin. Invest.* **92,** 2858–2866.

Saarialho-Kere, U. K., Crouch, E. C., and Parks, W. C. (1995). The matrix metalloproteinase matrilysin is constitutively expressed in adult human exocrine epithelium. *J. Invest. Dermatol.* **105,** 190–196.

Saarialho-Kere, U. K., Vaalamo, M., Puolakkainen, P., Airola, K., Parks, W. C., and Karjalainen-Lindsberg, M.-L. (1996). Enhanced expression of matrilysin, collagenase, and stromelysin-1 in gastrointestinal ulcers. *Am. J. Pathol.* **148,** 519–526.

Sang, Q. A., Bodden, M. K., and Windsor, L. J. (1996). Activation of human progelatinase A by collagenase and matrilysin: Activation of procollagenase by matrilysin. *J. Protein Chem.* **15,** 243–253.

Sang, Q.-X., Birkedal-Hansen, H., and van Wart, H.E. (1995). Proteolytic and nonproteolytic activation of human neutrophil progelatinase B. *Biochim. Biophys. Acta* **1251,** 99–108.

Sasaki, T., Mann, K., Murphy, G., Chu, M. L., and Timpl, R. (1996). Different susceptibilities of fibulin-1 and fibulin-2 to cleavage by matrix metalloproteinases and other tissue proteases. *Eur. J. Biochem.* **240,** 427–434.

Schorpp, M., Mattei, M.-G., Herr, I., Gack, S., Schaper, J., and Angel, P. (1995). Structural organization and chromosomal localization of the mouse collagenase type I gene. *Biochem. J.* **308,** 211–217.

Sellers, A., and Woessner, J. F., Jr. (1980). The extraction of a neutral metalloproteinase from the involuting rat uterus, and its action on cartilage proteoglycan. *Biochem. J.* **189,** 521–531.

Shapiro, S. D., Griffin, G. L., Gilbert, D. J., Jenkins, N. A., Copeland, N. G., Welgus, H. G., Senior, R. M., and Ley, T. J. (1992). Molecular cloning, chromosomal localization, and bacterial expression of a murine macrophage metalloelastase. *J. Biol. Chem.* **267,** 4664–4671.

Sires, U. I., Griffin, G. L., Broekelmann, T. J., Mecham, R. P., Murphy, G., Chung, A. E., Welgus, H. G., and Senior, R. M. (1993). Degradation of entactin by matrix metalloproteinases. Susceptibility to matrilysin and identification of cleavage sites. *J. Biol. Chem.* **268,** 2069–2074.

Sires, U. I., Murphy, G., Baragi, V. M., Fliszar, C. J., Welgus, H. G., and Senior, R. M. (1994). Matrilysin is much more efficient than other matrix metalloproteinases in the proteolytic inactivation of α_1-antitrypsin. *Biochem. Biophys. Res. Commun.* **204,** 613–620.

Siri, A., Knauper, V., Veirana, N., Caocci, F., Murphy, G., and Zardi, L. (1995). Different susceptibility of small and large human tenascin-C isoforms to degradation by matrix metalloproteinases. *J. Biol. Chem.* **270,** 8650–8654.

Smith, M. M., Shi, L., and Navre, M. (1995). Rapid identification of highly active and selective substrates for stromelysin and matrilysin using bacteriophage peptide display libraries. *J. Biol. Chem.* **270,** 6440–6449.

Soler, D., Nomizu, T., Brown, W. E., Chen, M., Ye, Q.-Z., Van Wart, H. E., and Auld, D. S. (1994). Zinc content of promatrilysin, matrilysin and the stromelysin catalytic domain. *Biochem. Biophys. Res. Commun.* **201,** 917–923.

Sundareshan, P., Koster, J. J., Nagle, R. B., and Bowden, G. T. (1997). Coordinated expression of matrilysin during TPA-induced apoptosis of LNCaP cells: Two parallel processes affected by TPA. *Cancer Lett.* **113,** 17–24.

Vaalamo, M., Weckroth, M., Puolakkainen, P., Kere, J., Saarinen, P., Lauharanta, J., and Saarialho-Kere, U. K. (1996). Patterns of matrix metalloproteinase and TIMP-1 expression in chronic and normally healing human cutaneous wounds. *Br. J. Dermatol.* **135,** 52–59.

Vettraino, I. M., Roby, J., Tolley, T., and Parks, W. C. (1996). Collagenase-1, stromelysin-1, and matrilysin are expressed within the placenta during multiple stages of human pregnancy. *Placenta* **17,** 557–563.

Von Bredow, D. C., Nagle, R. B., Bowden, G. T., and Cress, A. E. (1995). Degradation of fibronectin fibrils by matrilysin and characterization of the degradation products. *Exp. Cell Res.* **221,** 83–91.

Wallon, U. M., Shassetz, L. R., Cress, A. E., Bowden, G. T., and Gerner, E. W. (1994). Polyamine-dependent expression of the matrix metalloproteinase matrilysin in a human colon cancer-derived cell line. *Mol. Carcinogen.* **11,** 138–144.

Wang, H., Rodgers, W. H., Chmell, M. J., Svitek, C., and Schwartz, H. S. (1995). Osteosarcoma oncogene expression detected by *in situ* hybridization. *J. Orthop. Res.* **13,** 671–678.

Welch, A. R., Holman, C. M., Huber, M., Brenner, M. C., Browner, M. F., and van Wart, H. E. (1996). Understanding the P1' specificity of the matrix metalloproteinases: Effect of S1' pocket mutations in matrilysin and stromelysin-1. *Biochemistry* **35,** 10103–10109.

Wilson, C. L., and Matrisian, L. M. (1996). Matrilysin: An epithelial matrix metalloproteinase with potentially novel functions. *Int. J. Biochem. Cell Biol.* **28,** 123–136.

Wilson, C. L., Heppner, K. J., Rudolph, L. A., and Matrisian, L. M. (1995). The metalloproteinase matrilysin is preferentially expressed by epithelial cells in a tissue-restricted pattern in the mouse. *Mol. Biol. Cell* **6,** 851–869.

Wilson, C. L., Heppner, K. J., Labosky, P. A., Hogan, B. L. M., and Matrisian, L. M. (1997). Intestinal tumorigenesis is suppressed in mice lacking the metalloproteinase matrilysin. *Proc. Natl. Acad. Sci. USA* **94,** 1402–1407.

Witty, J. P., McDonnell, S., Newell, K., Cannon, P., Navre, M., Tressler, R., and Matrisian, L. M. (1994). Modulation of matrilysin levels in colon carcinoma cell lines affects tumorigenicity *in vivo. Cancer Res.* **54,** 4805–4812.

Woessner, J. F., Jr. (1995). Matrilysin. *Meth. Enzymol.* **248,** 485–495.

Woessner, J. F., Jr. (1996). Regulation of matrilysin in the rat uterus. *Biochem. Cell Biol.* **74,** 777–784.

Woessner, J. F., Jr., and Taplin, C. (1988). Purification and properties of a small latent matrix metalloproteinase of the rat uterus. *J. Biol. Chem.* **263,** 16918–16925.

Wolf, C., Rouyer, N., Lutz, Y., Adida, C., Loriot, M., Bellocq, J.-P., Chambon, P., and Basset, P. (1993). Stromelysin 3 belongs to a subgroup of proteinases expressed in breast carcinoma fibroblastic cells and possibly implicated in tumor progression. *Proc. Natl. Acad. Sci. USA* **90,** 1843–1847.

Wolf, K., Sandner, P., Kurtz, A., and Moll, W. (1996). Messenger ribonucleic acid levels of collagenase (MMP-13) and matrilysin (MMP-7) in virgin, pregnant, and postpartum uterus and cervix of rat. *Endocrinology* **137,** 5429–5434.

Wright, J. H., McDonnell, S., Portella, G., Bowden, G. T., Balmain, A., and Matrisian, L. M. (1994). A switch from stromal to tumor cell expression of stromelysin-1 mRNA associated with the conversion of squamous to spindle carcimonas during mouse skin tumor progression. *Mol. Carcinogen.* **10,** 207–215.

Yamamoto, H., Itoh, R., Hinoda, Y., Senota, A., Yoshimoto, M., Nakamura, H., Imai, K., and Yachi, A. (1994). Expression of matrilysin mRNA in colorectal adenomas and its induction by truncated fibronectin. *Biochem. Biophys. Res. Commun.* **201,** 657–664.

Yamamoto, H., Itoh, R. Senota, A. Adachi, Y., Yoshimoto, M., Endoh, T., Hinoda, Y., Yachi, A., and Imai, K. (1995a). Expression of matrix metalloproteinase matrilysin (MMP-7) was induced by activated Ki-ras via AP-1 activation in SW1417 colon cancer cells. *J. Clin. Lab Anal.* **9,** 297–301.

Yamamoto, H., Itoh, F., Hinoda, Y., and Imai, K. (1995b). Suppression of matrilysin inhibits colon cancer cell invasion *in vitro. Int. J. Cancer* **61,** 218–222.

Yoshimoto, M., Itoh, F., Yamamoto, H., Hinoda, Y., Imai, K., and Yachi, A. (1993). Expression of MMP-7 (pump-1) mRNA in human colorectal cancers. *Int. J. Cancer* **54,** 614–618.

Zhang, Z., Winyard, P. G., Chidwick, K., Murphy, G., Wardell, M., Carrell, R. W., and Blake, D. R. (1994). Proteolysis of human native and oxidised alpha 1–proteinase inhibitor by matrilysin and stromelysin. *Biochim. Biophys. Acta* **1199,** 224–228.

Zhu, C., and Woessner, J. F., Jr. (1991). A tissue inhibitor of metalloproteinases and α-macroglobulins in the ovulating rat ovary: Possible regulators of collagen matrix breakdown. *Biol. Reprod.* **45,** 334–342.

Macrophage Elastase (MMP-12)

Steven D. Shapiro* and Robert M. Senior†

*Departments of Medicine and Cell Biology, and †Department of Medicine, Washington University School of Medicine at Barnes-Jewish Hospital, St. Louis, Missouri 63110

I. Introduction

Macrophage elastase (MMP-12) is either one of the oldest or newest members of the matrix metalloproteinase (MMP) family depending on one's perspective. In 1975 Werb and Gordon identified elastolytic activity in mouse peritoneal macrophage conditioned media. Subsequently in 1981, Banda and Werb purified a 22-kDa metal-dependent proteinase that was responsible for this activity. For a number of years thereafter interest in the enzyme languished, in part because of minimal interest in elastin and uncertainty about the status of the enzyme. However, the status of the enzyme as a distinct member of the MMP family was unsettled in 1992 when the murine cDNA (MME) was cloned from a murine macrophage (P388D1) library and found to have characteristic MMP features, but only 33–48% amino acid homology with other matrix metalloproteinases (Shapiro *et al.*, 1992). The human orthologue of macrophage elastase (HME) was then cloned from a cDNA library derived from human alveolar macrophages of a cigarette smoker (Shapiro *et al.*, 1993b). The cDNAs for human and murine macrophage elastase have 74% homology; there is 64% identity between the enzymes at the amino acid level.

The name *macrophage elastase* incorporates both the principal cellular source of this enzyme and the fact that it is able to degrade insoluble elastin. Because macrophages can express several elastolytic protein-

Matrix Metalloproteinases

ases other than macrophage elastase, including cysteine proteinases (cathepsins S and L) as well as metalloproteinases (particularly MMP-9 and to a lesser degree MMP-7 and perhaps MMP-3), macrophage elastase activity and macrophage elastase should not be considered interchangeable. In this context, rabbit and guinea pig alveolar macrophages in culture release elastolytic activity (Banda *et al.*, 1995), presumably due to macrophage elastase, but the enzymatic basis of the elastase activity from these sources has not in fact been established. Moreover, as discussed later, macrophage elastase cleaves a broad range of substrates in addition to elastin.

Macrophage elastase has typical MMP features. It is a matrix-degrading enzyme, its activity is dependent on coordination of zinc, and it is inhibited by tissue inhibitors of metalloproteinases (TIMPs), and it shares structural similarity to other MMPs. Macrophage elastase is composed of a proenzyme, catalytic, and C-terminal domain (Fig. 1), sharing the highest degree of homology to MMP-1 and MMP-3 (the human enzymes are 49% identical at the amino acid level). The gene for human macrophage elastase is part of the MMP gene cluster on chromosome 11q22.2/22.3. Macrophage elastase is most similar to the stromelysins (MMP-3 and MMP-10) and matrilysin (MMP-7) with respect to their capability to hydrolyze a broad variety of extracellular matrix (ECM) components and other proteins. Macrophage elastase is unique with respect to its predominantly macrophage-specific pattern of expression and perhaps the ability to readily shed its C-terminal domain on processing (see Section III). On a molar basis it is clearly the most active MMP against elastin. The participation of MME in

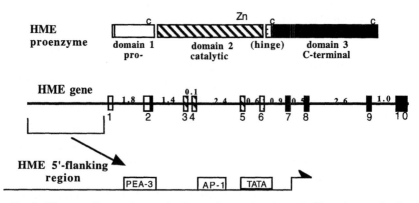

Fig. 1. Diagram of genomic organization and promoter elements. Domain organization of HME protein (top), genomic organization showing exons (boxes) and introns (lines) (middle), and promoter *cis*-elements (bottom).

(patho)biology is starting to become appreciated. Recently, MME has been shown to play a central role in the pathogenesis of pulmonary emphysema (Hautamaki *et al.*, 1997) and atherosclerotic lesions (Carmeleit *et al.*, 1997). In addition, MME may have beneficial functions such as modulating macrophage influx to inflammatory sites and limiting growth of tumor metastases (Dong *et al.*, 1997).

II. THE GENE FOR MACROPHAGE ELASTASE AND ITS EXPRESSION

The human macrophage elastase gene is 13 kb and consists of 10 exons and 9 introns. It is located on chromosome 11q22.2-22.3 (Belaaouaj *et al.*, 1995), sites also for the genes of the interstitial collagenases (MMP-1, MMP-8, and MMP-13), stromelysins (MMP-1 and MMP-10), and matrilysin. The murine macrophage elastase gene is on chromosome 9 (Shapiro *et al.*, 1992).

The 5′-flanking promoter region of the macrophage elastase gene contains activator protein 1 (AP-1) and polyomaviurs enhancer A-binding protein 3 (PEA-3) sites, similar to many other MMPs. Whether these or other *cis* acting elements regulate macrophage elastase responsiveness to biological agents has not been determined. Expression of macrophage elastase is induced in cultured macrophages by LPS, zymosan, GMCSF (Kumar *et al.*, 1996), hyaluronan fragments (Horton and Noble, personal communication), and surfactant protein D (Malone, Crouch, and Shapiro, unpublished). *In vivo,* macrophage elastase production is induced either directly or indirectly by cigarette smoke (Hautamaki *et al.*, 1997). Like other macrophage MMPs, macrophage elastase is down-regulated by dexamethasone, interferon γ (IFN-γ), and interleukin 4 (IL-4). The biological consequences of macrophage elastase regulation by this diverse group of agents remain to be determined.

Macrophage elastase is produced by activated tissue macrophages in the peritoneum, the lung, the central nervous system, and perhaps other tissues. Peripheral blood monocytes possess serine proteinases (cathepsin G and neutrophil elastase) and they produce matrilysin. However, most MMPs, including macrophage elastase, are not expressed by immature monocytes. Upon differentiation to tissue macrophages, these cells lose their serine proteinase profile and gain the capacity to produce macrophage elastase, MMP-1, MMP-3, MMP-9, MT-MMP-1 (Campbell *et al.*, 1991). The molecular basis for this transition and the basis for macrophage-specific expression of macrophage elastase are unknown.

III. ACTIVATION AND PROCESSING

In early studies (Banda and Werb, 1981), the basis for the marked increase in activity during purification was unclear; it was thought perhaps to be the result of separation from an inhibitor. Human macrophage elastase cDNA codes for a proenzyme of 54 kDa (Shapiro *et al.*, 1993b). The proenzyme has the typical signal peptide, proenzyme domain, catalytic domain, and C-terminal hemopexin-like domain found in all MMPs except matrilysin. During purification, both the propiece and the C-terminal domain become removed, leading to an active enzyme of 22 kDa. Murine macrophage elastase codes for a protein of 53 kDa and, like the human enzyme, it also undergoes processing to an active 22-kDa form during purification (Shapiro *et al.*, 1992). Purified 45-kDa MME autoactivates to the 22-kDa form. This process is markedly accelerated in the presence of trypsin, plasmin, neutrophil elastase, and several MMPs, most notably stromelysin-1. Processing is a stepwise process with several cleavages within the C-terminal domain before cleavage at the hinge region resulting in loss of the entire C-terminal domain (Januscz *et al.*, 1998).

Carmeleit and colleagues have demonstrated that urokinase-like plasminogen activator (uPA)-mediated plasmin generation appears to be required for macrophage MMP activation *in vivo* (Carmeleit *et al.*, 1997). Mice deficient in uPA, but not tissue-type plasminogen activator (tPA)-deficient or wild type mice, were unable to activate and process macrophage MMPs in culture, including MME, MMP-13, MMP-3, and MMP-9. Progeny of ApoE−/− mice crossed to uPA−/− mice are protected from atherosclerotic (micro)aneurysmal lesions associated with matrix destruction and macrophage infiltration. ApoE−/− and crosses of ApoE−/− X tPA−/− are not protected from these lesions.

IV. Protein Preparation

Murine macrophage elastase is present in conditioned medium of cultures of normal peritoneal macrophages and murine macrophage cell lines (Werb and Gordon, 1975), for example, P388D1 cells (available from the American Type Culture Collection, Rockville, MD). Gel filtration and ion exchange chromatography (Banda and Werb, 1981), affinity chromatography over κ-elastin linked to agarose, and heparin-agarose chromatography followed by gel filtration over ACA-54 (Shapiro *et al.*, 1992) have been used as purification procedures. Recombinant murine and human macrophage elastases can be generated in *Escherichia coli* after transformation with full-length macrophage elastase cDNAs that have been ligated into pET vectors (Shapiro *et al.*, 1992, 1993b).

V. Catalytic Features

As with the enzymatic activities of other matrix metalloproteinases, macrophage elastase has an optimum pH at 8.0 with considerable activity at pH 7.4. Calcium is required for activity; addition of Ca^{2+} to a final concentration of 5 mM in calcium-free Tris buffer restores full elastase activity.

A. Amino Acid Specificity

Cleavage fragments of elastin (Banda and Werb, 1981) and the insulin B-chain (Kettner and Shaw, 1981) produced by MME show leucine in the P_1' position. A single cleavage in α_1-antitrypsin (α_1-AT) occurs between Pro 357 and Met 358 after it is exposed to macrophage elastase. However, if Met 358 is oxidized, the cleavage occurs between Phe[352] and Leu[353] (Banda et al., 1988). Interestingly, human macrophage elastase cleaves α_1-AT at Phe[352]–Leu[353] as well as Glu[199]–Val[200], and cleaves α_1-AT with at least one order of magnitude more efficiency than any other MMP (Gronski et al., 1997).

Using synthetic octapeptides differing for the P_1' position, it was shown that HME has a preference for Leu, but can tolerate a variety of large and small residues at the P_1' position (Gronski et al., 1997). In fact, HME is the only MMP that can accommodate Arg in the P_1' position. This is consistent with the prediction that macrophage elastase should have a deep S_1' pocket based on its predicted amino acid sequence.

To identify the peptide bonds in elastin that are susceptible to elastolytic MMPs, insoluble bovine ligament elastin was subjected to partial hydrolysis by macrophage elastase (Mecham et al., 1997). Parallel studies were done with the 92-kDa gelatinase, neutrophil elastase, and thermolysin. To eliminate free N terminals in the elastin substrate prior to exposure to the enzymes, the elastin was blocked with dinitrofluorobenzene and blockage was confirmed by sequence analysis, which yielded no product. By sequencing the residual insoluble elastin rather than solubilized elastin peptides, Mecham and colleagues (1997) identified the earliest peptide bonds hydrolyzed because there was no requirement for sufficient hydrolysis to liberate peptides. There was a highly significant difference in amino acids in the P_1' position between 92-kDa gelatinase, macrophage elastase, and serine elastases. The 92-kDa gelatinase and thermolysin had a strong preference for bulky aliphatic side chains, whereas the serine elastases preferred nonbulky amino acids. In contrast, MME and HME had a marked preference for bonds with leucine in the P_1' position. Cleavage at these sites generates much more "knicking" of elastin than other proteinases prior to solubilization.

B. Hydrolysis of Extracellular Matrix Components

Macrophage elastase cleaves a number of matrix proteins besides elastin, including basement membrane components including entactin, laminin-1, fibronectin, various proteoglycans, and type IV collagen (Gronski et al., 1997). Macrophage elastase does not cleave interstitial collagens. Macrophages from mice completely lacking MME as a result of a targeted disruption of the gene cannot penetrate reconstituted basement membranes or degrade insoluble elastin (Shipley et al., 1996).

C. Hydrolysis against Other Non-ECM Proteins

As mentioned earlier, macrophage elastase cleaves α_1-AT, casein, and the insulin B-chain. Other substrates include fibrinogen and mouse immunoglobulins (Banda et al., 1983). Recently, HME has been shown to degrade myelin basic protein and activate latent tumor necrosis factor α (TNF-α) (Chandler et al., 1996). To a greater degree than other MMPs, macrophage elastase can cleave plasminogen yielding the anti-angiogenic fragment, angiostatin, consisting of kringle regions 1–4 (Cornelius et al., 1997). Considering that macrophage elastase degrades many ECM components and other proteins, one might question naming the enzyme for its elastolytic activity.

VI. Inhibition

TIMP-1 and α_2-macroglobulin inhibit macrophage elastase stoichiometrically. Detailed kinetics of inhibition have not been performed. Because enzymatic activity requires zinc in the catalytic domain, the enzyme can be reversibly inactivated by zinc chelators such a 1,10-phenanthroline. Various hydroxymates are also effective inhibitors, including BB94 and other nonselective MMP inhibitors.

VII. Biological Aspects

Macrophage elastase has been associated with several macrophage-mediated destructive diseases in humans and murine models of disease. These diseases include the leading causes of death in the United States: atherosclerosis, cancer, and chronic obstructive pulmonary disease. Use of MME-deficient mice has uncovered both pathological and physiological functions of macrophage elastase.

Atherosclerosis is a chronic inflammatory process whereby plaques are formed in the intimal layer of the vessel wall as a result of accumulation of lipid-laden macrophages, smooth muscle cells, lipids, and ECM. Plaques may become unstable and rupture, triggering intravascular thrombosis and clinical symptoms of tissue ischemia. Alternatively, the atherosclerotic vessel wall may dilate due to destruction of the media, leading to aneurysm formation and rupture of the weakened vessel wall. Recently, the plasminogen and metalloproteinase systems have been implicated in the pathogenesis of myocardial infarction and aneurysm formation by their association with atherosclerotic arteries. We detected HME and matrilysin by *in situ* hybridization in atherosclerotic lesions, particularly in macrophages localized to the "shoulder regions" of plaques, which are most susceptible to fissuring leading to acute myocardial infarctions (Halpert *et al.*, 1996). In addition, we found prominent expression of HME in macrophages associated with elastic fiber disruption in specimens of human abdominal aortic aneurysms (Thompson, unpublished).

However, the pathophysiological relevance and causal roles for proteinases in atherosclerotic lesions remain undefined. To investigate the contribution of individual proteins in atherosclerosis, mouse models of atherosclerosis can be used, particularly mice with a targeted disruption of apolipoprotein E gene (ApoE−/−). These mice have a delayed clearance of lipoproteins, and when fed a Western diet develop serum cholesterol levels of 1400–2000 mg/dL, and develop fatty streaks progressing to fibrous plaques at branch points of major vessels. This is associated with macrophage recruitment causing disruption of the medial external elastic lamina (EEL) and microaneurysm formation. Complex lesions with hemorrhage and fibrosis have yet to be observed.

To investigate the role of plasminogen activators (tPA and uPA), Carmeleit and Collen crossed ApoE−/− mice with uPA−/− and tPA−/− mice. Interestingly, ApoE−/− X uPA−/− mice (but not ApoE−/− or ApoE−/− X tPA−/−) were protected from macrophage-mediated destruction of medial external elastic lamina (EEL) and microaneurysm formation (Carmeleit *et al.*, 1997). It appears that local production of plasmin (by uPA) is required to activate pro-MME and perhaps other MMPs. In the absence of plasmin, macrophages line up but do not penetrate or disrupt the EEL (Fig. 2). This is consistent with findings that macrophages of MME−/− mice cannot penetrate basement membranes or degrade elastin (Shipley *et al.*, 1996). These results suggest that plasmin is required for MMP activation in this model, and that MMPs, particularly MME, are responsible for matrix destruction and macrophage infiltration associated with atherosclerotic microaneurysm formation and potentially plaque rupture.

In *cancer,* MMPs are believed to promote tumor progression by (1) enhancing growth via angiogenesis, (2) disrupting local tissue architecture to allow tumor growth, and (3) breaking down basement membrane barriers to allow metastatic spread. Much correlative evidence exists that demonstrates an association of MMP expression in tumors with an invasive phenotype. While some MMPs, such as matrilysin, collagenase-3, and gelatinase A, are expressed by tumors themselves,

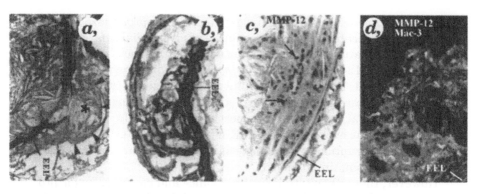

FIG. 2. Association of MME (MMP-12) expression with elastic lamina disruption and microaneurysm formation in ApoE−/− mice (but not ApoE−/− X uPA−/− mice). (a) Verhoff–von Gieson (VG) staining (elastic membranes black) of ApoE−/− aorta. EEL, external elastic lamina; *, pseudo-microaneurysm with arrows denoting outside border of adventitial cap. (b) VG staining of aortas from ApoE−/− X uPA−/− mice. Note that EEL is intact without medial destruction. (c) Immunostain for MME (arrow) in ApoE−/− mice. (d) Confocal microscopy in ApoE−/− mice. Colocalization of marcophages and MME appears white. Note that MME is expressed in macrophages infiltrating between degraded elastic lamina. (With permission from Shapiro et al., 1997, Nature Genetics 17(4), 439–444.)

most other MMPs are predominantly produced by surrounding stromal host cells in response to cytokines released by the tumors. Once produced, the MMPs appear to promote tumor progression. For example, gelatinase A was shown to bind through its C-terminal domain to endothelial cell $\alpha_v\beta_3$ promoting tumor angiogenesis (Brooks et al., 1996). Application of hydroxamates, specific MMP zinc chelating agents, is effective in inhibiting growth of several primary tumors and metastases in animal models (Brown and Givazzi, 1995). Inhibitors of this type are currently undergoing clinical trials in the United Kingdom and United States with the aim that MMP inhibiton will be cytostatic to tumors by inhibiting angiogenesis and tissue invasive capacity. Moreover, these agents could inhibit metastases as well.

MMP inhibitors appear to be most effective in models using nude mice and least effective in inhibiting growth of tumors with tumor-associated macrophages. This may be due to unique properties of macrophage elastase. Unlike other MMPs, macropahge elastase is expressed exclusively in host macrophages. A study of MMP expression in human breast cancer demonstrated HME as the single MMP expressed in macrophages associated with tumors (MacDougall and Matrisian, 1995). MME has recently been implicated as the enzyme responsible for maintenance of dormancy in lung metastases in the murine Lewis lung cell carcinoma model (Dong *et al.*, 1997). It has long been recognized that removal of certain primary tumors, in both experimental models and clinical practice, can be followed by rapid growth of dormant metastases. To help explain this finding, O'Reilly and Folkman subcutaneously implanted murine Lewis lung carcinoma cells (LLC) into mice. Two weeks following resection of the primary tumors dormant "micrometastases" had grown to gross metastases (O'Reilly *et al.*, 1994). From the urine, they isolated an angiogenesis inhibitor termed *angiostatin*, a fragment of plasminogen containing kringle regions 1–4. Application of angiostatin prevented growth of lung metastases following resection. They surmised that primary tumors generate angiostatin which is lost upon resection resulting in tumor growth. Subsequently it was demonstrated that generation of angiostatin in the LLC model was not caused by the tumor cells but was associated with the presence of macrophages in the primary tumor. Furthermore, angiostatin activity (inhibition of endothelial cell proliferation) was correlated with the presence of EDTA-inhibitable elastolytic activity, presumably MME (Dong *et al.*, 1997).

Using mice deficient in MME by gene targeting, we have demonstrated directly the requirement for MME in preventing growth of lung metastases in the LLC model (Grisolano and Shapiro, unpublished observations). While primary tumors and "micrometastases" are unaffected by the presence of MME, MME−/− mice have markedly more gross lung metastases. We also found by Western blot analysis that MME and angiostatin are induced in lung tissue after implantation of primary LLC cells and both are down-regulated following removal of the primary tumor.

We also found that macrophage elastase generates angiostatin from plasminogen, resulting in inhibition of endothelial cell proliferation and "tube formation" (Cornelius *et al.*, 1998). Together these findings lead us to suspect that MME-induced generation of angiostatin may be responsible for tumor growth inhibition, although other possibilities have not been ruled out. We believe that local production of MME by macrophages in the lung, induced by soluble factors from the primary tumor (perhaps GMCSF), is responsible for the effects on metastases growth.

These data suggest that macrophage elastase prevents tumor metastases growth, while other MMPs promote tumor progression. Our hypothesis is that MMPs spatially associated with the neovasculature, and capable of interacting with endothelial cells, may promote tumor angiogenesis. In contrast, macrophage elastase expressed by macrophages around the tumor comes in contact with and cleaves plasminogen and perhaps other molecules such as thrombospondin (Ts) or type XVIII collagen generating anti-angiogenic molecules; angiostatin, Ts fragments, and endostatin, respectively. Additionally, macrophage elastase may have more potent proteolytic capacity and higher binding affinity for elastin and other matrix components than other MMPs, perhaps further enhancing this property. We also speculate that macrophage elastase, which, unlike other MMPs, processes its C-terminal domain during activation, would be unable to bind to endothelial cells. The free C terminal may even compete with "pro-angiogenic" MMPs for endothelial cell binding. Further work is required to prove or disprove these theories, which might influence drug design of MMP inhibitors to maximize their anti-tumor effects.

Pulmonary emphysema is a major component of the morbidity and mortality of chronic obstructive pulmonary disease (COPD), a condition that afflicts more than 14 million persons in the United States and has become the fourth leading cause of death. Given the large increase in smoking in many foreign countries, COPD will become a larger worldwide problem in the ensuing years.

Emphysema is defined as enlargement of peripheral air spaces of the lung including respiratory bronchioles, alveolar ducts, and alveoli, accompanied by destruction of the walls of these structures. Inherited deficiency of α_1-AT, the primary inhibitor of neutrophil elastase, predisposes individuals to early-onset emphysema, and intrapulmonary instillation of elastolytic enzymes in experimental animals causes emphysema. Together, these findings led to the elastase : antielastase hypothesis for the pathogenesis of emphysema, which has been the prevailing hypothesis for more than 30 years. However, macrophages, not neutrophils, are the most abundant defense cell in the lung both under normal conditions and in the lungs of smokers. The capacity of macrophages to degrade elastin and, hence, to contribute to emphysema, was controversial until Chapman and Stone (1984) identified elastolytic cysteine proteinases. Subsequently, we found elastolytic MMPs produced by alveolar macrophages. In our studies the macrophage elastolytic activity was inhibited by the TIMPs, therefore we focused on the role of MMPs, particularly macrophage elastase, in the pathogenesis of emphysema.

Macrophage elastase, nearly undetectable in normal macrophages, is expressed in human alveolar macrophages of cigarette smokers. We also detect HME by immunohistochemistry and *in situ* hybridization in macrophages in patients with emphysema, but not normal lung tissue. To determine directly the contribution of macrophage elastase to emphysema we (1) developed a murine model of cigarette smoke-induced emphysema, (2) generated macrophage elastase-deficient (MME−/−) mice by gene targeting, and (3) subjected MME−/− mice and wild type (MME+/+) littermates to chronic cigarette smoke expo-

MME +/+   MME -/-

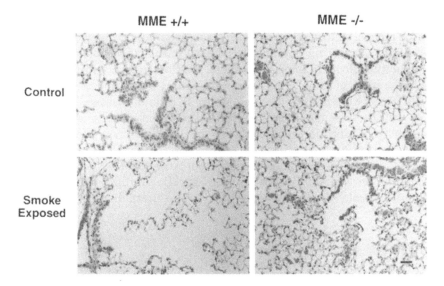

Control

Smoke
Exposed

FIG. 3. MME+/+, but not MME−/− mice, develop emphysema in response to cigarette smoke. Lungs from MME+/+ and MME−/− mice exposed to cigarette smoke for 6 months and age-matched controls were inflated by intratracheal administration of 10% formalin under constant pressure (25 cm H$_2$O). Following fixation, hematoxylin-eosin-stained midsagittal sections were prepared. Note that the MME+/+ smoke-exposed lung has centriacinar dilation of alveolar ducts compared to the control. MME−/− lung of smoke-exposed mice resembles control. Bar = 48 mm.

sure (Hautamaki *et al.*, 1997). We found that in contrast to MME+/+ mice, mice lacking MME (MME−/−) did not develop emphysema (Fig. 3). Surprisingly, MME−/− mice also failed to recruit macrophages into their lungs in response to cigarette smoke. Monthly intratracheal instillation of monocyte chemoattractant protein 1 to smoke-exposed MME−/− mice resulted in recruitment of MME−/− alveolar macrophages but failed to cause air-space enlargement. Thus, macrophage elastase is required for both macrophage accumulation and emphysema resulting from chronic inhalation of cigarette smoke. Our current working model is that cigarette smoke induces constitutive macrophages to produce MME, which cleaves elastin-generating fragments chemotactic for monocytes. This positive feedback loop perpetuates macrophage accumulation and lung destruction.

VIII. SUMMARY

The unusual history of macrophage elastase helps explain why it is perhaps the least studied MMP with respect to structure and biochemistry, but perhaps the most advanced with respect to biological function. In the mouse, this enzyme is critical to macrophage proteolysis and the use of gene targeting has uncovered both pathological roles including destructive effects in emphysema and atherosclerosis, and physiological roles such as tumor growth inhibition and regulation of inflammation. Translation of findings from mouse to human disease depends on how well the disease models replicate the human conditions and the similarity of enzyme profile between species. We know that HME is associated with these diseases, but as opposed to the mouse, other macrophage MMPs may also be of importance (MMP-9, and perhaps MMP-7 in particular). Our interpretation is that findings in mice reflect the critical role of macrophage proteolysis in these disease processes.

REFERENCES

1. Banda M. J., and Werb Z. (1981). Mouse macrophage elastase. Purification and characterization as a metalloproteinase. *Biochem. J.* **193**, 589–605.
2. Banda, M. J., Clark, E. J., and Werb, Z. (1983). Selective proteolysis of immunoglobulins by mouse macrophage elastase. *J. Exp. Med.* **157**, 1184–1196.
3. Banda, M. J., Clark, E. J., and Werb, Z. (1985). Regulation of alpha-1-proteinase inhibitor function by rabbit alveolar macrophages. Evidence for proteolytic rather than oxidative inactivation. *J. Clin. Invest.* **75**, 1758–1762.
4. Banda, M. J., Rice, A. G., Griffin, G. L., and Senior, R. M. (1988). Alpha-1-proteinase inhibitor is a neutrophil chemoattractant after proteolytic inactivation by macrophage elastase. *J. Biol. Chem.* **263**, 4481–4484.
5. Belaaouaj, A., Shipley, J. M., Kobayashi, D. K., Zimonji, D. B., Popescu, N., Silverman, G., and Shapiro, S. D. (1995). Human macrophage metalloelastase. Genomic organization, chromosomal location, gene linkage, and tissue-specific expression. *J. Biol. Chem.* **270**, 14568–14575.
6. Brooks, P. C., Strombla, S., Sanders, L. C., Schalscha, T. L., Aimes, R. T., Stetler-Stevenson, W., Quigley, J. F., and Cheresh, D. A. (1996). Localization of matrix metalloproteinase MMP-2 to the surface of invasive cells by interaction with integrin avb3. *Cell* **85**, 683–693.
6A. Brown, P. D., and Givazzi, R. (1995). Matrix metalloproteinase inhibition: A review of anti-tumor activity. *Ann. Oncol.* **6**, 967–974.
7. Campbell, E. J., Cury, J. D., Shapiro, S. D., Goldberg, G. I., and Welgus, H. G. (1991). Neutral proteinases of human mononuclear phagocytes—cellular-differentiation markedly alters cell phenotype for serine proteinases, metalloproteinases, and tissue inhibitor of metalloproteinases. *J. Immunol.* **146**, 1286–1293.
8. Carmeleit, P., Moons, L., Lijnen, R., Crawley, J., Tipping, P., Drew, A., Eeckhout, Y., Shapiro, S. D., Lupu, F., and Collen, D. (1997). Plasmin predisposes to atherosclerotic aneurysm formation by activation of matrix metalloproteinases. *Nature Genetics* **17**, 439–444.
9. Chandler, S., Cossins, J., Lury, J., and Wells, G. (1996). Macrophage metalloelastase degrades matrix and myelin proteins and processes a tumor necrosis factor-alpha fusion protein. *Biochem. Biophys. Res. Commun.* **228**, 421–429.
10. Chapman, H. A. Jr., and Stone, O. L. (1984). Comparison of live human neutrophil and alveolar macrophage elastolytic activity *in vitro*. Relative resistance of macrophage

elastolytic activity to serum and alveolar proteinase inhibitors. *J. Clin. Invest.* **74,** 1693–1700.

11. Cornelius, L. A., Nehring, L., Klein, B., Pierce, R., Bolinski, M., Welgus, H. G., and Shapiro, S. D. (1997). Generation of angiostatin by matrix metalloproteinases: Effects on neovascularization. *Clin. Invest.* In press.
12. Dong, Z., Kuma, R., Yan, X., and Fidler, I. J. (1997). Macrophage-derived metalloelastase is responsible for the generation of angiostatin in Lewis lung carcinoma. *Cell* **88,** 801–810.
13. Gronski, T. J., Martin, R., Kobayashi, D. K., Walsh, B. C., Holman, M. C., Van Wart, H. E., and Shapiro, S. D. (1997). Hydrolysis of a broad spectrum of extracellular matrix proteins by human macrophage elastase. *J. Biol. Chem.* **272,** 12189–12194.
14. Halpert, I., Sires, U. I., Roby, J. D., Potter-Perigo, S., Wight, T. N., Shapiro, S. D., Welgus, H. G., Wickline, S. A., and Parks, W. C. (1996). Matrilysin is expressed in lipid-laden macrophages at sites of potential rupture in atherosclerotic lesions and localizes to areas of versican deposition, a proteoglycan substrate for the enzyme. *Proc. Natl. Acad. Sci. USA* **93,** 9748–9753.
15. Hautamaki, R. D., Kobayashi, D. K., Senior, R. M., and Shapiro, S. D. (1997). Requirement for macrophage elastase for cigarette smoke-induced emphysema in mice. *Science* **277,** 2002–2004.
16. Januscz, M., Travis, W., and Shapiro, S. D. (1997). Macrophage elastase is responsible for macrophage proteolytic activity: Activation and processing. *J. Biol. Chem.* In press.
17. Kettner, C., and Shaw, E. (1981). The specificity of macrophage elastase on the insulin B-chain. *Biochem. J.* **195,** 369–372.
18. Kumar, R., Dong, Z., and Fidler, I. J. (1996). Differential regulation of metalloelastase activity in murine peritoneal macrophages by granulocyte-macrophage colony-stimulating factor and macrophage colony-stimulating factor. *J. Immunol.* **157,** 5104–5111.
19. MacDougall, J. R., and Matrisian, L. M. (1995). Contributions of tumor and stromal matrix metalloproteinases to tumor progression, invasion and metastasis. *Cancer Metastasis Rev.* **14,** 351–362.
20. Mecham, R. P., Broekelmann, T. C., Fliszar, C., Shapiro, S. D., Welgus, H. G., and Senior R. M. (1997). Elastin degradation by matrix metalloproteinases. *J. Biol. Chem.* **272,** 18071–18076.
21. O'Reilly, M., Holmgren, L., Shin, Y., Chen, C., Rosenthal, R., Moses, M., Lane, W., Ca, Y., Sag, E., and Folkman, J. (1994). Angiostatin: A novel angiogenesis inhibitor that mediates the suppression of metastases by a Lewis lung carcinoma. *Cell* **79,** 315–328.
22. Shapiro, S. D., Griffin, G., Gilbert, D. J., Jenkins, N. A., Copeland, N. G., Welgus, H. G., Senior, R. M., and Ley, T. J. (1992). Molecular cloning, chromosomal localization and bacterial expression of a novel murine macrophage metalloelastase. *J. Biol. Chem.* **267,** 4664–4671.
23. Shapiro, S. D., Doyle, G. A., Ley, T. J., Parks, W. C., Welgus, H. G., (1993a). Molecular mechanisms regulating the production of collagenase and TIMP in U937 cells: Evidence for involvement of delayed transcriptional activation and enhanced mRNA stability. *Biochemistry* **32,** 4286–4292.
24. Shapiro, S. D., Kobayashi, D. and Ley, T. (1993b). Cloning and characterization of a unique elastolytic metalloproteinase produced by human alveolar macrophages. *J. Biol. Chem.* **268,** 23824–23829.
25. Shipley, J. M., Wesselschmidt, R. L., Kobayashi, D. K., Ley, T. J., and Shapiro, S. D. (1996). Metalloelastase is required for macrophage-mediated proteolysis and matrix invasion in mice. *Proc. Natl. Acad. Sci. USA* **93,** 3942–3946.
26. Werb, Z., and Gordon, S. (1975). Elastase secretion by stimulated macrophages. *J. Exp. Med.* **142,** 361–377.

Membrane-Type Matrix Metalloproteinases and Cell Surface-Associated Activation Cascades for Matrix Metalloproteinases

Vera Knäuper and Gillian Murphy

Strangeways Research Laboratory, Cambridge, and School of Biological Sciences, University of East Anglia, Norwich NR4 7TJ, United Kingdom

I. Introduction

A growing body of evidence suggests that matrix metalloproteinases (MMPs) initiate degradation of the extracellular matrix (ECM) during remodeling of connective tissues. A clear understanding of the mechanisms governing the regulation of their activity during physiological processes should give further insights into the uncontrolled remodeling that occurs in degradative pathologies. Regulation of the MMPs occurs at the level of gene expression, with precise spatial and temporal compartmentalization of both synthesis and secretion. Most soluble MMPs

Matrix Metalloproteinases

are secreted into the pericellular and extracellular environment as inactive proenzymes and an important level of regulation of their enzymatic activity is the activation of the respective proenzymes by proteolysis. The potential for association of MMPs with the cell surface or ECM components constrains their relationship with substrates, activators, and inhibitors, acting as further levels for the regulation of MMP activity. This localized activity has led to the concept that these enzymes play a role in the determination of cell phenotype by the modulation of cell–matrix interactions (Basbaum and Werb, 1996).

The structural basis of latency of proMMPs is due to the presence of an N-terminal pro-domain that protrudes into the active site of the enzyme (Freimark et al., 1994; Becker et al., 1995). The conserved sequence PRCGVPD within the propeptide domain is a major determinant of latency since the free cysteine residue of this sequence motif is coordinated to the catalytic zinc. The residues making up the start of the catalytic domain are not accessible to proteolytic enzymes because the first α-helix of the propeptide domain blocks this site of the molecule. Detailed studies on the activation mechanisms of various proMMPs by endoproteinases have revealed that the propeptide domain is cleaved sequentially, with the initial cleavage typically taking place in a "bait" region located in a readily accessible site between the first and second α-helix in the propeptide domain. This cleavage destabilizes the interaction of the propeptide with the catalytic domain. Further proteolytic cleavages are then possible and the final step is often a bimolecular autoproteolytic event (Imai et al., 1996; Nagase et al., 1991; Nagase, 1997; Knäuper et al., 1993, 1996). The fully active enzyme is generated by the release of the PRCGVPD sequence motif from the active site cleft ("cysteine switch" mechanism: Springman et al., 1990), and its proteolytic removal coincides with a structural rearrangement of the first six amino acid residues in the mature active from. The fully active MMPs have an N-terminal Tyr or Phe residue, which forms a salt bridge with a conserved Asp residue. It is believed that this salt bridge is vital for high specific enzymatic activity (Benbow et al., 1996). This is particularly important for the triple helicase activity of the collagenases against native collagens (Reinemer et al., 1994; Knäuper et al., 1993; Suzuki et al., 1990).

II. Potential Activation Routes for Matrix Metalloproteinases

In the case of some MMPs, the residues of the bait region of the propeptide can be cleaved at least in vitro by a number of proteinases of different specificities, including trypsin, tryptase, chymotrypsin, cathepsin G, plasmin, plasma kallikrein, neutrophil elastase, and thermolysin (Table I; Nagase et al., 1991; Nagase, 1997). Activation by plasmin, which has been extensively studied biochemically (Fig. 1), is also thought to be of potential physiological importance. A number of cell model studies have demonstrated that MMP activation could occur at the cell surface through the urokinase-like plasminogen activator (uPA)/uPA receptor (uPAR)/plasminogen cascade for plasmin generation (Mignatti et al., 1986; Murphy et al., 1992a). Macrophages from uPA−/− mice are unable to activate MMP-3, MMP-9, MMP-12, or MMP-13 in a plasminogen-dependent manner, whereas cells from wild type or tissue-type plasminogen activator (tPA) −/− mice can. Furthermore, atherosclerotic plaques from apolipoprotein A (ApoE) −/− : uPA−/− mice showed far less matrix degradation at sites of macrophage accumulation relative to those from apoE−/− or apoE−/− : tPA−/− mice, implicating the abrogation of MMP activity (Carmeliet et al., 1997). Inhibitors of uPA, such as the plasminogen activator inhibitors (PAIs 1–4), and of plasmin, such as α_2-antiplasmin, may play important roles in regulating MMP activation by this route (Saksela and Rifkin, 1988).

Following cleavage within the propeptide bait region, secondary cleavages yielding the fully active MMP are usually catalyzed by MMPs. Complex activation cascades may be established as active MMPs are able to process other proenzymes within the family, either in the bait

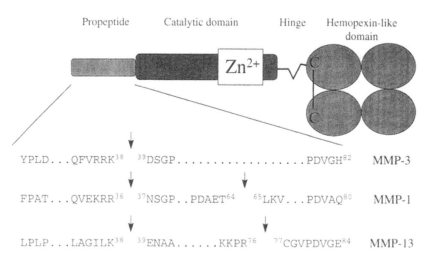

FIG. 1. Activation of MMP-1, -3, and -13 by plasmin. The identifiable sites at which plasmin cleaves the propeptide of stromelysin-1 (MMP-3; Nagase et al., 1990), collagenase-1 (MMP-1; Suzuki et al., 1990), and collagenase-3 (MMP-13; Knäuper et al., 1996) are shown by arrows. The C-terminal residue shown in each case is the site of the subsequent cleavage required to generate the fully active enzyme.

region or at secondary sites, to generate their active counterparts (Table I). These MMP activation cascades could be tightly regulated by the natural inhibitors of the MMPs (TIMPs 1–4) (Gavrilovic *et al.*, 1987; Ward *et al.*, 1991; DeClerck *et al.*, 1991; Benbow *et al.*, 1996; Knäuper *et al.*, 1996).

The recent discovery of four membrane-associated MMPs (membrane-type, MT-MMPs) has strengthened the concept of pericellular activation cascade mechanisms for the MMPs. It has been shown that MT1-MMP, MT2-MMP, and MT3-MMP can activate progelatinase A, which has a propeptide that is not generally susceptible to proteolytic initiation of activation by serine proteinases (Okada *et al.*, 1990). The emphasis in this chapter is on reviewing the mechanisms by which the MT-MMPs modulate the activation of progelatinase A (proMMP-2) and procollagenase-3 (proMMP-13) and the potential interaction with other activation cascades as well as the consequences for pericellular matrix turnover.

III. Structure of Membrane-Type Matrix Metalloproteinases

To date, four MT-MMPs have been cloned by RT-PCR-based homology screening and sequenced: MT1-MMP (MMP-14), MT2-MMP (MMP-

TABLE I

ACTIVATION OF MMPS

Human MMP	No.	Exogenous activators	Activating
Collagenase-1	1	Plasmin, kallikrein, chymase, MMP-3	MMP-2
Collagenase-2	8	Plasmin, MMP-3, MMP-10	Not known
Collagenase-3	13	Plasmin, MMP-2, MMP-3, MMP-14	MMP-2, MMP-9
Stromelysins-1, -2	3, 10	Plasmin, kallikrein, chymase, tryptase, elastase, cathepsin G	MMP-1, MMP-9, MMP-8, MMP-13
Stromelysin-3	11	Furin	Not known
Matrilysin	7	Plasmin, MMP-3	MMP-3
Metalloelastase	12	Not known	Not known
Gelatinase A	2	MMP-1, MMP-7, MMP-14, MMP-15, MMP-16	MMP-9, MMP-13
Gelatinase B	9	Plasmin, neutrophil elastase, MMP-3, MMP-2, MMP-13	Not known
MT1-MMP	14	Furin (transmembrane deletion mutants only), plasmin	MMP-2, MMP-13
MMT2-MMP	15	Not known	MMP-2
MT3-MMP	16	Not known	MMP-2
MT4-MMP	17	Not known	Not known
MMP-19	19	Not known	Not known

15), MT3-MMP (MMP-16), and MT4-MMP (MMP-17). The enzymes are closely related to each other, sharing 50–30% sequence homology and a common multidomain structure (Sato *et al.*, 1994; Will and Hinzmann, 1995; Takino *et al.*, 1995; Puente *et al.*, 1996). Their primary structure is composed of a signal peptide, a propeptide, a catalytic domain that contains the HEXXHXXGXXH consensus motif for zinc binding, a hinge sequence motif of varying length that connects the catalytic domain with the C-terminal hemopexin-like domain, and a transmembrane spanning hydrophobic sequence motif followed by a short cytoplasmic tail (Fig. 2).

Common to all MT-MMPs is the insertion of a potential furin/prohormone convertase cleavage site at the end of the propeptide domain, which is also conserved in stromelysin-3, a soluble member of the MMP family. With the exception of MT4-MMP, they also have an insertion of 8-amino-acid residues within the catalytic domain, 43-amino-acid residues downstream of the start of this domain (Fig. 3). The function of this insertion is not clear to date, but it could possibly determine some aspects of the substrate specificity of the enzymes or be responsible for the impaired TIMP-1 binding of MT1-MMP and MT2-MMP.

The hinge region of these enzymes is very variable in size, ranging from 34- to 65-amino-acid residues and it is not known whether this

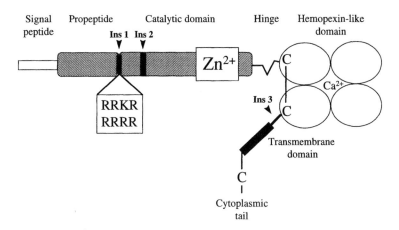

FIG. 2. The MT-MMPs show the typical domain structure of cleavable signal and propeptide sequences, a Zn^{2+} and Ca^{2+} binding catalytic domain connected by a short flexible "hinge" sequence to a disulfide bonded C-terminal domain. A furin cleavable site (Ins1) in the propeptide and an 8-amino-acid insert in the catalytic domain (Ins2; see Fig. 3) are common to all the known MT-MMPs (with the exception of MT4-MMP, which does not have Ins2). Unique to the subfamily is a C-terminal transmembrane domain and a very short cytoplasmic tail.

PYAYIREG	human MT1-MMP
PYAYIREG	mouse MT1-MMP
EYAYIRDG	rabbit MT1-MMP
PYEDIRLR	human MT2-MMP
SYDDIRLR	mouse MT2-MMP
PYSELENG	human MT3-MMP
PYIELENG	chicken MT3-MMP
--------	human MT4-MMP

FIG. 3. The sequences of insert 2 in the MT-MMPs. An insertion of 8-amino-acid residues is found within the catalytic domain of all the MT-MMPs, with the exception of MT4-MMP. Its function is not yet known.

sequence motif is important for mediating substrate specificity as has been demonstrated for the collagenases (Hirose *et al.*, 1993; Knäuper *et al.*, 1997b). The hemopexin-like domain shows all the characteristics earlier demonstrated for the other MMPs, including conservation of the two cysteine residues that form a disulfide bond (Murphy and Knäuper, 1997). Due to the high percentage of homology of this domain in the different MMPs it can be assumed that the overall structure will be very similar to those already solved by X-ray crystallography, such as fibroblast collagenase, gelatinase A, and collagenase-3 (Li *et al.*, 1995; Libson *et al.*, 1995; Gohlke *et al.*, 1996; Gomis-Rüth *et al.*, 1996).

The MT-MMPs contain a further insertion at the end of the hemopexin domain, just preceding the hydrophobic transmembrane spanning sequence motif, which is also variable in size. The cytoplasmic tail at the carboxyl end of the MT-MMPs contains an additional conserved cysteine residue flanked by tyrosine and serine residues, which may be potential phosphorylation sites. However, these residues are absent in MT4-MMP, which emphasizes that this enzyme is less conserved within the subfamily (Puente *et al.*, 1996).

IV. CELLULAR SOURCES OF MT-MMPs AND REGULATION OF EXPRESSION

Most soluble members of the MMP family are not expressed to a high level in normal adult tissues and their expression is very tightly regulated at the transcriptional level by growth factors and cytokines. However, the mRNAs for the MT-MMPs are detectable in extracts of such tissues as assessed by Northern blot analysis (Fig. 4). Kinoh *et al.* (1996) have described prominent expression of MT1-MMP and TIMP-2 mRNAs in bone and perichondrium during mouse embryo skeletogenesis. These observations have recently been confirmed, but with intri-

	brain	colon	heart	intestine	kidney	leukocytes	liver	lung	muscle	ovary	pancreas	placenta	prostate	spleen	testis
MT1-MMP	-	-	-	+	+	-	-	+	-	+	-	+	+	+	-
MT2-MMP	-	+	+	+	+	-	+	+	+	-	+	+	-	-	+
MT3-MMP	+	-	-	-	-	-	-	+	-	-	-	+	-	-	-
MT4-MMP	+	+	-	+	-	+	-	-	-	+	-	-	-	-	+

FIG. 4. Expression of the mRNA for the MT-MMP in normal human tissues. Northern blot analyses of the mRNA levels in tissues are summarized according to the data of Will and Hinzmann (1995) for MT1-MMP and MT2-MMP, Takino *et al.* (1995) for MT3-MMP, and Puente *et al.* (1996) for MT4-MMP. + denotes stronger expression than +.

guing differences in MT1-MMP and TIMP-2 distribution at sites of endochondral ossification (S. Apte, personal communication), who also found co-expression of both proteins in tendons, ligaments, muscle, and joint capsules of the embryo as well as in the media of arteries, during bladder development and in specific urogenital organs. The MT1-MMP expression pattern during mouse embryogenesis was quite distinct from that described for the chicken MT3-MMP homologue during chick development (Yang *et al.*, 1996).

A number of reports of regulation of MT1-MMP mRNA by cytokines have now been made, notably tumor necrosis factor α (TNF-α) up-regulation in synovial cells and induction of mRNA by TNF-α, interleukin 1β (IL-1β), epidermal growth factor (EGF), and basic fibroblast growth factor (bFGF) in human embryonic lung fibroblasts (Migita *et al.*, 1996; Lohi *et al.*, 1996). In chick embryo fibroblasts the mRNA levels for an MT-MMP (most closely related to MT3-MMP) were only modestly increased by bFGF or TNF-α and not affected by IL-1β or retinoic acid (Yang *et al.*, 1996). To date, no knowledge is available on the regulatory elements of the MT-MMP promoter sequences, and elucidation of the mechanisms regulating the gene expression of the MT-MMPs will certainly be the topic of future research.

The tetravalent lectin concanavalin A (con A) induces MT1-MMP activity in some cell types, including fibroblasts, and is generally accompanied by induction of progelatinase A activation (Atkinson *et al.*, 1995; Lohi *et al.*, 1996; Gilles *et al.*, 1996). It is now clear that con A mediates the up-regulation of MT1-MMP in a c-Ras dependent manner (Thant *et al.*, 1997). However, con A could also be responsible for cross-linking of cell surface MT1-MMP in a form of "capping" or concentration of the enzyme, thereby facilitating rapid activation of progelatinase A. In some cell types treatment with phorbol esters can also modulate MT1-MMP and progelatinase A processing (Foda *et al.*, 1996; Lohi *et al.*, 1996). Recent studies have provided evidence that MT1-MMP synthesis is regulated by the cytoskeleton (Tomasek *et al.*, 1997). Fibroblasts grown in a collagen substratum produce low levels of MT1-MMP and synthesis is up-regulated in relaxed lattices. This effect is mediated by the lack of stress fibers in those cells, and treatment of fibroblasts with cytochalasin D, which disrupts stress fibers, leads to the initiation of increased MT1-MMP mRNA levels and to progelatinase A activation (Ailenberg and Silverman, 1996; Tomasek *et al.*, 1997).

V. Activation of MT-MMPs

MT-MMPs contain a basic sequence motif (Ins1 in Fig. 2) at the C-terminal end of the propeptide domain that is a potential recognition site for prohormone convertases, suggesting that these enzymes might be processed by subtilisin-related mammalian endoproteases. To date, a number of proteases have been described that could be physiological activators of MT-MMPs, namely, proprotein convertase 2 (PC2), PC3/PC1, PC4, furin/PC, PC 4, PC 5/6 and 7/8 (Smeekens, 1993; Taylor *et al.*, 1997). Thus, MT-MMPs could potentially be processed either within the regulated or the constitutive secretory pathway. The first studies attempting to address this question demonstrated that MT1-MMP transmembrane deletion mutants were processed by co-expressed furin/PC (Pei and Weiss, 1996) as had been demonstrated earlier for stromelysin-3 (Pei and Weiss, 1995). This processing was inhibited by co-expression of a mutant α_1-proteinase inhibitor (α_1-Pittsburgh), which is known to inhibit furin/PC (Pei and Weiss, 1996; Sato *et al.*, 1996a), thus providing evidence that furin/PC is a candidate for activation of MT1-MMP. Furthermore, mutagenesis of the basic furin/PC processing motif and expression of these mutants in COS cells revealed that transmembrane domain deleted MT1-MMP was secreted into the culture medium as a proenzyme, thereby implicating furin/PC in intracellular processing of this membrane-associated MMP. However, in the case of the expression of recombinant full-length membrane-bound MT1-MMP, co-expression of furin/PC in COS cells had no effect on the molecular mass (63 kDa) of the membrane-associated proteinase, and it was suggested that the full-length enzyme might not be processed by furin/PC (Cao *et al.*, 1996). Furthermore, co-expression of α_1-Pittsburgh had no effect on the activation of progelatinase A by membrane-associated

MT1-MMP in transfected COS cells, implying that the membrane bound MT1-MMP was fully functional and active. In addition, mutagenesis of the furin/PC recognition site in the full-length MT1-MMP molecule did not abrogate activation of progelatinase A, as would have been expected from the earlier studies using transmembrane deletion mutants. Thus furin/PC-induced activation of membrane-bound MT1-MMP is not a prerequisite for progelatinase A activation. Intriguingly, Cao et al. (1996) also found that the 63-kDa MT1-MMP expressed did not bind to α_2-macroglobulin, implying that the species responsible for progelatinase A cleavage may not be readily detectable by the techniques utilized.

Recently, it was domonstrated that soluble proMT1-MMP can be activated by human plasmin in vitro by cleavage immediately downstream of Arg[108] and Arg[111] in the basic furin/PC recognition site (Okumura et al., 1997). These results suggest that proMT1-MMP could be transported to the plasma membrane where the proenzyme is extracellularly activated by membrane-associated plasmin. However, there is no evidence for plasmin activation of cell-associated MT1-MMP, in that the activation of progelatinase A by cell-bound MT1-MMP is not inhibited by addition of plasmin inhibitors in cellular activation experiments (S. Cowell, V. Knäuper, and G. Murphy, unpublished data). Because the amino terminal sequence of native active membrane-bound MT1-MMP is Tyr[112], and this amino terminus was only partially generated by plasmin-mediated cleavage of proMT1-MMP, this suggests that activation in vivo is unlikely to be mediated by plasmin. Studies using MMP inhibitors, including TIMP-2 and -3 and peptide hydroxamates, suggest that the activation process is intracellular and involves MMP activity (H. Stanton, S. Atkinson, and G. Murphy, unpublished observation; Lee et al., 1997). In conclusion, the physiological activator of membrane-associated MT-MMP has not yet been discovered, but it is clearly critical to the initiation of MMP activation cascades (see later discussion).

VI. ACTIVATION OF PROGELATINASE A BY MT1-MMP AND MT2-MMP

A. Activation in Solution

Progelatinase A is activated by soluble MT1-MMP and MT2-MMP in a two-step activation mechanism analogous to the activation of many other MMPs. The initial cleavage is observed at the Asn[37]–Leu[38] peptide bond and is due to either MT1-MMP- or MT2-MMP-mediated proteolysis in a region of the progelatinase A propeptide domain that is solvent exposed (Butler et al., 1997; Will et al., 1996). The secondary cleavage event is due to autoproteolytic cleavage, since an inactive progelatinase A mutant (proE^{375}–A gelatinase A) is only processed by MT1-MMP or MT2-MMP to the Asn^{37}–Leu^{38} intermediate form. Processing by MT1-MMP or MT2-MMP was inhibited by either TIMP-2 or TIMP-3 in a concentration-dependent fashion, and TIMP-1 was not effective under the same conditions. Determination of the rate constants for MT1-MMP or MT2-MMP inhibition by the TIMPs revealed that TIMP-1 has very low association binding constants with both MT-MMPs, whereas TIMP-2 and TIMP-3 bind rapidly (Will et al., 1996). These results suggest that activation of progelatinase A by MT1-MMP or MT2-MMP may be regulated by TIMP-2 and TIMP-3.

B. Activation in Cellular Systems and the Role of TIMP-2

Progelatinase A activation by cells expressing MT1-MMP at their surface also involves a two-step activation mechanism as described earlier for the soluble system, the secondary cleavage event being autoproteolytic (Atkinson et al., 1995; Will et al., 1996). The process appears to involve binding of the proenzyme to an MT1-MMP/TIMP-2 complex ("receptor") on the cell surface through interaction between the C-terminal domain of the enzyme and the C-terminal domain of TIMP-2 (Fig. 5) (Murphy et al., 1992b; Ward et al., 1994; Strongin et al., 1993, 1995; Atkinson et al., 1995). By establishing a trimolecular complex, consisting of MT1-MMP/TIMP-2/progelatinase A as demonstrated by cross-linking experiments (Strongin et al., 1995), the components are concentrated on the cell surface. Processing of progelatinase A to the Leu^{38} intermediate form may then be initiated by an adjacent free and active MT1-MMP molecule (Fig. 5). This initial cleavage event destabilizes the structure of the progelatinase A propeptide domain, and autoproteolysis then proceeds in a gelatinase A concentration-dependent manner, which releases the rest of the propeptide domain and fully active gelatinase A. In cell culture studies, the enzyme concentrations in solution are low and deletion of either the progelatinase A C-terminal domain or the transmembrane domain of MT1-MMP abolishes progelatinase A activation, emphasizing that the binding mechanism involving the MT1-MMP/TIMP-2 complex on the cell surface acts as a concentration mechanism that is crucial for the efficiency of activation (Murphy et al., 1992b; Ward et al., 1994; Atkinson et al., 1995; Cao et al., 1995; Sato et al., 1996b).

Addition of small amounts of TIMP-2 to cells expressing MT1-MMP can enhance progelatinase A activation, because this increases the concentration of the MT1-MMP/TIMP-2 receptor for progelatinase A on the cell surface (Strongin et al., 1995; Cao et al., 1996). However,

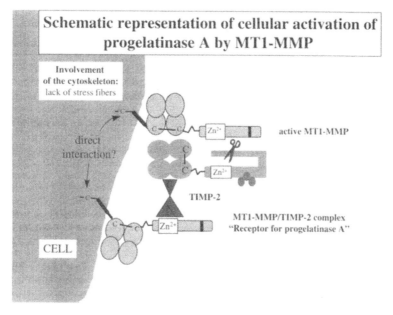

Schematic representation of cellular activation of progelatinase A by MT1-MMP

Involvement of the cytoskeleton: lack of stress fibers

direct interaction?

active MT1-MMP

TIMP-2

MT1-MMP/TIMP-2 complex "Receptor for progelatinase A"

CELL

FIG. 5. A scheme for the cellular activation of progelatinase A by MT1-MMP. Based on the data from a number of laboratories, it has been proposed that the processed form of MT1-MMP is displayed at the cell surface in part as a TIMP-2 complex. The complexed form is not a functional enzyme because the catalytic site is occupied by interactions with the N-terminal domain of TIMP-2. The C-terminal domain of TIMP-2 is free and can bind to the C-terminal domain of progelatinase A. There is potential for dimerization of MT1-MMP through either the free cysteine residue in the catalytic domain or the cytoplasmic tail. The cytoplasmic domain may be involved in interactions with cellular signaling mechanisms, for example, those mediated by the cytoskeleton. This "receptor" mechanism serves to sequester and concentrate progelatinase A at the cell surface. Cleavage of the propeptide of progelatinase A by TIMP-2-free active MT1-MMP on the cell surface initiates autolytic activation.

at high TIMP-2 concentrations all the MT1-MMP molecules on the cell surface are complexed with TIMP-2, and although progelatinase A binding occurs, no active MT1-MMP remains to initiate processing. This suggests that progelatinase A activation is regulated by the amount of TIMP-2 secreted by MT1-MMP expressing cells as well as by the extent of MT1-MMP activation. In addition, if high concentrations of TIMP-3 are present, progelatinase A activation is strongly inhibited. In contrast, TIMP-1 is not efficient in preventing progelatinase A activation, since this inhibitor does not inhibit MT1-MMP (Will *et al.,* 1996). Since MT1-MMP seems to be activated intracellularly, the formation of TIMP-2 complexes and the activation of progelatinase A may occur intracellularly to some extent (R. Hembry and G. Murphy, unpublished results; Lee *et al.*, 1997).

VII. Activation of Procollagenase-3 by MT1-MMP

A. *Activation in Solution*

Procollagenase-3 is also activated by MT1-MMP in a two-step activation mechanism. Initial cleavage by MT1-MMP is observed at the Gly^{35}–Ile^{36} peptide bond within the propeptide domain, followed by a secondary cleavage event thereby releasing the rest of the propeptide domain (Tyr^{85} N terminus) (Knäuper *et al.*, 1996). Inhibition studies using TIMP-2, TIMP-3, and synthetic metalloproteinase inhibitors revealed that processing of procollagenase-3 was abrogated by these MMP inhibitors. In contrast, TIMP-1 was less efficient in inhibiting procollagenase-3 processing, because hydrolysis of the Gly^{35}–Ile^{36} peptide bond still occurred although processing to the Tyr^{85} active collagenase-3 was prevented. These results indicate that the initial cleavage within the propeptide domain of procollagenase-3 is mediated by MT1-MMP, since TIMP-1 is a very poor inhibitor of this enzyme as discussed earlier. It can also be concluded that the secondary cleavage event may be autoproteolytic, since TIMP-1 is an excellent inhibitor of collagenase-3, although direct evidence for this hypothesis is still lacking (Knäuper *et al.*, 1997a). A C-terminal deletion mutant of procollagenase-3 and a chimeric procollagenase-3 mutant constructed from N-terminal collagenase-3 and C-terminal MMP-19 were also activated by MT1-MMP in solution, thereby demonstrating that at high enzyme concentrations activation does not require C-terminal domain interactions (Knäuper *et al.*, 1996; V. Knäuper and G. Murphy, unpublished results).

B. *Activation in Cellular Systems*

Activation of procollagenase-3 in cellular model systems, such as con A stimulated fibroblast monolayers or HT1080 cells transfected with wild type MT1-MMP, can also be observed, and analysis of the activation products by Western blotting revealed that this also involves a minimum of two sequential propeptide cleavage events, thereby confirming the data obtained in solution using purified MT1-MMP (Knäuper *et al.*, 1996; S. Cowell, V. Knäuper, and G. Murphy, unpublished results). However, the situation in these cellular model systems is more complicated, since the cells produce progelatinase A, which is activated by MT1-MMP. In turn, active gelatinase A can activate procollagenase-3 (Knäuper *et al.*, 1996). From inhibitor studies using the homologous TIMPs 1–3 it is clear that the initial cleavage of procollagenase-3 in these cellular model systems is mediated by MT1-MMP.

Again, TIMP-1 was unable to prevent procollagenase-3 processing, indicating that the first cleavage event in this system is due to MT1-MMP. If gelatinase A were responsible for the activation of procollagenase-3, processing to the active form would be inhibited by TIMP-1, since this inhibitor has fast association rate constants for gelatinase A inhibition.

TIMP-2 and TIMP-3 were efficient inhibitors of procollagenase-3 activation in line with their efficient inhibition of MT1-MMP. Initial mechanistic studies to elucidate the functions of the different domains of procollagenase-3 in cellular activation revealed that processing is dependent on the presence of the C-terminal hemopexin-like domain, because a C-terminal deletion mutant or chimeric enzyme constructed from N-terminal collagenase-3 and C-terminal MMP-19 (Pendás *et al.*, 1997) is not processed by cells expressing MT1-MMP (V. Knäuper, A. M. Pendás, C. López-Otin, and G. Murphy, unpublished results). The mechanism of procollagenase-3 activation by cells expressing MT1-MMP shows the characteristics demonstrated earlier for progelatinase A processing in cellular model systems, but it is not at all clear whether this process also requires the presence of TIMP-2. Our current data show that collagenase-3 does not interact in kinetic studies with the C-terminal domain of free TIMP-2 (Knäuper *et al.*, 1997a) and therefore binding to cell surfaces may be mediated by a profoundly different mechanism when compared with progelatinase A binding to the MT1-MMP/TIMP-2 receptor. Thus, further detailed studies are required to solve these questions and understand this interesting mechanism of activation.

VIII. Substrate Specificity of MT-MMPs

Knowledge of the substrate specificity of the MT-MMPs has largely been gained through the preparation of soluble transmembrane deletion mutants in mammalian and bacterial expression systems, with the inherent assumption that membrane localization has no effect on the enzymes' properties. Although the initial analysis of MT1-MMP, MT2-MMP, and MT3-MMP revealed that these membrane-associated MMPs are physiological activators of progelatinase A and collagenase-3 (MT1-MMP) there is now increasing evidence that these enzymes hydrolyze a much broader range of ECM components. A number of elegant studies have provided evidence that MT1-MMP and MT2-MMP are potent proteinases and both enzymes have been shown to degrade denatured interstitial collagens, cartilage aggrecan, perlecan, fibulin-1 and -2, fibronectin, vitronectin, nidogen, large tenascin-C, and laminin-1, as well as proTNF-α fusion protein (Ohuchi et al., 1997; Imai et al., 1996; Pei and Weiss, 1996; d'Ortho et al., 1997). In case the of MT1-MMP, cleavage of triple helical interstitial collagens (type I, II and III) into characteristic ¾ and ¼ fragments has been demonstrated, indicating that MT1-MMP shares some characteristics with the collagenases (Ohuchi et al., 1997; d'Ortho et al., 1997). MT1-MMP preferentially hydrolyzes type I relative to type II and III collagen (Ohuchi et al., 1997). However, the enzyme was five- to sevenfold less efficient than fibroblast collagenase (MMP-1) in hydrolyzing these triple helical substrates. C-terminal deletion mutants of MT1-MMP lost the ability to hydrolyze the triple helical interstitial collagens, thus demonstrating that the mechanism of action involves the hemopexin-like C-terminal domain, as earlier demonstrated for the three homologous human collagenases (Clark and Cawston, 1989; Knäuper et al., 1993, 1997a). These results indicate that MT-MMPs can regulate the integrity of the extracellular matrix through two different mechanisms: (1) indirectly through activation of either progelatinase A or procollagenase-3 and (2) directly through hydrolysis of a large number of ECM components.

IX. Summary

The restriction of extracellular matrix proteolysis to the discrete pericellular environment in normal turnover is determined by complex mechanisms for the sequestration and activation of degradative proteinases and their subsequent inhibition. In the case of MMPs, the production of active forms by the action of cell-associated plasmin and of the MT-MMPs is clearly critical to the initiation of complex activation cascades of MMPs (Fig. 6). Plasmin has been found to initiate the activation of MMPs such as collagenase-1 and stromelysin-1, which bind to collagens. MT-MMPs, as exemplified by MT1-MMP, appear to act as receptors for gelatinase A, through the binding of TIMP-2, and can therefore both initiate and potentiate activation of this enzyme, as well as abrogate activation if TIMP-2 is expressed in excess of MT-MMPs. Comparable mechanisms may occur in the case of collagenase-3 and gelatinase B. Other cell surface proteinases may be involved, such as seprase (Pineiro-Sanchez et al., 1997) or the mammalian reprolysins or ADAMs (Wolfsberg et al., 1995), although direct evidence for this hypothesis is still lacking. Cell surface activation of MMPs allows precise focusing of ECM hydrolysis, which is vital for spatial control of turnover of these macromolecules. This is furthermore tightly regu-

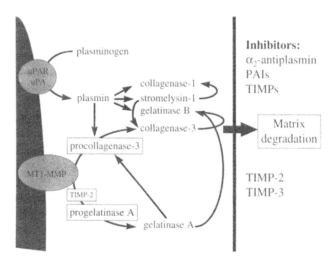

FIG. 6. Cell surface associated activation cascades for matrix metalloproteinases. The extracellular activation of most proMMPs is limited to the pericellular environment where membrane-associated proteinases can function in a relatively inhibitor depleted environment. Key initiators of the MMP activation cascades are (1) MT1–3 MMPs, which are themselves activated intracellularly and (2) uPA, which is activated by plasmin and bound to its specific cell surface receptor. The generation of partially active or active MMPs allows a cascade of autolytic and heterolytic cleavages to generate fully active enzymes. The efficiency of these interactions is dependent on mechanisms for concentrating the MMPs, for example, progelatinase A on the MT1-MMP/TIMP-2 receptor (see Fig. 5) and stromelysin and other MMPs through binding to ECM proteins.

lated by the local concentrations of PAIs, α_2-antiplasmin, and TIMPs in the pericellular environment of cells, which control the activity of uPA, plasmin, and MT-MMPs. Strikingly, where the levels of active MMPs are high, PAIs and α_2-antiplasmin are susceptible to inactivation by MMPs which could lead to elevated levels of serine proteinase activity at the cell surface, thereby allowing more efficient activation of those proMMPs which are susceptible to plasmin. Both the cell membrane and the ECM may modulate the activation and the activity of MMPs in other ways (Rice and Banda, 1995; Emonard and Hornebeck, 1997) that have yet to be elucidated.

ACKNOWLEDGMENTS

Our work is supported by grants from the Arthritis and Rheumatism Council, the Medical Research Council, Wellcome Trust, Nuffield Foundation, and Biomed-2 Programme of the European Union. We thank all members of the MMP community for their contribution to the understanding of the matrixins in extracellular connective tissue turnover. We furthermore express our gratitude to all our collaborators and colleagues for their support.

REFERENCES

Ailenberg, M., and Silverman, M. (1996). Cellular activation of mesangial gelatinase A by cytochalasin D is accompanied by enhanced mRNA expression of both gelatinase A and its membrane-associated gelatinase A activator (MT-MMP). *Biochem. J* **313**, 879–884.

Atkinson, S. J., Crabbe, T., Cowell, S., Ward, R. V., Butler, M. J., Sato, H., Seiki, M., Reynolds, J. J., and Murphy, G. (1995). Intermolecular autolytic cleavage can contribute to the activation of progelatinase A by cell membranes. *J. Biol. Chem.* **270**, 30479–30485.

Basbaum, C. B., and Werb, Z. (1996). Focalized proteolysis: Spatial and temporal regulation of extracellular matrix degradation at the cell surface. *Curr. Opin. Cell Biol.* **8**, 731–738.

Becker, J. W., Marcy, A. I., Rokosz, L. L., Axel, M. G., Burbaum, J. J., Fitzgerald, P. M. D., Cameron, P. M., Esser, C. K., Hagmann, W. K., Hermes, J. D., and Springer, J. P. (1995). Stromelysin-1: Three-dimensional structure of the inhibited catalytic domain and of the C-truncated proenzyme. *Protein Sci.* **4**, 1996–1976.

Benbow, U., Buttice J. G., Nagase, H., and Kurkinen, M. (1996). Characterization of the 46-kDa intermediates of matrix metalloproteinase 3 (stromelysin 1) obtained by site-directed mutation of phenylalanine 83. *J. Biol. Chem.* **271**, 10715–10722.

Butler, G. S., Will, H., Atkinson, S. J., and Murphy, G. (1997). Membrane-type-2 matrix metalloproteinase can initiate the processing of progelatinase A and is regulated by the tissue inhibitors of metalloproteinases. *Eur. J. Biochem.* **244**, 653–657.

Cao, J., Sata, H., Takino, T., and Seiki, M. (1995). The C-terminal region of membrane type matrix metalloproteinase is a functional transmembrane domain required for pro-gelatinase A activation. *J. Biol. Chem.* **270**, 801–805.

Cao, J., Rehemtulla, A., Bahou, W., and Zucker, S. (1996). Membrane type matrix metalloproteinase 1 activates pro-gelatinase A without furin cleavage of the N-terminal domain. *J. Biol. Chem.* **271**, 30174–30180.

Carmeliet, P., Lupu, F., Lijnen, R., Lemaitre, V., Rabinovitch, M., Tipping, P., Drew, A., Crawley, J., Eeckhout, Y., Shapiro, S., Moons, L., and Collen, D. (1997). Role of urokinase-type plasminogen activator (u-PA)-mediated activation of matrix metalloproteinases (MMP) MMP-3, MMP-9, MMP-12 and MMP-13 in atherosclerotic aneurysm in apolipoprotein E deficient mice. Submitted for publication.

Clark, I., and Cawston, T. E. (1989). Fragments of human fibroblast collagenase. Purification and characterisation. *Biochem. J.* **263**, 201–206.

DeClerck, Y. A., Yean, T.-D., Lu, H. S., Ting, J., and Langley, K. E. (1991). Inhibition of autoproteolytic activation of interstitial procollagenase by recombinant metalloproteinase inhibitor MI/TIMP-2. *J. Biol. Chem.* **266**, 3893–3899.

d'Ortho, M.-P., Will, H., Atkinson, S., Butler, G., Messent, A., Gavrilovic, J., Smith, B., Timpl, R., Zardi, L., and Murphy, G. (1998). Membrane-type matrix metalloproteinases 1 and 2 (MT1-MMP and MT2-MMP) exhibit a broad spectrum proteolytic capacity comparable to many matrix metalloproteinases. *Eur. J. Biochem.* In press.

Emonard, H., and Hornebeck, W. (1997). Binding of 92 kDa and 72 kDa progelatinases to insoluble elastin modulates their proteolytic activation. *Biol. Chem. Hoppe Seyler* **378**, 265–271.

Foda, H. D., George, S., Conner, C., Drews, M., Tompkins, D. C., and Zucker, S. (1996). Activation of human umbilical vein endothelial cell progelatinase A by phorbol myristate acetate: A protein kinase C-dependent mechanism involving a membrane-type matrix metalloproteinase. *Lab. Invest.* **74**, 538–545.

Freimark, B. D., Feeser, W. S., and Rosenfeld, S. A. (1994). Multiple sites of the propeptide region of human stromelysin-1 are required for maintaining a latent form of the enzyme. *J. Biol. Chem.* **269**, 26982–26987.

Gavrilovic, J., Hembry, R. M., Reynolds, J. J., and Murphy, G. (1987). Tissue inhibitor of metalloproteinases (TIMP) regulates extracellular type I collagen degradation by chondrocytes and endothelial cells. *J. Cell Sci.* **87**, 357–362.

Gilles, C., Polette, M., Piette, J., Munaut, C., Thompson, E. W., Bierembaut, P., and Foidart, J. M. (1996). High level of MT-MMP expression is associated with invasiveness of cervical cancer cells. *Int. J. Cancer* **65**, 209–213.

Gohlke, U., Gomis-Rüth, F. X., Crabbe, T., Murphy, G., Docherty, A. J. P., and Bode, W. (1996). The C-terminal (haemopexin-like) domain structure of human gelatinase A (MMP2): Structural implications for its function. *FEBS Lett.* **378**, 126–130.

Gomis-Rüth, F. X., Gohlke, U., Betz, M., Knäuper, V., Murphy, G., López-Otín, C., and Bode, W. (1996). The helping hand of collagenase-3 (MMP-13): 2.7 Å crystal structure of its C-terminal haemopexin-like domain. *J. Mol. Biol.* **264**, 556–566.

Hirose, T., Patterson, C., Pourmotabbed, T., Mainardi, C. L., and Hasty, K. A. (1993). Structure–function relationship of human neutrophil collagenase: Identification of regions responsible for substrate specificity and general proteinase activity. *Proc. Natl. Acad. Sci. USA* **90**, 2569–2573.

Imai, K., Ohuchi, E., Aoki, T., Nomura, H., Fujii, Y., Sato, H., Seiki, M., and Okada, Y. (1996). Membrane-type matrix metalloproteinase 1 is a gelatinolytic enzyme and is secreted in a complex with tissue inhibitor of metalloproteinases 2. *Cancer Res.* **56**, 2707–2710.

Kinoh, H., Sato, H., Tsunezuka, Y., Takino, T., Kawashima, A., Okada, Y., and Seiki, M. (1996). MT-MMP, the cell surface activator of proMMP-2 (pro-gelatinase A), is expressed with its substrate in mouse tissue during embryogenesis. *J. Cell Sci.* **109**, 953–959.

Knäuper, V., Osthues, A., DeClerck, Y. A., Langley, K. E., Bläser, J., and Tschesche, H. (1993). Fragmentation of human polymorphonuclear-leucocyte collagenase. *Biochem. J.* **291**, 847–854.

Knäuper, V., Will, H., López-Otin, C., Smith, B., Atkinson, S. J., Stanton, H., Hembry, R. M., and Murphy, G. (1996). Cellular mechanisms for human procollagenase-3 (MMP-13) activation—evidence that MT1-MMP (MMP-14) and gelatinase A (MMP-2) are able to generate active enzyme. *J. Biol. Chem.* **271**, 17124–17131.

Knäuper, V., Cowell, S., Smith, B., López-Otin, C., O'Shea, M., Morris, H., Zardi, L., and Murphy, G. (1997a). The role of the C-terminal domain of human collagenase-3 (MMP-13) in the activation of procollagenase-3, substrate specificity, and tissue inhibitor of metalloproteinase interaction. *J. Biol. Chem.* **272**, 7608–7616.

Knäuper, V., Docherty, A. J. P., Smith, B., Tschesche, H., and Murphy, G. (1997b). Analysis of the contribution of the hinge region of human neutrophil collagenase (HNC, MMP-8) to stability and collagenolytic activity by alanine scanning mutagenesis. *FEBS Lett.* **405**, 60–64.

Lee, A. I., Akers, K. T., Collier, M., Li, L., Eisen, A. Z., and Seltzer, J. L. (1997). Intracellular activation of gelatinase A (72-kDa type IV collagenase) by normal fibroblasts. *Proc. Natl. Acad. Sci. USA* **94**, 4424–4429.

Li, J., Brick, P., O'Hare, M. C., Skarzynski, T., Lloyd, L. F., Curry, V. A., Clark, I. M., Bigg, H. F., Hazleman, B. L., Cawston, T. E., and Blow, D. M. (1995). Structure of full-length porcine synovial collagenase reveals a C-terminal domain containing a calcium-linked, four-bladed β-propeller. *Structure* **3**, 541–549.

Libson, A. M., Gittis, A. G., Collier, I. E., Marmer, B. L., Goldberg, G. I., and Lattman, E. E. (1995). Crystal structure of the haemopexin-like C-terminal domain of gelatinase A. *Nature Struct. Biol.* **2**, 938–942.

Lohi, J., Lehti, K., Westermarck, J., Kähäri, V. M., and Keski-Oja, J. (1996). Regulation of membrane-type matrix metalloproteinase-1 expression by growth factors and phorbol 12-myristate 13-acetate. *Eur. J. Biochem.* **239**, 239–247.

Migita, K., Eguchi, K., Kawabe, Y., Inchinose, Y., Tsukada, T., Aoyahi, T., Nakamura, H., and Nagataki, S. (1996). TNF-α-mediated expression of membrane-type matrix metalloproteinase in rheumatoid synovial fibroblasts. *Immunology* **89**, 553–557.

Mignatti, P., Robbins, E., and Rifkin, D. E. (1986). Tumor invasion through the human amniotic membrane: Requirement for a proteinase cascade. *Cell* **47**, 487–498.

Murphy, G. and Knäuper, V. (1997). Relating matrix metalloproteinase structure to function: Why the "hemopexin" domain? *Matrix Biol.* **15**, 511–518.

Murphy, G., Atkinson, S., Ward, R., Gavrilovic, J., and Reynolds, J. J. (1992a). The role of plasminogen activators in the regulation of connective tissue metalloproteinases. *Ann. NY Acad. Sci.* **667**, 1–12.

Murphy, G., Willenbrock, F., Ward, R. V., Cockett, M. I., Eaton, D., and Docherty, A. J. P. (1992b). The C-terminal domain of 72 kDa gelatinase A is not required for catalysis, but is essential for membrane activation and modulates interactions with tissue inhibitors of metalloproteinases. *Biochem. J.* **283**, 637–641.

Nagase, H. (1997). Activation mechanisms of matrix metalloproteinases. *Biol. Chem. Hoppe Seyer* **378**, 151–160.

Nagase, H., Enghild, J. J., Suzuki, K., and Salvesen, G. (1990). Stepwise activation mechanisms of the precursor of matrix metalloproteinases 3 (stromelysin) by proteinases and (4-aminophenyl)mercuric acetate. *Biochemistry* **29**, 5783–5789.

Nagase, H., Ogata, Y., Suzuki, K., Enghild, J. J., and Salvesen, G. (1991). Substrate specificities and activation mechanisms of matrix metalloproteinases. *Biochem. Soc. Trans.* **19**, 715–718.

Ohuchi, E., Imai, K., Fujii, Y., Sato, H., Seiki, M., and Okada, Y. (1997). Membrane type 1 matrix metalloproteinase digests interstitial collagens and other extracellular matrix macromolecules. *J. Biol. Chem.* **272**, 2446–2451.

Okada, Y., Morodomi, T., Enghild, J. J., Suzuki, K., Yasui, A., Nakanishi, I., Salvesen, G., and Nagase, H. (1990). Matrix metalloproteinase 2 from human rheumatoid synovial fibroblasts. Purification and activation of the precursor and enzymic properties. *Eur. J. Biochem.* **194**, 721–730.

Okumura, Y., Sato, H., Seiki, M., and Kido, H. (1997). Proteolytic activation of the precursor of membrane type 1 matrix metalloproteinase by human plasmin—a possible cell surface activator. *FEBS Lett.* **402**, 181–184.

Pei, D., and Weiss, S. J. (1995). Furin-dependent intracellular activation of the human stromelysin-3 zymogen. *Nature* **375**, 244–247.

Pei, D. Q., and Weiss, S. J. (1996). Transmembrane-deletion mutants of the membrane-type matrix metalloproteinase-1 process progelatinase A and express intrinsic matrix-degrading activity. *J. Biol. Chem.* **271**, 9135–9140.

Pendás, A. M., Knäuper, V., Puente, X. S., Llano, E., Mattei, M. G., Apte, S., Murphy, G., and López-Otín, C. (1997). Identification and characterization of a novel human matrix metalloproteinase with unique structural characteristics, chromosomal location, and tissue distribution. *J. Biol. Chem.* **272**, 4281–4286.

Pineiro-Sanchez, M. L., Goldstein, L. A., Dodt, J., Howard, L., Yeh, Y., and Chen, W. T. (1997). Identification of the 170-kDa melanoma membrane-bound gelatinase (seprase) as a serine integral membrane protease. *J. Biol. Chem.* **272**, 7595–7601.

Puente, X. S., Pendás, A. M., Llano, E., Velasco, G., and López-Otín, C. (1996). Molecular cloning of a novel membrane-type matrix metalloproteinase from a human breast carcinoma. *Cancer Res.* **56**, 944–949.

Reinemer, P., Grams, F., Humber, R., Kleine, T., Schnierer, S., Piper, M., Tschesche, H., and Bode, W. (1994). Structural implications for the role of the N terminus in the 'superactivation' of collagenases. *FEBS Lett.* **338,** 227–233.

Rice, A., and Banda, M. J. (1995). Neutrophil elastase processing of gelatinase A is mediated by extracellular matrix. *Biochemistry* **34,** 9249–9256.

Saksela, O., and Rifkin, D. B. (1988). Cell-associated plasminogen activation: Regulation and physiological functions. *Ann. Rev. Cell Biol.* **4,** 93–126.

Sato, H., Takino, T., Okada, Y., Cao, J., Shinagawa, A., Yamamoto, E., and Seiki, M. (1994). A matrix metalloproteinase expressed on the surface of invasive tumor cells. *Nature* **370,** 61–65.

Sato, H., Kinoshita, T., Takino, T., Nakayama, K., and Seiki, M. (1996a). Activation of a recombinant membrane type 1-matrix metalloproteinase (MT1-MMP) by furin and its interaction with tissue inhibitor of metalloproteinases (TIMP)-2. *FEBS Lett.* **393,** 101–104.

Sato, H., Takino, T., Kinoshita, T., Imai, K., Okada, Y., Stevenson, W. G. S., and Seiki, M. (1996b). Cell surface binding and activation of gelatinase A induced by expression of membrane-type-1-matrix metalloproteinase (MT1-MMP). *FEBS Lett.* **385,** 238–240.

Smeekens, S. P. (1993). Processing of protein precursors by a novel family of subtilisin-related mammalian endoproteases. *Bio/Technology* **11,** 182–186.

Springman, E. B., Angleton, E. L., Birkedal-Hansen, H., and Van Wart, H. E. (1990). Multiple modes of activation of latent fibrolast collagenase: Evidence for the role of a Cys^{73} active-size zinc complex in latency and a "cysteine switch" mechanism for activation. *Proc. Natl. Acad. Sci. USA* **87,** 364–368.

Strongin, A. Y., Marmer, B. L., Grant, G. A., and Goldberg, G. I. (1993). Plasma membrane-dependent activation of the 72-kDa type IV collagenase is prevented by complex formation with TIMP-2. *J. Biol. Chem.* **268,** 14033–14039.

Strongin, A. Y., Collier, I., Bannikov, G., Marmer, B. L., Grant, G. A., and Goldberg, G. I. (1995). Mechanism of cell surface activation of 72-kDa type IV collagenase. Isolation of the activated form of the membrane metalloprotease. *J. Biol. Chem.* **270,** 5331–5338.

Suzuki, K., Enghild, J. J., Morodomi, T., Salvesen, G., and Nagase, H. (1990). Mechanisms of activation of tissue procollagenase by matrix metalloproteinase 3 (stromelysin). *Biochemistry* **29,** 10261–10270.

Takino, T., Sato, H., Shinagawa, A., and Seiki, M. (1995). Identification of the second membrane-type matrix metalloproteinase (MT-MMP-2) gene from a human placenta cDNA library—MT-MMPs form a unique membrane-type subclass in the MMP family. *J. Biol. Chem.* **270,** 23013–23020.

Taylor, N., Shennan, K. I. J., Cutler, D. F., and Docherty, K. (1997). Mutations within the propeptide, the primary cleavage site or the catalytic site, or deletion of C-terminal sequences, prevents secretion of proPC2 from transfected COS-7 cells. *Biochem. J.* **321,** 367–373.

Thant, A. A., Serbulea, M., Kikkawa, F., Liu, E., Tomoda, Y., and Hamaguchi, M. (1979). c-Ras is required for the activation of the matrix metalloproteinases by concanavalin A in 3Y1 cells. *FEBS Lett.* **406,** 28–30.

Tomasek, J. J., Halliday, N. L., Updike, D. L., Ahern-Moore, J. S., Vu, T. K. H., Liu, R. W., and Howard, E. W. (1997). Gelatinase A activation is regulated by the organization of the polymerized actin cytoskeleton. *J. Biol. Chem.* **272,** 7482–7487.

Ward, R. V., Atkinson, S. J., Slocombe, P. M., Docherty, A. J. P., Reynolds, J. J., and Murphy, G. (1991). Tissue inhibitor of metalloproteinases-2 inhibits the activation of 72 kDa progelatinase by fibroblast membranes. *Biochim. Biophys. Acta* **1079,** 242–246.

Ward, R. V., Atkinson, S. J., Reynolds, J. J., and Murphy, G. (1994). Cell surface-mediated activation of progelatinase A: Demonstration of the involvement of the C-terminal domain of progelatinase A in cell surface binding and activation of progelatinase A by primary fibroblasts. *Biochem. J.* **304,** 263–269.

Will, H., and Hinzmann, B. (1995). cDNA sequence and mRNA tissue distribution of a novel human matrix metalloproteinase with a potential transmembrane segment. *Eur. J. Biochem.* **231,** 602–608.

Will, H., Atkinson, S. J., Butler, G. S., Smith, B., and Murphy, G. (1996). The soluble catalytic domain of membrane type 1 matrix metalloproteinase cleaves the propeptide of progelatinase A and initiates autoproteolytic activation—regulation by TIMP-2 and TIMP-3. *J. Biol. Chem.* **271,** 17119–17123.

Wolfsberg, T. G., Straight, P. D., Gerena, R. L., Huovila, A.-P. J., Primakoff, P., Myles, D. G., and White, J. M. (1995). ADAM, a widely distributed and developmentally regulated gene family encoding membrane proteins with A Disintegrin And Metalloprotease domain. *Dev. Biol.* **169,** 378–383.

Yang, M. Z., Hayashi, K., Hayashi, M., Fujii, J. T., and Kurkinen, M. (1996). Cloning and developmental expression of a membrane-type matrix metalloproteinase from chicken *J. Biol. Chem.* **271,** 25548–25554.

Substrate Specificity and Mechanisms of Substrate Recognition of the Matrix Metalloproteinases

Vera Imper* and Harold E. Van Wart†

*Faculty of Pharmacy and Biochemistry, University of Zagreb, 10000 Zagreb, Croatia; and †Inflammatory Diseases Unit, S3-1, Roche Bioscience, Palo Alto, California 94304

I. INTRODUCTION

The significance of the action of any enzyme is generally related to the substrate(s) that it acts on and the physiological consequences of these actions. Accordingly, the matrix metalloproteinase (MMP) family has been recognized as a class of enzymes that plays a critical role in extracellular matrix (ECM) turnover and remodeling based on the ability of individual members to hydrolyze the major protein components of the ECM (Birkedal-Hansen *et al.*, 1993). It has become apparent, however, that the MMP family members have a broader substrate specificity than was originally recognized. Thus, detailed studies of these enzymes have shown they they also act on non-ECM substrates and, consequently, play a more diverse role in physiology than originally believed. For example, MMPs can activate other MMP zymogens (proMMPs) by hydrolyzing propeptide bonds, modulate the activities of other proteinases by degrading their macromolecular inhibitors, activate growth factors by degrading their neutralizing binding proteins, and affect a variety of cell functions by destroying cell–matrix binding interactions.

The goal of this article is to present an overview of the substrate specificity of the MMPs. Whenever possible, the potential physiological consequences of MMP-mediated proteolysis are considered. In approaching this topic, substrate specificity is considered at several levels. A necessary but not sufficient criterion for hydrolysis is that the substrate contain a hydrolyzable bond that conforms to the *sequence specificity* of the MMP. This is defined as the specificity of the MMP toward short peptides that do not adopt specific secondary or tertiary structures. On binding to MMPs, these peptides adapt their conformations to conform to the structure of the enzyme surface. The degree of complementarity between the subsites on the peptide substrate (P_n and P_n') and the enzyme (S_n and S_n'), as defined by both steric and energetic criteria, determines the free energy of binding. Good substrates are those that are able to utilize this binding energy to lower the activation energy for peptide bond hydrolysis. The kinetic parameter that quantitates specificity is k_{cat}/K_M. For the purposes of general discussion, sequence specificity can be considered to be a description of which residues the enzyme "prefers" in each subsite.

When considering substrate specificity toward a protein substrate, the effect of conformation of potential scissile bonds must be superimposed on sequence specificity. Thus, not all sequences that are hydrolyzed in unordered peptides are hydrolyzed when present in proteins either because they are buried in the protein interior, or are accessible but presented in a conformation that is not suitable for binding in the active site of the enzyme. For example, the tightly wound triple helical conformation of collagens protects many potentially cleavable sites from hydrolysis by proteases. In certain cases, the reverse can be true. For example, the k_{cat}/K_M value for the hydrolysis of the bait region of α_2-macroglobulin is considerably larger than that for a peptide with the identical sequence. In this case, it is likely that the protein presents that target sequence in a conformation that is more favorable for binding than that of the unordered peptide with the matching sequence.

Sequence specificity, whether in the context of short peptides or of target sequences in proteins, is a concept that is restricted to interactions of the substrate with the active site. This is defined as that region of the surface of the enzyme that binds the residues of the substrate that are proximal to the scissile bond and contains the catalytic machinery used to affect bond hydrolysis. A second feature that can affect substrate specificity and recognition involves the interaction of substrates with *exosites* on the enzyme. These are sites distant from the active site that recognize features of the substrate distant from the scissile bond. A well-known example of a proteinase that contains exosites is thrombin. The evidence for exosites in the MMPs is considered later. The diversity in the domain structures of the MMPs coupled with their different protein substrate specificities suggests that they may contain substrate-specific exosites.

At the time this article was written there were 14 members of the human MMP family in the published literature. The specificities of all of these will be considered, although little is known about the more recently discovered members. These enzymes are defined here and will subsequently be referred to only by abbreviation: Collagenase-1 (fibroblast collagenase, MMP-1 or Col-1), collagenase-2 (neutrophil collagenase, MMP-8 or Col-2), collagenase-3 (MMP-13 or Col-3), stromelysin-1 (proteoglycanase, MMP-3 or Str-1), stromelysin-2 (MMP-10 or Str-2), stromelysin-3 (MMP-11 or Str-3), gelatinase A (72-kDa gelatinase, MMP-2 or Gel A), gelatinase B (92-kDa gelatinase, MMP-9 or Gel B), matrilysin (PUMP-1, MMP-7 or Mat), macrophage metalloelastase (MMP-12 or MME), and four membrane-type MMPs (MMP-14, -15, -16, and -17 referred to as MT1-MMP, MT2-MMP, MT3-MMP, and MT4-MMP, respectively).

II.   PEPTIDE SEQUENCE SPECIFICITY

A.   Active Site Architecture of MMPs Revealed from Crystal Structures

The crystal structures of the catalytic domains of Col-1 (Borkakoti *et al.*, 1994; Lovejoy *et al.*, 1994a, 1994b; Spurlino *et al.*, 1994), Col-2 (Bode *et al.*, 1994; Stams *et al.*, 1994; Grams *et al.*, 1995), Col-3 (Browner *et al.*, in preparation), Str-1 (Becker *et al.*, 1995; Dhanaraj *et al.*, 1996), and Mat (Browner *et al.*, 1995; Welch *et al.*, in preparation) have been solved. The overall secondary and tertiary structures of these enzymes are well conserved. This allows homology models of the catalytic domains of the other MMPs to be constructed. Based on the solved structures and homology models, several generalizations about the architecture of the active sites of the different MMPs can be drawn. In particular, it is possible to provide a general description of the S_n and S_n' subsites on the enzymes.

Insights into substrate binding to the S_n side of the active site come from the structures of Pro–Leu–Gly–NHOH bound to Col-2 (Bode *et al.*, 1994) and of phthalyl–(Gly–PO$_2$H)–(CH$_2$–DL–Leu)–Trp–NH–benzyl bound to Mat (Welch *et al.*, in preparation). Both of these inhibitors have Gly in subsite P_1. Both crystal structures show the absence of a well-defined S_1 subsite on the enzyme, accounting for the inability of the MMPs to bind substrates with large side chains in this subsite (see later discussion). The Col-2 structure contains an inhibitor with Leu and Pro in subsites P_2 and P_3, respectively. The P_2 Leu residue nestles into a shallow groove lined by His-201, Ala-206, and His-207 (this numbering of residues assumes that residue 79 is the first residue of the catalytic domain). The P_3 Pro residue binds in a small cleft defined by the side chains of His-162, Phe-164, and Ser-151. In general, the S_n side of the active sites of MMPs consists of poorly defined subsites.

Much more data are available concerning the S_n' side of the active sites of the MMPs. The S_1' subsite is the most well-defined subsite in these enzymes and consists of a hydrophobic pocket of variable depth. The amino acid residues that contribute side chains to this pocket are at positions 214, 215, 218, 238, and 240 (Welch *et al.*, in preparation; this numbering assumes that residue 100 is the first residue of the catalytic domain). The residues present at these sites in the 14 human MMPs are shown in Fig. 1. Two types of S_1' architecture can be distinguished from the available X-ray structures. The first consists of a shallower pocket whose bottom is defined by the side chain of residue 214. Col-1 has an Arg residue in this position and falls into this category.

The second type of S_1' architecture consists of a deep hydrophobic pocket in which the side chain of residue 214 points away from the P_1'

```
Mat          ...LYAATHEL........MYPTYGN...
Col-1        ...HRVAAHEL........MYPSYTF...
Str-3        ...LQVAAHEF........MSAFYTF...
MME          ...FLTAVHEI........MFPTYKY...
Str-1        ...FLVAAHEI........MYPLYHS...
Str-2        ...FLVAAHEL........MYPLYNS...
Col-2        ...FLVAAHEF........MYPNYAF...
Col-3        ...FLVAAHEF........MFPIYTY...
Gel A        ...FLVAAHEF........MAPIYTY...
Gel B        ...FLVAAHEF........MYPMYRF...
MT-1 MMP     ...FLVAVHEL........MAPFYQW...
MT-2 MMP     ...FLVAVHEL........MAPFYQW...
MT-3 MMP     ...FLVAVHEL........MAPFYQY...
MT-4 MMP     ...FLVAVHEF........MRPYYGG...
                   |              |
                  213            236
```

FIG. 1. Amino acid residues (bold typeface) contributing side chains to the P_1' subsite of Mat and, by analogy, to the other 13 human MMPs.

side chain of bound inhibitors. Str-1, Col-2, and Col-3 all have Leu at position 214 and are examples of this class. Mat has a Tyr residue at position 214 that would be expected to endow it with the shallower pocket shape. However, it constitutes an exception to this classification scheme because its S_1' subsite structure can change in response to inhibitor binding. Inhibitors with large S_1' side chains induce movement of the surface loop consisting of residues 240–249 that stabilize the S_1' pocket conformation, resulting in the deep pocket architecture. It is not only the depth of the S_1' pocket that influences specificity, but also subtle interactions between pocket residues with portions of the inhibitor. The S_1' pocket architecture of the other 10 MMPs is predicted to be of the deep pocket type.

Subsites S_2' and S_3' are also poorly defined. They are flat recesses that bind the substrate in a sheet conformation via four hydrogen bonds. The P_2' side chain of inhibitors often points away from the enzyme, while the P_3' residues interact adventitiously with residues in a shallow pocket.

B. Specificity toward Peptide Substrates

The specificities of a number of the MMPs toward peptide substrates have been examined in several studies. The most systematic data come from three studies in which the k_{cat}/K_M values for the hydrolysis of a series of 60 oligopeptides containing different amino acids in the P_4 through P_4' positions by Col-1 and Col-2 (Netzel-Arnett et al., 1991a), Mat, Gel A and Gel B (Netzel-Arnett et al., 1993), and Str-1 (Netzel-Arnett, Fields, Nagase, and Van Wart, unpublished data) were quantitated. These data are summarized in Tables I through VII. The reference peptide for these studies is Gly–Pro–Gln–Gly–Ile–Ala–Gly–Gln, which contains the four amino acid residues surrounding the scissile bond in the collagenase cleavage site in the $\alpha 1(I)$-chain of collagen. The k_{cat}/K_M value for the hydrolysis of this peptide by each MMP is assigned the value of 100%. The other peptides listed differ from this peptide either by being truncated from the N- or C-terminal direction (Table I) or by having a single amino acid substitution in one of the P_3 through P_4' positions (Tables II, III, IV, VI, and VII). A separate set of peptides has been used to study the P_1' specificity of MME and MT1-MMP (Table V).

Successive N- and C-terminal truncation of the reference peptide shows that the rate of hydrolysis drops off sharply for all six MMPs if the peptide is shortened beyond the central P_3–P_3' hexapeptide (Table I). Only a few substitutions at subsite P_3 have been investigated. Replacement of Pro in this position by Ala, Leu, or Asn leads to a worse

TABLE I

EFFECT OF C- AND N-TERMINAL ELONGATION ON THE HYDROLYSIS OF PEPTIDES BY SIX
HUMAN MMPs

Peptide	Activity (%)					
	Col-1	Col-2	Str-1	Mat	Gel A	Gel B
↓						
Gly–Pro–Gln–Gly–Ile–Ala–Gly–Gln	100	100	100	100	100	100
Gly–Pro–Gln–Gly–Ile–Ala–Gly	68	54	64	99	60	93
Gly–Pro–Gln–Gly–Ile–Ala	13	<5.0	34	10	<5.0	<5.0
Gly–Pro–Gln–Gly–Ile	<5.0	<5.0	<5.0	<5.0	<5.0	<5.0
Pro–Gln–Gly–Ile–Ala–Gly–Gln	150	100	30	43	62	65
Gln–Gly–Ile–Ala–Gly–Gln	7.3	<5.0	26	8.3	<5.0	<5.0
Gly–Ile–Ala–Gly–Gln	<5.0	<5.0	<5.0	<5.0	<5.0	<5.0

substrate for all of the MMPs for which data have been collected (Table II). Replacement of Gln in position P_2 by the hydrophobic residues Leu, Met, and Tyr increases the rate of hydrolysis, whereas Hyp and Asp lower the rates for all five MMPs (Table II). Arg and Val produce mixed effects and are preferred by Str-1, Gel A, and Gel B.

The effect of substitutions at subsite P_1 varies markedly for the six MMPs (Table III). Of the 11 amino acid substitutions examined for Gly, all with the exception of Ala decrease the hydrolysis rates for the two gelatinases. Thus, Gel A and Gel B tolerate only small amino acids

TABLE II

P_3 AND P_2 SUBSITE SPECIFICITIES OF SIX HUMAN MMPs

Peptide	Activity (%)					
	Col-1	Col-2	Str-1	Mat	Gel A	Gel B
↓						
Gly–Pro–Gln–Gly–Ile–Ala–Gly–Gln	100	100	100	100	100	100
Gly–Ala–Gln–Gly–Ile–Ala–Gly–Gln	50	23	62	—	22	9.4
Gly–Leu–Gln–Gly–Ile–Ala–Gly–Gln	14	27	75	—	10	—
Gly–Asn–Gln–Gly–Ile–Ala–Gly–Gln	17	46	68	25	60	<5.0
Gly–Pro–Leu–Gly–Ile–Ala–Gly–Gln	150	270	190	420	330	290
Gly–Pro–Hyp–Gly–Ile–Ala–Gly–Gln	11	15	83	17	32	15
Gly–Pro–Arg–Gly–Ile–Ala–Gly–Gln	17	31	96	13	160	83
Gly–Pro–Asp–Gly–Ile–Ala–Gly–Gln	30	44	89	7.0	11	10
Gly–Pro–Val–Gly–Ile–Ala–Gly–Gln	32	29	160	57	130	105
Gly–Pro–Met–Gly–Ile–Ala–Gly–Gln	160	160	120	400	120	180
Gly–Pro–Tyr–Gly–Ile–Ala–Gly–Gln	110	110	230	240	200	150

TABLE III

P₁ SUBSITE SPECIFICITY OF SIX HUMAN MMPs

	Activity (%)					
Peptide	Col-1	Col-2	Str-1	Mat	Gel A	Gel B
↓						
Gly–Pro–Gln–Gly–Ile–Ala–Gly–Gln	100	100	100	100	100	100
Gly–Pro–Gln–Met–Ile–Ala–Gly–Gln	200	140	110	150	22	12
Gly–Pro–Gln–Glu–Ile–Ala–Gly–Gln	28	330	190	170	15	29
Gly–Pro–Gln–Tyr–Ile–Ala–Gly–Gln	130	180	68	34	58	30
Gly–Pro–Gln–Ala–Ile–Ala–Gly–Gln	640	330	300	530	96	110
Gly–Pro–Gln–Pro–Ile–Ala–Gly–Gln	260	190	170	140	32	46
Gly–Pro–Gln–Gln–Ile–Ala–Gly–Gln	37	150	140	180	25	13
Gly–Pro–Gln–Phe–Ile–Ala–Gly–Gln	95	170	68	63	15	26
Gly–Pro–Gln–Leu–Ile–Ala–Gly–Gln	27	54	170	49	21	8.8
Gly–Pro–Gln–Val–Ile–Ala–Gly–Gln	5.5	7.9	32	<5.0	<5.0	<5.0
Gly–Pro–Gln–Arg–Ile–Ala–Gly–Gln	<5.0	27	<5.0	—	38	—
Gly–Pro–Gln–His–Ile–Ala–Gly–Gln	160	50	87	—	65	44

at this position. The response of the other four MMPs is more varied with some substitutions increasing and others decreasing the hydrolysis rates. Substitution of Met, Ala, and Pro for Gly increases the rates for all four enzymes, whereas Arg and Val produce uniform decreases. The most preferred residue in this position is Ala. The other substitutions produce mixed changes for the different MMPs. The fact that residues with markedly different sizes and polarities (Glu, Tyr, Ala, Pro, Gln, Leu, and His) can produce increases in rates for individual enzymes indicates that these residues can probably make a variety of different types of interactions with the S_1 subsite of these MMPs.

Replacement of the Ile residue in subsite P_1' by 11 different positively or negatively charged, aliphatic or aromatic hydrophobic amino acid residues generally produces reductions in rate (Table IV). Pro, Glu, Val, Gln, Ser, and Arg markedly lower the rates for all six MMPs, whereas Phe and Trp cause reductions for all of these enzymes except Mat. Substitutions that produce mixed changes for these MMPs include Leu, Tyr, and Met. The two special preferences that are apparent are that of Col-2 for Tyr and of Mat for Leu. The ability of Str-1 to tolerate Trp and Phe in this position is also noteworthy. The basis for the preferences of Str-1 for aromatic residues and the preference of Mat for Leu in subsite P_1' have been investigated in detail by producing engineered enzymes in which the residues in the S_1' pocket were mutated (Welch et al., 1996). These studies clearly showed that the identity

TABLE IV

P₁' SUBSITE SPECIFICITY OF SIX HUMAN MMPS

	Activity (%)					
Peptide	Col-1	Col-2	Str-1	Mat	Gel A	Gel B
Gly–Pro–Gln–Gly–Ile–Ala–Gly–Gln ↓	100	100	100	100	100	100
Gly–Pro–Gln–Gly–Trp–Ala–Gly–Gln	<5.0	48	120	<5.0	<5.0	<5.0
Gly–Pro–Gln–Gly–Pro–Ala–Gly–Gln	<5.0	<5.0	<5.0	<5.0	<5.0	<5.0
Gly–Pro–Gln–Gly–Glu–Ala–Gly–Gln	<5.0	<5.0	<5.0	8.0	<5.0	<5.0
Gly–Pro–Gln–Gly–Leu–Ala–Gly–Gln	130	180	110	430	88	80
Gly–Pro–Gln–Gly–Tyr–Ala–Gly–Gln	45	480	150	21	50	96
Gly–Pro–Gln–Gly–Phe–Ala–Gly–Gln	<5.0	46	140	24	55	24
Gly–Pro–Gln–Gly–Met–Ala–Gly–Gln	110	100	60	89	230	180
Gly–Pro–Gln–Gly–Val–Ala–Gly–Gln	9.1	9.0	53	17	30	25
Gly–Pro–Gln–Gly–Gln–Ala–Gly–Gln	28	10	38	<5.0	34	20
Gly–Pro–Gln–Gly–Ser–Ala–Gly–Gln	5.9	<5.0	45	5.5	15	<5.0
Gly–Pro–Gln–Gly–Arg–Ala–Gly–Gln	<5.0	<5.0	<5.0	<5.0	<5.0	<5.0

of the residue in position 214 plays a dominant role in shaping the substrate specificity of these enzymes.

A separate set of fluorogenic peptides has been used to examine the P₁' specificity of MME (Gronski *et al.*, 1997) and MT1-MMP (Imper *et al.*, unpublished data) (Table V). The results for MME show that substitution of seven different residues for Leu all produce lower rates. The next best residues are Phe and Lys. For MT1-MMP, the aromatic

TABLE V

P₁' SUBSITE SPECIFICITY OF HUMAN MME AND MT1-MMP

	Activity (%)	
Peptide	MME	MT1-MMP
Dnp–Arg–Pro–Leu–Ala–Leu–Trp–Arg–Ser–NH₂ ↓	100	100
Dnp–Arg–Pro–Leu–Ala–Arg–Trp–Arg–Ser–NH₂	7.8	<5.0
Dnp–Arg–Pro–Leu–Ala–Gln–Trp–Arg–Ser–NH₂	0.78	<5.0[a]
Dnp–Arg–Pro–Leu–Ala–Lys–Trp–Arg–Ser–NH₂	25	10
Dnp–Arg–Pro–Leu–Ala–Phe–Trp–Arg–Ser–NH₂	25	3.9
Dnp–Arg–Pro–Leu–Ala–Ser–Trp–Arg–Ser–NH₂	5.1	11[b]
Dnp–Arg–Pro–Leu–Ala–Trp–Trp–Arg–Ser–NH₂	13	85
Dnp–Arg–Pro–Leu–Ala–Tyr–Trp–Arg–Ser–NH₂	14	110

[a] Peptide was hydrolyzed only at the Gln–Trp bond.
[b] Peptide was also hydrolyzed at the Ser–Trp bone at 2/3 of this rate.

residues Tyr and Trp are approximately as good as Leu, whereas the other substitutions all produce significant decreases in rate. Interestingly, replacement of Leu in this subsite by Gln causes MT1-MMP to hydrolyze exclusively the Gln–Trp bond instead of the Ala–Gln bond. In a similar fashion, replacement of Leu by Ser results in hydrolysis at both the Ala–Ser and Ser–Trp bonds.

The P_2' specificity of six MMPs is shown in Table VI. In contrast to subsite P_1', four of the six substitutions produce faster rates of hydrolysis for all six enzymes. In spite of the fact that X-ray structures show that the P_2' amino acid side chain points away from the enzyme surface, substitutions at this position produce the largest increases in rates at any subsite. For example, substitution of Ala by Trp increases specificity by 8.3-fold for Col-2. It is interesting that the MMPs like the bulky hydrophobic Phe, Trp, and Leu residues at this site, but also like Arg. Hyp and Glu are generally not well tolerated.

A limited amount of information on the P_3' and P_4' specificities of the MMPs is shown in Table VII. Replacement of Gly in this subsite by Ala or Ser increases the hydrolysis rates for all six MMPs. The other three substitutions (Val, Arg, and Met) are favorable for Str-1 and Mat, but detrimental for the other four enzymes. The marked preferences of Str-1 for Met and Arg, and of Mat for Met in this position, are noteworthy. Substitution of the Gln residue in subsite P_4' by four different residues generally has a small effect on the hydrolysis rates, suggesting that this position on the substrate is not interacting with the enzyme.

Other studies of the action of MMPs on synthetic peptides have been reported including studies of Col-1 (Weingarten *et al.*, 1985), Gel A (Seltzer *et al.*, 1989, 1990), and Str-1 (Niedzwiecki *et al.*, 1992). The

TABLE VI
P_2' Subsite Specificity of Six Human MMPs

Peptide	Activity (%)					
	Col-1	Col-2	Str-1	Mat	Gel A	Gel B
Gly–Pro–Gln–Gly–Ile–Ala–Gly–Gln	100	100	100	100	100	100
Gly–Pro–Gln–Gly–Ile–Phe–Gly–Gln	430	310	130	140	310	390
Gly–Pro–Gln–Gly–Ile–Trp–Gly–Gln	730	830	280	330	330	240
Gly–Pro–Gln–Gly–Ile–Leu–Gly–Gln	210	400	280	250	400	240
Gly–Pro–Gln–Gly–Ile–Hyp–Gly–Gln	<5.0	<5.0	42	8.0	32	11
Gly–Pro–Gln–Gly–Ile–Arg–Gly–Gln	180	170	250	270	180	200
Gly–Pro–Gln–Gly–Ile–Glu–Gly–Gln	35	58	58	86	85	130

TABLE VII

P₃' AND P₄' SUBSITE SPECIFICITIES OF SIX HUMAN MMPs

TABLE VII

P_3' AND P_4' SUBSITE SPECIFICITIES OF SIX HUMAN MMPs

Peptide	Activity (%)					
	Col-1	Col-2	Str-1	Mat	Gel A	Gel B
Gly–Pro–Gln–Gly–Ile–Ala–Gly–Gln (↓)	100	100	100	100	100	100
Gly–Pro–Gln–Gly–Ile–Ala–Val–Gln	77	56	170	170	26	49
Gly–Pro–Gln–Gly–Ile–Ala–A̅r̅g̅–Gln	55	33	490	220	35	45
Gly–Pro–Gln–Gly–Ile–Ala–M̅e̅t̅–Gln	41	33	810	450	40	35
Gly–Pro–Gln–Gly–Ile–Ala–A̅l̅a̅–Gln	220	120	280	300	180	140
Gly–Pro–Gln–Gly–Ile–Ala–S̅e̅r̅–Gln	130	130	230	150	350	130
Gly–Pro–Gln–Gly–Ile–Ala–G̅l̅y̅–Thr	160	140	72	100	59	160
Gly–Pro–Gln–Gly–Ile–Ala–Gly–H̅i̅s̅	91	140	110	87	150	120
Gly–Pro–Gln–Gly–Ile–Ala–Gly–A̅l̅a̅	86	110	130	91	85	110
Gly–Pro–Gln–Gly–Ile–Ala–Gly–P̅r̅o̅	77	48	98	—	—	—
Gly–Pro–Gln–Gly–Ile–Ala–Gly–p̅N̅A̅	130	110	130	70	54	51

specificities of Str-1 and Mat have also been studied using a bacterio-phage peptide display library (Smith *et al.*, 1995). The results of these studies are generally in good agreement with the data presented in Tables I through VII. The results of these sequence specificity studies have been used to design optimized chromogenic peptide substrates for *in vitro* assays (Teahan *et al.*, 1989; Netzel-Arnett *et al.*, 1991b; Nagase *et al.*, 1994; Smith *et al.*, 1995).

III. PROTEIN SUBSTRATE SPECIFICITY

In the next sections, attention is turned to the protein substrates of the MMPs. To facilitate discussion, the known protein substrates of each MMP are tabulated in Table VIII. The information in this table has been taken from Sang and Douglas (1996), but has been updated with the additional citations shown at the bottom. This table does not include proMMP substrates because these are considered separately in a later section. In many cases, the locations of the MMP cleavages in these proteins are known. However, no attempt has been made here to catalog these. The reader is referred elsewhere for such information (Birkedal-Hansen *et al.*, 1993). In all cases where the cleavage sequence has been identified, it has followed from the sequence specificty of the MMP revealed from the peptide sequence studies discussed earlier.

A. *Extracellular Matrix Substrates*

The MMP family has gained its current notoriety by virtue of its collective ability to hydrolyze all of the major protein components of

TABLE VIII
PROTEIN SUBSTRATES OF MMPs

MMP	Substrate
Col-1	Collagens I,[a] II,[a] III,[a] VII,[a] and X,[a] gelatins,[a] aggrecan,[a] entactin,[a] tenascin,[a] serum amyloid-A,[b] insulin-like growth factor binding protein-3[c]
Col-2	Collagens, I,[a] II,[a] and III,[a] gelatins,[a] fibronectin,[a] laminin,[a] aggrecan,[a] elastin,[a] α_1-proteinase inhibitor,[d] α_1-antichymotrypsin[d]
Col-3	Collagens I,[e] II,[e] and III,[e] gelatins,[e] α_1-antichymotrypsin,[e] plasminogen activator inhibitor[e]
Gel A	Collagens I,[a] IV,[a] V,[a] VII,[a] and X,[a] gelatins,[a] fibronectin,[a] laminin,[a] aggrecan,[a] elastin,[a] galectin-3,[f] insulin-like growth factor binding protein-3[c]
Gel B	Collagens IV[a] and V,[a] gelatins,[a] elastin,[a] entactin,[a] aggrecan,[a] α_1-proteinase inhibitor,[d] galectin-3,[f] angiostatin[g]
Str-1	Collagens II,[a] IV,[a] IX,[a] X,[a] and XI,[a] gelatins,[a] laminin,[a] fibronectin,[a] elastin,[a] tenascin,[a] aggrecan,[a] serum amyloid-A,[b] insulin-like growth factor binding protein-3[c]
Str-2	Collagen IV,[a] laminin,[a] fibronectin,[a] elastin,[a] aggrecan[a]
Str-3	Fibronectin,[a] laminin,[a] aggrecan,[a] α_1-proteinase inhibitor[h]
Mat	Collagen IV,[a] gelatins,[a] laminin,[a] fibronectin,[a] entactin,[a] elastin,[a] aggrecan,[a] tenascin,[a] angiostatin[g]
MME	Collagen IV,[i] elastin,[i] fibronectin,[i] laminin,[i] entactin,[i] chondroitin sulfate,[i] heparin sulfate,[i] α_1-proteinase inhibitor,[i] α_1-antichymotrypsin[j]
MT1-MMP	Collagens I,[k] II,[k] and III,[k] gelatins,[k] proteoglycan,[k] fibronectin,[k,l] vitronectin,[k,l] laminin,[k,l] α_1-proteinase inhibitor,[k] dermatin sulfate,[l] elastin[l]

[a] Referenced in Sang and Douglas, 1996. [b] Mitchell et al., 1993. [c] Fowlkes et al., 1994. [d] Desrochers et al., 1992. [e] Knauper et al., 1996. [f] Ochieng et al., 1994. [g] Patterson and Sang, in press. [h] Pei et al., 1994. [i] Gronski et al., 1997. [j] Dong et al., 1997. [k] Ohuchi et al., 1997. [l] Pei & Weiss, 1996.

the ECM. This includes many of the various collagen types, elastin, proteoglycans, and several matrix glycoproteins such as laminin, fibronectin, tenascin, etc. These activities are believed to endow the MMPs with the ability to catabolize the ECM in both normal and pathological biological processes.

An important question related to the substrate specificity of the MMPs is whether certain members of the family have evolved to specifically hydrolyze certain ECM substrates. This question is perhaps most relevant to the collagenases that degrade the interstitial collagens (types I, II, and III), which comprise the majority of the matrix protein in most tissues and which are known to be particularly resistant to proteolysis. Indeed, the ability of Col-1, Col-2, and Col-3 to hydrolyze these collagens within the triple helical domain at an appreciable rate has led not only to their being named collagenases, but also to the view that these enzymes are the ones responsible for collagen turnover *in vivo*.

Another interesting question is whether the different collagenases have specific roles in the turnover of the three types of interstitial collagens. It is known that the three collagenases have different preferences for the three types of interstitial collagens (Welgus *et al.*, 1981; Hasty *et al.*, 1987; Mallya *et al.*, 1990; Knauper *et al.*, 1996; Mitchell *et al.*, 1996). A role for Col-1 in the turnover of types I and III collagens in soft tissues seems reasonable, since this enzyme is made by resident fibroblasts and is quite active against both of these collagens. In contrast, Col-1 has minimal activity against type II collagen, which is found in cartilage, intravertebral disk, and vitreous humor. Interestingly, Col-3 is an excellent type II collagenase and is made by chondrocytes (Mitchell *et al.*, 1996), suggesting that Col-3 is the enzyme that is responsible for the degradation of the type II collagen matrix in cartilage. The ability of Col-3 selective collagenase inhibitors to block this process in live cartilage samples lends support to this view (Billinghurst *et al.*, 1997; Eugui *et al.*, in press).

Although a role for these three collagenases in collagen turnover has become widely accepted, a number of observations suggest that the picture may be more complicated. Certain MMPs such as Str-1 (Wu *et al.*, 1991) can hydrolyze the interstitial collagens in their telopeptide regions, an event that could contribute to collagenolysis. However, this is not true collagenase activity and is probably not sufficient in itself to enable the enzyme to catabolize collagen *in vivo*. Recently, the TIMP-free form of Gel A has been reported to be able to cleave soluble type I collagen into characteristic TC^A and TC^B fragments at a rate similar to that of Col-1 (Aimes and Quigley, 1995). Interestingly, the TIMP-free form of Gel B was found to be unable to perform this reaction under the same conditions. Other workers have found that the TIMP-free forms of both Gel A and Gel B are interstitial collagenases (Kleiner, Stetler-Stevenson, and Birkedal-Hansen, personal communication). Most recently, Ohuchi and associates (1997) have shown that MT1-MMP acts as a collagenase toward types I, II, and III collagens. Much more work needs to be carried out to quantitate these activities and to assess whether any of them are physiologically relevant.

Assigning specific roles for MMPs in the degradation of other matrix substrates is even more difficult. Elastin is also an extremely protease-resistant substrate. Accordingly, the ability of MME to hydrolyze this substrate suggests that this is a physiologically relevant activity. Evidence in support of this concept comes from the recent report that

MME-deficient mice do not develop smoking-induced emphysema (Hautamaki et al., 1997), a condition brought about by lung elastin degradation. Gel A and Gel B have long been known to degrade type IV collagen found in basement membranes and it has been suggested that this activity enables cells to traverse these membranes (Stetler-Stevensen et al., 1993). However, many other MMPs also possess this activity. Str-1 has long been known to degrade proteoglycans, an activity that led to the enzyme being referred to as proteoglycanase. Because the enzyme is upregulated in arthritic joints (Okada et al., 1992), a leading role for this enzyme in aggrecan turnover has been postulated. However, the observations that other MMPs found in arthritic cartilage (Col-1, Col-2, Gel A, Gel B) can also degrade aggrecan and the revelation that a distinct non-MMP aggrecanase activity exists (Lohmander et al., 1993) cast doubt on the singular importance of Str-1 in cartilage proteoglycan turnover.

B. Nonextracellular Matrix Protein Substrates

An intriguing possibility is that the MMPs have activities toward non-ECM substrates that are of physiological significance and that these enzymes play a broader role in physiology than simply catabolizing ECM macromolecules. Several of the substrates listed in Table VIII suggest that this may be the case. The ability of Col-2, Col-3, Gel B, Str-3, MT1-MMP, and MME to degrade and inactivate serpins such as α_1-proteinase inhibitor and α_1-antichymotrypsin (Desrochers et al., 1992; Pei et al., 1994; Gronski et al., 1997; Ohuchi et al., 1997) suggests that they may play a role in modulating the activity of serine proteases. This could be of significance in complex inflammatory conditions where both serine proteases and MMPs are simultaneously expressed. The ability of these classes of enzymes to degrade and inactivate each other's natural inhibitors is noteworthy.

Col-1, Gel A, and Str-1 have all been reported to degrade and inactivate insulin-like growth factor binding protein-3 (IGFBP-3) (Fowlkes et al., 1994). This protein binds and keeps latent the potent mitogenic insulin-like growth factors (IGFs) in tissue. The hydrolysis of IGFBP-3 by these MMPs may be a mechanism by which IGFs are released in tissue so that they can interact with cells. Thus, the action of these MMPs may modulate cell growth and proliferation through hydrolysis of this non-ECM substrate.

Angiostatin is a 38-kDa fragment of plasminogen that selectively inhibits the proliferation of endothelial cells and is a potent inhibitor of angiogenesis. In a recent study (Dong et al., 1997), it was shown that the generation of angiostatin in vivo requires the presence of macrophages and is correlated with the up-regulation of MME. This is consistent with the observation that pancreatic elastase can generate angiostatin from plasminogen *in vitro*. Patterson and Sang (1997) have subsequently demonstrated that Mat and Gel B can also generate angiostatin from plasminogen *in vitro*. Thus, these MMPs may be modulating the physiological levels of angiostatin and its subsequent action on endothelial cells.

Col-1 and Str-1 have also been shown to hydrolyze the acute phase reactant serum amyloid A (Mitchell *et al.*, 1993). This protein can serve as a paracrine/autocrine factor that induces Col-1 in fibroblasts. Its hydrolysis by Col-1 could serve as a feedback mechanism to turn off expression during inflammatory conditions. In another study, Gel A and Gel B have been shown to be able to hydrolyze the cell-surface carbohydrate binding protein galectin-3 (Ochieng *et al.*, 1994). This protein is thought to modulate the growth and differentiation of cells and its hydrolysis by the gelatinases has been hypothesized to modulate these biological activities. It seems likely that such roles for the MMPs will continue to be discovered as their actions against more of these non-ECM substrates are investigated.

C. ProMMP Activation

Another potentially important function of MMPs is to activate other MMP zymogens (proMMPs) extracellularly. With the exception of the MT MMPs, all of these enzymes are soluble and are secreted into the extracellular space as zymogens that are kept inactive by their propeptides. The active sites of all of the MMPs contain a catalytic zinc atom that is complexed to three histidine ligands and water. The propeptides of all of the MMPs contain a highly conserved region, which contains a cysteine residue that substitutes for water as the fourth ligand in the zymogen. This converts the zinc from a catalytic zinc in the active enzyme to a structural zinc in the zymogen (Vallee and Auld, 1992). The propeptides of the MMPs make this interaction possible by folding into a conformation that allows the propeptide cysteine residue to ligate to the zinc and by forming noncovalent interactions with the catalytic domain that stabilize this conformation.

The activation of the pro-MMPs requires that the propeptide cysteine residue dissociate from the zinc atom and that the propeptide uncover the active site. This has been referred to as the "cysteine switch" mechanism of activation (Springman *et al.*, 1990; Van Wart and Birkedal-Hansen, 1990). An efficient way to achieve these activating events is by proteolytic removal of sections of the propeptide, a process that destabilizes the conformation of the propeptide, abolishes interactions with the catalytic domain, and leads to dissociation of the cysteine switch residue from the zinc. It is this initial proteolysis that starts the activation process and has been shown in the case of many of the pro-MMPs to be carried out by other active MMPs *in vitro.*

The proteolytic processing of the propeptide during activation often occurs in several steps that shorten the propeptide sequentially. Some of these proteolytic events, particularly the last cleavage, are autolytic. In all cases, the final result is the proteolytic removal of the cysteine switch region and permanent activation of the enzyme. The reader is referred elsewhere for a listing of the locations of the cleavage sites in the propeptides of the various proMMPs (Birkedal-Hansen *et al.,* 1993; Sang *et al.,* 1995, 1996; Sang and Douglas, 1996; Nagase, 1996). As noted earlier for the other non-ECM substrates, the bonds that are hydrolyzed generally conform to the sequence specificities found with peptide substrates.

A question that remains unresolved is the physiological relevance of MMP-mediated proMMP activation. In most cases, these observations have been *in vitro* and are of questionable *in vivo* relevance. A notable exception that deserves special mention is that of the activation of pro-Gel A by MT-MMPs. The discovery of this process began with the observation of a cell-surface-dependent activation of pro-Gel A in fibroblasts (Ward *et al.,* 1991) and tumor cells (Brown *et al.,* 1990; Strongin *et al.,* 1995) on treatment with TPA or concanavilin A. This activation event has been attributed to the action of MT MMPs whose transmembrane domains localize their action to the cell surface. Both MT1-MMP (Sato *et al.,* 1994) and MT3-MMP (Takino *et al.,* 1995) have been shown to catalyze this activation in cell-based systems. Thus, this subclass of the MMP family may have evolved the specific function of activating pro-Gel A, which in turn could participate in activation cascades involving other proMMPs.

IV. Mechanisms of Substrate Recognition

In this section, we consider substrate recognition events that involve domains other than the catalytic domain. The domain structures of the MMPs are shown in Fig. 2. All members of the family possess the propeptide and catalytic domains. However, the members of the family are distinguished by the presence or absence of the hemopexin-like, fibronectin-like, $\alpha 1(V)$-like and transmembrane domains. The functions of the catalytic and propeptide domains were discussed earlier. The roles of the other domains remain a subject of intense interest. In this section, we consider potential substrate recognition roles for some of these domains.

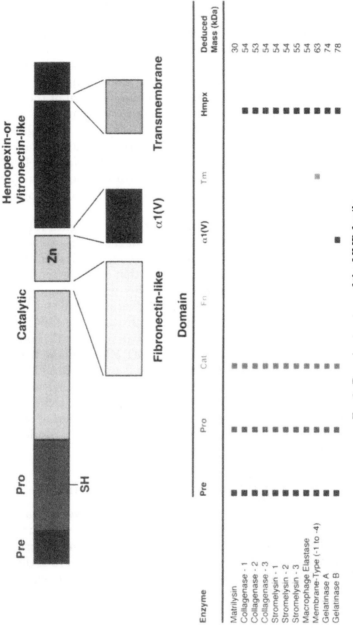

Fig. 2. Domain structure of the MMP family.

The best example of a substrate recognition role for a noncatalytic domain is that of the hemopexin-like domain of the collagenases in the recognition of their collagen substrates. Col-1, Col-2, and Col-3 all cleave the interstitial collagen types I, II, and III by making a single initial scission across all three α-chains at a characteristic locus to produce a 3/4 length N-terminal TC^A fragment and a 1/4 C-terminal TC^B fragment. There are two remarkable features of these reactions. First, the cleavage locus cannot be explained purely on the basis of sequence specificity, because the respective unordered gelatins are hydrolyzed at many other sites at similar rates. Moreover, peptides that mimic the collagenase cleavage site sequence are not hydrolyzed faster than those that mimic other sequences that are not cleaved in the intact collagens (Fields *et al.*, 1987). Thus, some conformational feature of the collagen in the susceptible region contributes to the specificity of these reactions. The second feature is that the catalytic domains of these collagenases are not able to catalyze these reactions (Clark and Cawston, 1989; Schnierer *et al.*, 1993; Knauper *et al.*, 1997). Thus, the hemopexin-like domain plays a role in orchestrating these reactions. Other MMPs that contain the hemopexin-like domain such as Str-1 are not able to hydrolyze these collagens in this manner, indicating that this domain functions differently in these other MMPs. Interestingly, the hemopexin-like domain of Str-1 endows it with the property to bind collagens, but this is not sufficient to make Str-1 a collagenase (Murphy *et al.*, 1992).

Based on these observations, it has been postulated that the collagenases are able to hydrolyze the interstitial collagens in this manner by virtue of a recognition event between the collagen triple helix and an exosite located on the hemopexin-like domains (Netzel-Arnett *et al.*, 1994). Evidence for the existence of such a site comes from experiments in which triple helical model peptides with the repeating [Gly–Pro–Hyp]$_n$ sequences have been shown to inhibit the hydrolysis of type I collagen by Col-1 and Col-2 (Table IX). These peptides do not inhibit by binding to the active site because they have no effect on the hydrolysis of peptide substrates. Moreover, the triple helical peptides do not have any effect on Str-1 or Mat, neither of which hydrolyzes triple helical collagens. Thus, the hemopexin-like domains of the collagenases appear to contain a recognition site for the collagen triple helix and binding of the interstitial collagens to this site is essential for the specific hydrolysis by the collagenases. Other workers have studied the hydrolysis of triple helical peptides containing the collagenase cleavage sequence by full-length Col-2 and its catalytic domain (Ottl *et al.*, 1996). They find that only the full-length enzyme containing the hemopexin-

TABLE IX

EFFECT OF TRIPLE HELICAL PEPTIDES ON THE ACTIVITIES
OF VARIOUS MMPs

MMP	Substrate	Peptide	IC_{50} (mM)
Col-1	Type I collagen	Bridge[a]	>10
		$[(GPHy)_7]_3$–bridge	0.8
		$[(GPHy)_{10}]_3$–bridge	0.1
	DNP-peptide[b]	Bridge[a]	>10
		$[(GPHy)_7]_3$–bridge	>5
		$[(GPHy)_{10}]_3$–bridge	>5
Col-2	Type I collagen	Bridge[a]	>10
		$[(GPHy)_7]_3$–bridge	0.6
		$[(GPHy)_{10}]_3$–bridge	0.08
	DNP-peptide[b]	Bridge[a]	>10
		$[(GPHy)_7]_3$–bridge	>5
		$[(GPHy)_{10}]_3$–bridge	>5
Mat	Casein	$[(GPHy)_{10}]_3$–bridge	>5
	DNP-peptide[b]	$[(GPHy)_{10}]_3$–bridge	>5
Str-1	Casein	$[(GPHy)_{10}]_3$–bridge	>5
	DNP-peptide[b]	$[(GPHy)_{10}]_3$–bridge	>5

[a] The bridge consists of three aminohexanoic acid units linked to the α- and ε-amino groups of Lys–Lys–Gly.

[b] DNP–Pro–Leu–Ala–Leu–Trp–Ala–Arg.

like domain is able to hydrolyze this substrate with kinetic parameters that mimic those for the collagens.

A number of other investigations have been carried out to study the role of the hemopexin-like domain in collagenase activity. A variety of chimeras have been prepared in which different catalytic and hemopexin-like domains have been fused with different linker regions (Murphy et al., 1992; Hirose et al., 1993; Sanchez-Lopez et al., 1993; Sankovic and Van Wart, unpublished data). Hybrids consisting of the catalytic domain of Col-1 and hemopexin-like domain of Str-1 (Murphy et al., 1992) or Str-2 (Sanchez-Lopez et al., 1993), and of the catalytic domain of Col-2 and the hemopexin-like domain of Str-1 (Hirose et al., 1993), all fail to exhibit collagenase activity. In contrast, chimeras in which the catalytic and hemopexin-like domains of Col-1 and Col-2 were exchanged produced species that retained collagenase activity (Sankovic and Van Wart, unpublished data). These experiments indicate that the hemopexin-like domains of the collagenases are unique in endowing collagenase activity to these enzymes and that the location of the splice site and size of the linker region are important. The region in the hemopexin-like domains of the collagenases responsible for this recognition has been traced to a 16-amino-acid region (Hirose *et al.*, 1993).

The X-ray structures of full-length porcine Col-1 (Li *et al.*, 1995) and of the hemopexin-like domains of Col-3 (Gomis-Ruth *et al.*, 1996) and Gel A (Gohlke *et al.*, 1996; Libson *et al.*, 1995) have been elucidated. These studies reveal that this domain is folded into a four-bladed β-propeller structure in which the four blades are arranged around a funnel-like tunnel. It is not immediately clear from these structures how the hemopexin-like domain of the collagenases interacts in conjunction with the catalytic domain to give these enzymes the ability to hydrolyze the interstitial collagens. It has been proposed that a number of small structural features act in concert to make this interaction possible (Gomis-Ruth *et al.*, 1996). The linker region that joins the catalytic and hemopexin-like domains in the collagenases is believed to be crucial to the correct juxtaposition of the two domains.

Much less is known about possible substrate recognition roles of the other MMP domains. Gel A and Gel B both contain three tandem copies of a 58-amino-acid type II fibronectin module that is inserted in front of the catalytic zinc binding site (Fig. 2). This domain gives these enzymes the ability to bind to a number of ECM proteins. The effect of deleting this domain on the biochemical and functional properties of Gel A has been investigated (Murphy *et al.*, 1994). The mutant enzyme lacking the fibronectin-like domain had the same peptidase activity as the wild type enzyme, but only 10% of the gelatinase activity. This implies that the fibronectin-like domain interacts with gelatin in some way that facilitates its gelatinase activity. It was also observed that the mutant lost the ability to bind to collagen, indicating that it is the fibronectin-like domain, and not the hemopexin-like domain, of this enzyme that confers collagen-binding activity. The ability of Gel A to bind to a number of ECM macromolecules through its fibronectin-like domain has been postulated to deposit the enzyme at sites where it can be utilized by migrating cells (Steffensen *et al.*, 1995).

The function of the transmembrane domain of the MT MMPs deserves mention in the context of substrate recognition. As mentioned earlier, the MT MMPs are believed to play a critical role in the activation of pro-Gel A at the cell surface. Although this domain functions to localize these enzymes to the cell surface rather than to interact directly with the pro-Gel A substrate, its presence is essential for the pericellular action of the activated Gel A. Thus, the transmembrane domain plays a key role in substrate localization and, therefore, to the overall recognition process. No information is available on the function of the α1(V)-domain of Gel B. Last, although outside of the scope of this chapter, it should be mentioned that the different MMP domains modulate the interactions of these enzymes with TIMPs.

REFERENCES

Aimes, R. T., and Quigley, J. P. (1995). Matrix metalloproteinase-2 is an interstitial collagenase. *J. Biol. Chem.* **270**, 5872–5876.

Becker, J. W., Marcy, A. I., Rokosz, L. L., Axel, M. G., Burbaum, J. J., Fitzgerald, P. M. D., Cameron, P. M., Esser, C. K., Hagman, J. D., and Springer, J. P. (1995). Stromelysin-1: Three-dimensional structure of the inhibited catalytic domain and of the C-truncated proenzyme. *Protein Sci.* **4**, 1966–1976.

Billinghurst, R. C., Dahlberg, L., Ionescu, M., Reiner, A., Bourne, R., Rorabeck, C., Mitchell, P., Hambor, J., Diekmann, O., Tschesche, H., Chen, J., Van Wart, H. E., and Poole, R. A. (1997). Enhanced cleavage of type II collagen by collagenases in osteoarthritic articular cartilage. *J. Clin. Invest.* **99**, 1534–1545.

Birkedal-Hansen, H., Moore, W. G. I., Bodden, M. K., Windsor, L. J., Birkedal-Hansen, B., De Carlo, A., and Engler, J. A. (1993). Matrix metalloproteinases: A review. *Crit. Rev. Oral Biol. Med.* **4**, 197–250.

Bode, W., Reinemer, P., Huber, R., Kleine, T., Schnierer, S., and Tschesche, H. (1994). The x-ray crystal structure of the catalytic domain of human neutrophil collagenase inhibited by a substrate analogue reveals the essentials for catalysis and specificity. *EMBO J.* **13**, 1263–1269.

Borkakoti, N., Winkler, F. K., Williams, D. H., D'Arcy, A., Broadhurst, M. J., Brown, P. A., Johnson, W. H., and Murray, E. J. (1994). Structure of the catalytic domain of human fibroblast collagenase complexed with an inhibitor. *Struct. Biol.* **1**, 106–110.

Brown, P. D., Levy, A. T., Margulies, I. M. K., Liotta, L. A., and Stetler-Stevenson, W. G. (1990). Independent expression and cellular processing of Mr 72,000 type IV collagenase and interstitial collagenase in human tumorigenic cell lines. *Cancer Res.* **50**, 6184–6191.

Browner, F. M., Smith, W. W., and Castelhano, A. L. (1995). Matrilysin-inhibitor complexes: Common themes among metalloproteases. *Biochemistry* **34**, 6602–6610.

Clark, I. M., and Cawston, T. E. (1989). Fragments of human fibroblast collagenase. *Biochem. J.,* **263**, 201–206.

Desrochers, P. E., Mookhtiar, K., Van Wart, H. E., Hasty, K. A., and Weiss, S. J. (1992). Proteolytic inactivation of α_1-proteinase inhibitor and α_1-antichymotrypsin by oxidatively activated human neutrophil metalloproteinases. *J. Biol. Chem.* **267**, 5005–5012.

Dhanaraj, V., Ye, Q. Z., Johnson, L. L., Hupe, D. J., Ortwine, D. F., Dubar, J. B., Rubin, J. R., Pavlovsly, A., Humblet, C. and Blundell, T. L. (1996). X-ray structure of a hydroxamate inhibitor complex of stromelysin catalytic domain and its comparison with members of the zinc metalloproteinase family. *Structure* **4**, 375–386.

Dong, Z., Kumar, R., Yang, X., and Fidler, I. J. (1997). Macrophage-derived metalloelastase is responsible for the generation of angiostatin in Lewis lung carcinoma. *Cell* **88**, 801–810.

Eugui, E. M., Mirkovich, A., De Lustro, B., Haller, J., Rouhafza, S., Kim, Y., Broka, C., Walker, K. A. M., Hendricks, R. T., Lollini, L., Martin, R., Van Wart, H. E., and Caulfield, J. P. A synthetic matrix metalloproteinase inhibitor prevents cartilage degradation *in vitro* and *in vivo. In* "TIMP96." In press.

Fields, G. B., Van Wart, H. E., and Birkedal-Hansen, H. (1987). Sequence specificity of human skin fibroblast collagenase. *J. Biol. Chem.* **262**, 6221–6226.

Fowlkes, J. L., Enghild, J. J., Suzuki, K., and Nagase, H. (1994). Matrix metalloproteinases degrade insulin-like growth factor-binding protein-3 in dermal fibroblast cultures. *J. Biol. Chem.* **269,** 25742–25746.

Gohlke, U., Gomis-Ruth, F. X., Crabbe, T., Murphy, G., Docherty, A. J. P., and Bode, W. (1996). The C-terminal (haemopexin-like) domain structure of human gelatinase A (MMP-2): Structural implications for its function. *FEBS Lett.* **378,** 126–130.

Gomis-Ruth, F. X., Gohlke, U., Betz, M., Knauper, V., Murphy, G., Lopez-Otin, C., and Bode, W. (1996). The helping hand of collagenase-3 (MMP-13): 2.7 A crystal structure of its C-terminal heamopexin-like domain. *J. Mol. Biol.* **264,** 556–566.

Grams, F., Reinemer, P., Powers, J. C., Kleine, T., Pieper, M., Tschesche, H., Huber, R., and Bode, W. (1995). X-ray structures of human neutrophil collagenase complexed with peptide hydroxamate and peptide thiol inhibitors. Implications for substrate binding and rational drug design. *Eur. J. Biochem.* **228,** 830–841.

Gronski, T. J., Martin, R. L., Kobayashi, D. K., Walsh, B. C., Holman, M. C., Huber, M., Van Wart, H. E., and Shapiro, S. D. (1997). Hydrolysis of a broad spectrum of extracellular matrix proteins by human macrophage elastase. *J. Biol. Chem.* **272,** 12189–12194.

Hasty, K. A., Jeffrey, J. J., Hibbs, M. S., and Welgus, H. G. (1987). The collagen substrate specificity of human neutrophil collagenase. *J. Biol. Chem.* **262,** 10048–10052.

Hautamaki, R. D., Kobayashi, D. K., Senior, R. M., and Shapiro, S. D. (1997). Requirement for macrophage elastase for cigarette-induced emphysema in mice. *Science* **277,** 2002–2004.

Hirose, T., Patterson, C., Pourmotabbed, T., Mainardi, C. L., and Hasty, K. A. (1993). Structure–function relationship of human neutrophil collagenase: Identification of regions responsible for substrate specificity and general proteinase activity. *Proc. Natl. Acad. Sci. USA* **90,** 2569–2573.

Knauper, V., Lopez-Otin, C., Smith, B., Knight, G., and Murphy, G. (1996). Biochemical characterization of human collagenase-3. *J. Biol. Chem.* **271,** 1544–1550.

Knauper, V., Cowell, S., Smith, B., Lopez-Otin, C., O'Shea, M., Morris, H., Zardi, L., and Murphy, G. (1997). The role of the C-terminal domain of human collagenase-3 (MMP-13) in the activation of procollagenase-3, substrate specificity, and tissue inhibitor of metalloproteinase interaction. *J. Biol. Chem.* **272,** 7608–7616.

Li, J., Brick, P., O'Hare, M. C., Skarzynski, T., Loyd, L. F., Curry, V. A., Clark, I. M., Bigg, H. F., Hazleman, B. L., Cawston, T. E., and Blow, D. M. (1995). Structure of full-length porcine synovial collagenase reveals a C-terminal domain containing a calcium-linked, four-bladed β-propeller. *Structure* **3,** 541–549.

Libson, A., Gittis, A., Collier, I., Marmer, B., Goldberg, G., and Lattman, E. E. (1995). Crystal structure of the hemopexin-like C-terminal domain of gelatinase A. *Nature Struct. Biol.* **2,** 938–942.

Lohmander, L. S., Neame, J. P., and Sandy, J. D. (1993). The structure of aggrecan fragments in human synovial fluid. *Arthritis Rheum.* **36,** 1214–1222.

Lovejoy, B., Hassell, A. M., Luther, M. A., Weigl, D., and Jordan, S. R. (1994a). Crystal structures of recombinant 19-kDa human fibroblast collagenase complexed to itself. *Biochemistry* **33,** 8207–8217.

Lovejoy, B., Cleasby, A., Hassell, A. M., Longley, K., Luther, M. A., Weigl, D., McGeehan, G., McElroy, A. B., Drewry, D., Lambert, M. H., and Jordan, S. R. (1994b). Structure of the catalytic domain of fibroblast collagenase complexed with an inhibitor. *Science* **263,** 375–378.

Mallya, S. K., Mookhtiar, K. A., Gao, Y., Brew, K., Dioszegi, M., Birkedal-Hansen, H., and Van Wart, H. E. (1990). Characterization of 58-kilodalton human neutrophil collagenase; comparison with human fibroblast collagenase. *Biochemistry* **29,** 10628–10634.

Mitchell, P. G., Magna, H. A., Reeves, L. M., Lopresti-Morrow, L. L., Yocum, S. A., Rosner, P. J., Geoghegan, K. F., and Hambor, J. E. (1996). Cloning, expression, and type II collagen activity of matrix metalloproteinase-13 from human osteoarthritic cartilage. *J. Clin. Invest.* **97,** 761–768.

Mitchell, T. I., Jeffrey, J. J., Palmiter, D. R., and Brinckerhoff, E. C. (1993). The acute phase reactant serum amyloid A (SAA3) is a novel substrate for degradation by the metalloproteinases collagenase and stromelysin. *Biochem. Biophys. Acta* **1156,** 245–254.

Murphy, G., Allan, J. A., Willenbrock, F., Cockett, M. I., O'Connell, J. P., and Docherty, A. J. P. (1992). The role of the C-terminal domain in collagenase and stromelysin specificity. *J. Biol. Chem.* **267,** 9612–9618.

Murphy, G., Nguyen, Q., Cockett, M. I., Atkinson, S. J., Allan, J. A., Knight, C. G., Willenbrock, F., and Docherty, A. J. P. (1994). Assessment of the role of the fibronectin-like domain of gelatinase A by analysis of a deletion mutant. *J. Biol. Chem.* **269,** 6632–6636.

Nagase, H. (1996). Matrix metalloproteinases. *In* "Zinc Metalloproteases in Health and Disease" (N. M. Hooper, ed.), pp. 153–204, Taylor & Francis, London.

Nagase, H., Fields, G. C., and Fields, G. B. (1994). Design and characterization of a fluorogenic substrate selectively hydrolyzed by stromelysin-1 (matrix metalloproteinase-3). *J. Biol. Chem.* **269,** 20952–20957.

Netzel-Arnett, S., Fields, G., Birkedal-Hansen, H., and Van Wart, H. E. (1991a). Sequence specificities of human fibroblast and neutrophil collagenases. *J. Biol. Chem.* **266,** 6747–6755.

Netzel-Arnett, S., Mallya, S., Nagase, H., Birkedal-Hansen, H., and Van Wart, H. E. (1991b). Continuously recording fluorescent assays optimized for five human matrix metalloproteinases. *Anal. Biochem.* **195,** 86–92.

Netzel-Arnett, S., Sang, Q.-X., Moore, W. G. I., Navre, M., Birkedal-Hansen, H., and Van Wart, H. E. (1993). Comparative sequence specificities of human 72- and 92-kDa gelatinases (type IV collagenases) and PUMP (matrilysin). *Biochemistry* **32,** 6427–6432.

Netzel-Arnett, S., Salari, A., Goli, U. B., and Van Wart, H. E. (1994). Evidence for a triple helix recognition site in the hemopexin-like domains of human fibroblast and neutrophil interstitial collagenases. *Ann. N.Y. Acad. Sci.* **732,** 22–30.

Niedzwiecki, L., Teahan, J., Harrison, R. K., and Stein, R. L. (1992) Substrate specificity of the human matrix metalloproteinase stromelysin and the development of continuous fluorometric assays. *Biochemistry* **31,** 12618–12623.

Ochieng, J., Fridman, R., Nangia-Makker, P., Kleiner, D. E., Liotta, L. A., Stetler-Stevenson, W. G., and Raz, A. (1994). Galectin-3 is a novel substrate for human matrix metalloproteinases 2 and 9. *Biochemistry* **33,** 14109–14114.

Ohuchi, E., Imai, K., Fujii, Y., Sato, H., Seiki, M., and Okada, Y. (1997). Membrane type 1 matrix metalloproteinase digests interstitial collagens and other extracellular matrix macromolecules. *J. Biol. Chem.* **272,** 2446–2451.

Okada, Y., Shinmei, M., Tanaka, O., Naka, K., Kimura, A., Nakanishi, I., Bayliss, M. T., Iwata, K., and Nagase, H. (1992). Localization of matrix metalloproteinase 3 (stromelysin) in osteoarthritic cartilage and synovium. *Lab. Invest.* **66,** 680–690.

Ottl, J., Battistuta, R., Pieper, M., Tschesche, H., Bode, W., Kuhn, K., and Moroder, L. (1996). Design and synthesis of heterotrimeric collagen peptides with a built-in cystine-knot. Models for collagen catabolism by matrix metalloproteases. *FEBS Lett.* **398,** 31–36.

Patterson, B. C., and Sang, Q.-X. A. Angiostatin-converting enzyme activities of human matrilysin (MMP-7) and gelatinase B/type IV collagenase (MMP-9). *J. Biol. Chem.* **272.** In press.

Pei, D., and Weiss, S. J. (1996). Transmembrane-deletion mutants of the membrane-type matrix metalloproteinase-1 process progelatinase A and express intrinsic matrix-degrading activity. *J. Biol. Chem.* **271,** 9135–9140.

Pei, D., Majmudar, G., and Weiss, J. S. (1994). Hydrolytic inactivation of a breast carcinoma cell-derived serpin by human stromelysin-3. *J. Biol.Chem.* **269**, 25849–25855.

Sanchez-Lopez, R., Alexander, C. M., Behrendtsen, O., Breathnach, R., and Werb, Z. (1993). Role of zinc-binding and hemopexin domain-encoded sequences in the substrate specificity of collagenase and stromelysin-2 as revealed by chimeric proteins. *J. Biol. Chem.* **268**, 7238–7247.

Sang, Q.-X., Birkedal-Hansen, H., and Van Wart, H. E. (1995). Proteolytic and nonproteolytic activation of human neutrophil progelatinase B. *Biochim. Biophys. Acta* **1251**, 99–108.

Sang, Q. A., and Douglas, A. D. (1996). Computational sequence analysis of matrix metalloproteinases. *J. Protein Chem.* **15**, 137–160.

Sang, Q. A., Bodden, M. K., and Windsor, L. J. (1996). Activation of human progelatinase A by collagenase and matrilysin: Activation of procollagenase by matrilysin. *J. Protein Chem.* **15**, 243–253.

Sato, H., Takino, T., Okada, Y., Cao, J., Shinagawa, A., Yamamoto, E., and Seiki, M. (1994). A matrix metalloproteinase expressed on the surface of invasive tumour cells. *Nature* **370**, 61–65.

Schnierer, S., Kleine, T., Gote, T., Hillemann, A., Knauper, V., and Tschesche, H. (1993). The recombinant catalytic domain of human neutrophil collagenase lacks type I collagen substrate specificity. *Biochem. Biophysi. Res. Commun.* **191**, 319–326.

Seltzer, J. L., Weingarten, H., Akers, K. T., Eschbach, M. L., Grant, G. A., and Eisen, A. Z. (1989). Cleavage specificity of type IV collagenase (gelatinase) from human skin. *J. Biol. Chem.* **264**, 19583–19586.

Seltzer, J. L., Akers, K. T., Weingarten, H., Grant, G. A., McCourt, D. W., and Eisen, A. Z. (1990). Cleavage specificity of human skin type IV collagenase (gelatinase). *J. Biol. Chem.* **265**, 20409–20413.

Smith, M. M., Shi, L., and Navre, M. (1995). Rapid identification of highly active and selective substrates for stromelysin and matrilysin using bacteriophage peptide display libraries. *J. Biol. Chem.* **270**, 6440–6449.

Springman, E. B., Angleton, E. L., Birkedal-Hansen, H., and Van Wart, H. E. (1990). Multiple modes of activation of latent human fibroblast collagenase: Evidence for the role of a Cys^{73} active-site zinc complex in latency and a "cysteine switch" mechanism for activation. *Proc. Natl. Acad. Sci. USA* **87**, 364–368.

Spurlino, J. C., Smallwood, A. M., Carlton, D. D., Banks, T. M., Vavra, K. J., Johnson, J. S., Cook, E. R., Falvo, J., Wahl, R. C., Pulvino, T. A., Wendoloski, J. J., and Smith, D. L. (1994). 1.56 Å structure of mature truncated human fibroblast collagenase. *Proteins Struct. Function Genet.* **19**, 98–109.

Stams, T., Spurlino, J. C., Smith, D. L., Wahl, R. C., Ho, T. F., Qoronfleh, M. W., Banks, T. M., and Rubin, B. (1994). Structure of human neutrophil collagenase reveals large S_1' specificity pocket. *Struct. Biol.* **1**, 119–123.

Steffensen, B., Wallon, U. M., and Overall, C. M. (1995). Extracellular matrix binding properties of recombinant fibronectin type II-like modules of human 72-kDa gelatinase type IV collagenase. *J. Biol. Chem.* **270**, 11555–11566.

Stetler-Stevenson, W. G., Aznavoorian, S., and Liotta, L. A. (1993). Tumor cell interactions with the extracellular matrix during invasion and metastasis. *Annu. Rev. Cell Biol.* **9**, 541–573.

Strongin, A. Y., Collier, I., Bannikov, G., Marmer, B. L., Grant, G. A., and Goldberg, G. I. (1995). Mechanism of cell surface activation of 72-kDa type IV collagenase. *J. Biol. Chem.* **270**, 5331–5338.

Takino, T., Sato, H., Shinagawa, A., and Seiki, M. (1995) Identification of the second membrane-type matrix metalloproteinase (MT-MMP-2) gene from a human placenta cDNA library. *J. Biol. Chem.* **270**, 23013–23020.

Teahan, J., Harrison, R., Izquierdo, M., and Stein, R. (1989). Substrate specificity of human fibroblast stromelysin. Hydrolysis of substance P and its analogues. *Biochemistry* **28,** 8497–8501.

Vallee, B. L., and Auld, D. S. (1992). Active zinc binding sites of zinc metalloenzymes. *Matrix* Suppl. **1,** 5–19.

Van Wart, H. E., and Birkedal-Hansen, H. (1990). The cysteine swich: A principle of regulation of metalloproteinase activity with potential applicability to the entire matrix metalloproteinase gene family. *Proc. Natl. Acad. Sci. USA* **87,** 5578–5582.

Ward, R. V., Atkinson, S. J., Slocombe, P. M., Docherty, A. J. P., Reynolds, J. J., and Murphy, G. (1991) Tissue inhibitor of metalloproteinase-2 inhibits the activation of 72 kDa progelatinase by fibroblast membranes. *Biochim. Biophys. Acta* **1079,** 242–246.

Weingarten, H., Martin, R., and Feder, J. (1985). Synthetic substrates of vertebrate collagenases. *Biochemistry* **24,** 6730–6734.

Welch, A. R., Holman, C. M., Huber, M., Brenner, M. C., Browner, F. M., and Van Wart, H. (1996). Understanding the P_1' specificity of the matrix metalloproteinases: Effect of S_1' pocket mutations in matrilysin and stromelysin-1. *Biochemistry* **35,** 10103–10109.

Welgus, H. G., Jeffrey, J. J., and Eisen, A. Z. (1981). The collagen substrate specificity of human skin fibroblast. *J. Biol. Chem.* **256,** 9511–9515.

Wu, J.-J., Lark, M. W., Chun, L. E., and Eyre, D. R. (1991). Sites of stromelysin cleavage in collagen types II, IX, X, and XI of cartilage. *J. Biol. Chem.* **266,** 5625–5628.

Synthetic Inhibitors of Matrix Metalloproteinases

Peter D. Brown

Department of Clinical Research, British Biotech Pharmaceuticals Ltd., Oxford OX4 5LY, United Kingdom

I. Introduction

In the past few years the matrix metalloproteinase (MMP) family has grown appreciably. At least 16 mammalian enzymes have now been identified and the range of potential substrates has grown in both breadth and diversity. In normal physiology MMPs are considered to play a central role in the formation and maintenance of mammalian tissue architecture. Finely regulated MMP activity is associated with processes of ovulation (Butler *et al.*, 1991), trophoblast invasion (Graham and Lala, 1991), skeletal (Gack *et al.*, 1995) and appendageal development (Karelina *et al.*, 1994), and mammary gland involution (Lund *et al.*, 1996). The turnover of collagen in the adult is slow, of the order of 50–300 days (Kivirikko, 1970), but again MMPs are probably responsible for this turnover as part of the maintenance of load-bearing tissues such as bones, joints, and tendons.

In these normal physiological processes MMP activity is controlled at several levels. Firstly, gene expression is often tightly regulated with most matrix metalloproteinases being expressed only when their activity is required. Secondly, matrix metalloproteinases are synthesized as latent proenzymes that require the proteolytic removal of a 10-kDa amino-terminal domain in order to become proteolytically active (Kleiner and Stetler-Stevenson, 1993). For most of the matrix metalloproteinases this occurs after secretion from the cell. In the case of stromelysin-3 and MT1-MMP the presence of a furin-processing motif (Arg–Xaa–Lys–Arg) suggests that these enzymes can be processed intracellularly in the Golgi vesicles (Pei et al., 1994; Pei and Weiss, 1995). Thirdly, the activated matrix metalloproteinases can be inhibited by endogenous proteinase inhibitors such as α_2-macroglobulin and more importantly the family of tissue inhibitors of metalloproteinases, TIMPs 1–4 (Stetler-Stevenson et al., 1989).

As might be expected, loss of control of MMP activity appears to have serious consequences, and aberrations in MMP expression have been associated with several diseases. The proposed pathogenic roles for matrix metalloproteinases include the destruction of cartilage and bone in rheumatoid arthritis and osteoarthritis (Cawston, 1996; O'Byrne et al., 1995), tissue breakdown and remodeling during invasive tumor growth and tumor angiogenesis (Liotta and Stetler-Stevenson, 1990), degradation of myelin-basic protein in neuroinflammatory diseases (Gijbels et al., 1992; Chandler et al., 1995), opening of the blood–brain barrier following brain injury (Rosenberg, 1995), increased matrix turnover in restenotic lesions (Strauss et al., 1996), loss of aortic wall strength in aneuryms (Thompson and Parks, 1996), and tissue degradation in gastric ulceration (Saarialho-Kere et al., 1996).

Although in many of these diseases the "defect" is more than simple excessive matrix degradation, MMP inhibition may be of therapeutic benefit. This has led to the development of programs to design and test synthetic MMP inhibitors. The first disease target was rheumatoid arthritis but the range of potential applications has broadened considerably. Rather than provide a comprehensive review of progress in each of these diseases, this article focuses on the development of synthetic MMP inhibitors for the treatment of cancer. Progress in this disease has now reached the stage of clinical testing and the first findings in patients can be discussed.

II. Design of Synthetic Inhibitors

The first synthetic MMP inhibitors were developed in the early 1980s. These molecules were pseudopeptide derivatives based on the structure of the collagen molecule at the site of initial cleavage by interstitial collagenase. Compounds designed from the Ile–Ala–Gly and Leu–Leu–Ala sequences on the right-hand side of the cleavage site have emerged as the most promising drugs, although left-hand side compounds and inhibitors bridging the cleavage site have also been developed. The structure of a generic right-hand side inhibitor is shown in Fig. 1. The inhibitor binds reversibly at the active site of the matrix metalloprotei-

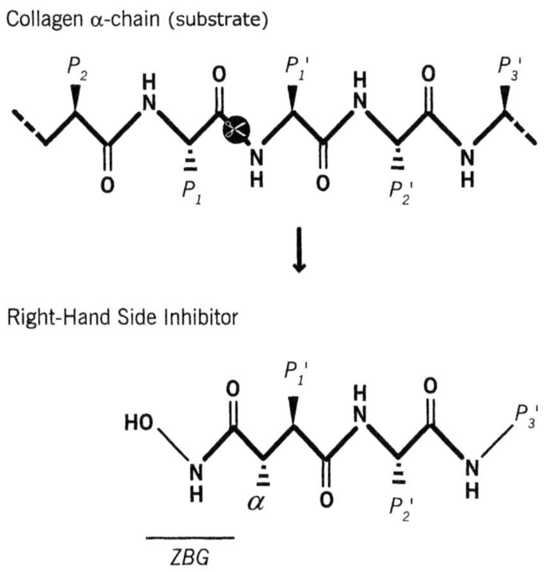

Collagen α-chain (substrate)

Right-Hand Side Inhibitor

ZBG

FIG. 1. Structure of a generic right-hand side inhibitor. Many of the inhibitors currently being studied are derived from the peptide structure of the α-chain of type I collagen at the point at which collagenase first cleaves the molecule (✂). The peptide backbone is retained although groups at the α, P_1', P_2', and P_3' positions are modified to give different inhibitory and physicochemical properties. The zinc-binding group (ZBG), in this case a hydroxamate (–CONHOH), binds the zinc atom in the active site of the MMP enzyme.

nase in a stereospecific manner. The zinc-binding group, in this case hydroxamic acid (–CONHOH), is then positioned to chelate the active site zinc ion. Modification of the stereochemistry of the molecule results in loss of inhibitory activity. Several zinc-binding groups have been tested including carboxylates, aminocarboxylates, sulphydryls, and derivatives of phosphorus acids (Beckett et al., 1996), but hydroxamates have proved to be the most useful and the majority of inhibitors currently in clinical testing contain this group.

Highly potent compounds with K_i values in the low nanomolar range were developed relatively easily. These compounds, typified by the hydroxamate batimastat (Fig. 2), showed broad specificity for members of the MMP family but displayed little detectable activity against other classes of metalloproteinase such as angiotensin-converting enzyme and enkephalinase (Brown and Giavazzi, 1995). Unfortunately, these early compounds showed poor oral bioavailability and therefore offered little advantage over the native TIMPs.

The next steps in the design of synthetic MMP inhibitors were made with two principal objectives: the development of compounds with im-

Batimastat (BB-94)

Marimastat (BB-2516)

Fɪɢ. 2. Structure of the hydroxamate-based inhibitors, batimastat and marimastat.

proved oral bioavailability and the development of compounds with selective inhibitory activity against individual matrix metalloproteinases. In both cases the design of new compounds was assisted by X-ray crystallography data on the three-dimensional structure of the collagenase active site (Grams et al., 1995). Marimastat (Fig. 2) was one of the first inhibitors to show improved oral bioavailability in both animals and man, differing from its predecessor batimastat in the group adjacent to the hydroxamate and the group at the P_2' position. In both positions a small substituent, -hydroxyl and -t-butyl, respectively, replaces a larger cyclic group. The precise reasons for marimastat's improved bioavailability are not clear but the substitutions made may reduce the compound's susceptibility to peptidases, improve absorption, or reduce first-pass metabolism (Beckett et al., 1996).

More selective compounds have also been developed. In practice, these compounds are not selective for one particular matrix metalloproteinase but instead show a selective loss of activity against one or more of the enzymes. A series of compounds described by Morphy et al. (1994) takes advantage of differences in the active site of gelatinase A and B that allow larger hydrophobic groups at the P_1' position. These compounds show greater than 1000-fold selectivity for gelatinases over interstitial collagenase but are also quite potent inhibitors of stromelysin-1. Unfortunately, these early selective inhibitors have not shown good bioavailability when given orally and CDP-845 (Celltech), a potent selective gelatinase inhibitor, has been withdrawn from clinical development partly for this reason. RO32-3555 (Roche) is a hydroxamate-based inhibitor with relatively weak activity against gelatinase A and stromelysin-1, but good activity against interstitial collagenase. This compound shows good activity when given orally in animal models of arthritis and is currently in clinical development as a treatment for rheumatoid arthritis (Wood *et al.*, 1996). It is expected that other orally bioavailable selective MMP inhibitors will follow RO32-3555 into the clinic.

Although hydroxamates make up most of the synthetic inhibitors in development, the availability of X-ray crystallographic structures and structure-based design has begun to yield various novel non-peptide-based inhibitors (Beckett, 1996). Low-molecular-weight MMP inhibitors have also been developed from natural products. Interestingly, the more potent natural inhibitors such as BE16627B (Banyu) (Naito *et al.*, 1993) and matlystatin B (Sankyo) (Tamaki *et al.*, 1995) are hydroxamates and are structurally very similar to the right-hand side pseudopeptide inhibitors obtained by rational substrate-based design (Fig. 3). Inhibitors have also been obtained through chemical modification of the tetracycline family of molecules where it has been possible to separate the antibiotic and protease inhibitory activities (Golub *et al.*, 1987).

III.   ROLE OF MATRIX METALLOPROTEINASES IN CANCER

The importance of proteolytic enzymes in facilitating invasive tumor growth was recognized some considerable time before these enzymes were isolated and characterized. Writing in 1949 Gersch and Catchpole postulated that "during rapid growth of a tumour, perhaps even as a

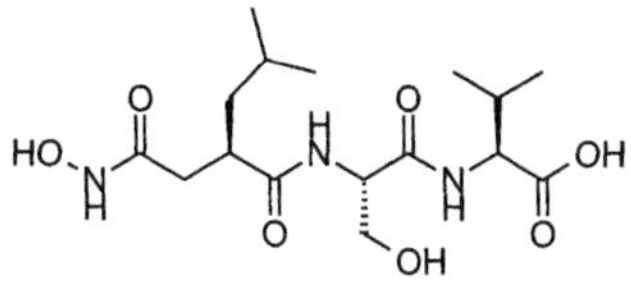

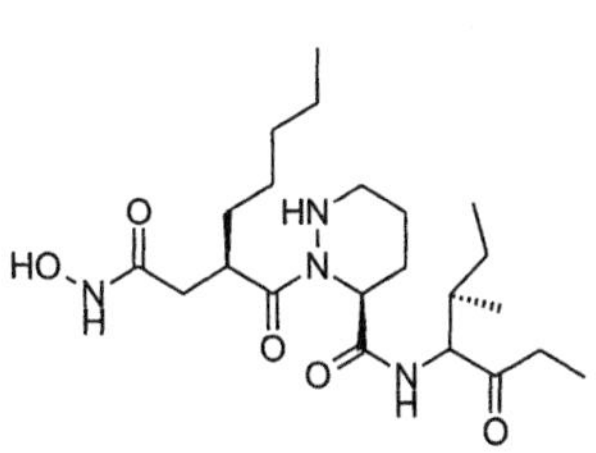

FIG. 3. Structure of natural MMP inhibitors BE16627B from *Streptomyces* and matlystatin B from *Actinomadura atramentaria*.

condition of tissue invasion by a neoplasm, the ground substance of neighboring connective tissue stroma, including the basement membrane of small blood vessels, may become more fluid as a result of the depolymerisation of the ground substance. This may be effected through the action of depolymerising enzymes possibly secreted by fibroblasts." The "depolymerizing" enzymes were subsequently identified as hyaluronidases, serine proteinases, and matrix metalloproteinases. The relative contribution of these and other families of enzymes such as the cysteine proteinases is still a matter of debate but patterns of expression in tumor tissue have been studied in some detail.

It is probably fair to say that when interstitial collagenase (Taylor *et al.*, 1970; Dresden *et al.*, 1972) and type IV collagenase (now termed gelatinase) (Liotta *et al.*, 1981) were first isolated from tumors it was generally considered that the enzymes were secreted by the tumor cells to degrade interstitial collagen and basal laminae to create space for invasive growth. The pattern of MMP expression is now known to be more complex and in many human malignancies it is characterized by the induction of metalloproteinase expression in "host" stromal cells (Basset *et al.*, 1994; Heppner *et al.*, 1996), including the fibroblasts described in Gersch and Catchpole's original postulate.

It is also clear that the MMP activity around a tumor is a feature of increased tissue remodeling as much as increased tissue degradation. Indeed, expression of both individual matrix metalloproteinases and their endogenous inhibitors (TIMPs) has been correlated with tumor progression and clinical outcome. This is perhaps most clearly demonstrated in colorectal cancer where high expression of gelatinase B mRNA has been shown to be associated with early relapse and poor survival (Zeng *et al.*, 1996) and high levels of TIMP-1 correlate with lymph node and distant metastatic spread (Zeng *et al.*, 1995). Other studies have reported an association between poor prognosis and high levels of expression of stromelysin-3 in breast cancer (Chenard *et al.*, 1996), activated gelatinase A in gastric cancer (Seir *et al.*, 1996), TIMP-1 in small cell lung cancer (Fong *et al.*, 1996), and TIMP-2 in bladder cancer (Grignon *et al.*, 1996).

The fact that high expression of both proteinase and inhibitor is associated with poor prognosis might seem surprising, but probably reflects the need for some regulation of the increased metalloproteinase activity. Clearly, to "succeed" a malignant neoplasm must do more than degrade its immediate environment. In effect, the tumor must "remodel" the local tissue to suit its own needs. The generation of a modified vasculature is perhaps the most obvious feature of this "remodeling," but associated with this must be the generation of supportive stromal tissues. To this extent the stroma acts as a "collaborator" with induction of proteinases and inhibitors by the adjacent tumor cells resulting in angiogenesis and invasive growth.

The growth of the MMP family has inevitably made the task to understand their role in tumor growth and spread more complex. The relative contribution of individual enzymes is of particular importance to the design of "selective" inhibitors. Gelatinase A, gelatinase B, stromelysin-3, matrilysin, and MT1-MMP seem to be associated most closely with the invasive phenotype although there are studies indicating a role for collagenase (Hewitt *et al.*, 1991; Murray *et al.*, 1996), stromelysin-1 (Gallegos *et al.*, 1995), and other enzymes. Analysis of the involvement of gelatinase B is complicated by the fact that in many tumors the source for this enzyme is inflammatory cells, which it can be argued are present in the tumor as part of the "host" defense. However, the association of gelatinase B (Zeng *et al.*, 1996) and, in some cases, inflammatory infiltrate (Leek *et al.*, 1996) with poor prognosis suggests a dual action.

It seems likely that some tumor types will depend more heavily on matrix metalloproteinases for their invasive growth than others and that even within a tumor type there may be a range of metalloproteinase involvement. This may also change as the tumor progresses to lymphatic and visceral metastases where the local stromal environment may result in a quite different metalloproteinase profile and dependency. The first indications of which tumor types will respond to MMP inhibition and the ways in which they respond have come from animal cancer models and more recently the first clinical trials. These are described in the following sections.

IV.   EFFECTS OF MATRIX METALLOPROTEINASE INHIBITORS IN CANCER MODELS

One of the first demonstrations of anticancer activity for synthetic MMP inhibitors was the finding that these compounds could block human fibrosarcoma cell invasion through reconstituted basement membranes and inhibit lung colonization by B16F10 mouse melanoma cells (Reich *et al.*, 1988). These results, together with the prevailing interest in molecular mechanisms of metastasis, led to a focus on these compounds as antimetastatic agents. However, a broader anticancer activity had already been proposed by, among others, Strauch (1972) who noted that "by reducing the rate of collagen degradation and simultaneously stimulating collagen biosynthesis, the body may be able to encapsulate the invading tumour in collagen structures and isolate it from adjoining tissue not yet affected. The invasion and metastases would thus be limited and the surgical removal or destruction of the

primary tumour would stand a better chance of successful outcome."
The concept of "encapsulation" or tumor containment can of course be
applied to either secondary or primary lesions and effects in both have
subsequently been demonstrated in animal models.

Mechanistically it has been proposed that MMP inhibitors could
inhibit tumor growth either by encouraging the development of fibrotic
tissue around the tumor, thereby preventing invasive growth, or by
inhibiting angiogenesis. Figure 4 shows a diagrammatic representation
of the possible effects of MMP inhibitors on tumor growth. Section A
shows an untreated tumor with a high proportion of viable tumor
cells and no detectable stromal tissue. The boundary of the tumor is
characterized by motile tumor cells invading adjacent tissue. The other
sections show three different possible treatment effects based on observations from experimental cancer models and the first clinical studies.
Section B shows the development of an enlarged necrotic core without
obvious signs of increased fibrotic tissue. Section C shows the development of fibrotic tissue around the tumor mass, as originally envisioned
by Strauch. Section D shows a third possible treatment effect in which
part of the cellular mass has been replaced by fibrotic tissue.

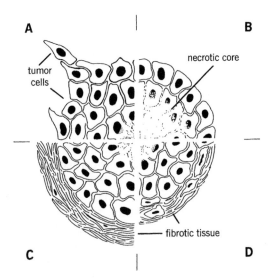

FIG. 4. A diagrammatic representation of the possible effects of MMP inhibition on a
solid tumor. Section A shows an untreated tumor with tumor cells moving into adjacent
tissue. Section B shows an enlarged necrotic core seen in some treated experimental
tumors. Section C shows a capsule of dense stromal tissue surrounding the tumor. Section
D shows fibrotic replacement of tumor with tumor cells showing signs of damage.

Two early studies with batimastat investigated the effect of the inhibitor on the growth of ascitic tumours in nude mice. In a study with human ovarian carcinoma xenografts, Davies et al., (1993) demonstrated almost complete inhibition of ascites development with a resulting five- to sixfold improvement in survival. The small nodules that did form in the batimastat-treated animals were avascular, noninvasive, and appeared to be composed primarily of fibrotic tissue surrounding small collections of tumor cells. In a similar study batimastat was shown to inhibit the development of human $C170HM_2$ colorectal carcinoma ascites (Watson et al., 1995, 1996). $C170HM_2$ is a liver-invasive colorectal carcinoma cell line and solid liver tumors were detected in both control and batimastat-treated animals. However, the batimastat-treated tumors were less frequent and smaller than the controls and were characterized by an enlarged necrotic center. This increase in necrosis may be the result of antiangiogenic activity. Alternatively, it is possible that in certain tumors constriction of invasive growth by MMP inhibitors results in increased interstitial pressure. This in turn leads to compression of blood vessels in the center of the tumor causing ischemia and subsequent necrosis (Jain, 1994; Folkman, 1996). There was no marked evidence of stromal development or encapsulation of the $C170HM_2$ tumors but it seems likely that their growth was still constrained by inhibition of invasion.

Synthetic MMP inhibitors have also been shown to reduce the growth rate of subcutaneously implanted tumors. The hydroxamate inhibitors GI168 and GI173 (GlaxoWellcome) both caused significant inhibition of subcutaneuos Mat Ly Lu rat prostate tumors in nude mice when given 3–10 days after implantation (Conway et al., 1996). Histological analysis revealed no obvious effect on tumour angiogenesis although, as with the $C170HM_2$ tumors, treatment was associated with increased tumor necrosis. Both inhibitors are generally broad-spectrum agents although GI168 is a relatively poor stromelysin-1 inhibitor and GI173 shows a preference for gelatinases. Studies with BE16627B, a natural MMP inhibitor with a broad spectrum of activity, have shown that it can inhibit the subcutaneous growth of HT-1080 human fibrosarcoma but not HCT116 human colon carcinoma (Naito et al., 1993). This is consistent with the observation that while HT-1080 cells secrete several matrix metalloproteinases, HCT116 cells show little detectable metalloproteinase activity in vitro.

Other studies have examined the anticancer activity of the inhibitors in settings that are of more relevance to the clinical situation than the simple subcutaneous models. In a model of human breast carcinoma Sledge et al. (1995) showed that regrowth of MDA-MB-435 tumor at the site of primary tumor resection could be significantly inhibited by batimastat, as could the number and size of lung metastases. Batimastat has also been shown to inhibit the growth of secondary lesions, in the form of lung metastases from B16-BL6 mouse melanoma (Chirivi *et al.*, 1994) and lymphatic metastases from HOSP rat mammary carcinoma (Eccles *et al.*, 1996). CT1746 (Celltech), a hydroxamate inhibitor with a similar range of activity to GI173, caused a marked inhibition in the growth and metastatic spread of Co-3 human colon carcinomas implanted orthotopically in the colon wall (An *et al.*, 1997). Similar activity in an orthotopic colon carcinoma model has also been demonstrated for batimastat (Wang *et al.*, 1994).

Although fibrotic changes have been observed in some models, in others there are no histopathological clues as to why the rate of tumor growth has been reduced. Inhibition of angiogenesis has been suggested as a mode of action and it is clear that synthetic MMP inhibitors can exert such activity in angiogenesis models. Batimastat was shown to reduce the angiogenic response *in vivo* to heparin-Matrigel implants to levels comparable to controls (Taraboletti *et al.*, 1995). In the same study it was shown that batimastat inhibited the invasion of human umbilical vein endothelial cells through Matrigel *in vitro,* but did not significantly alter endothelial cell proliferation, haptotaxis, or chemotaxis. The related compound, ilomastat (GM6001, Glycomed), inhibited angiogenesis in the chick chorioallantoic membrane assay and a rat corneal model (Galardy *et al.*, 1994). However, antiangiogenic activity in cancer models has been less easy to demonstrate. In studies where it has been examined there have been no clear changes in blood vessel density (Sledge *et al.*, 1995; Conway *et al.*, 1996) but in these models the tumors that do develop may be resistant to the effects of metalloproteinase inhibitors.

A general feature of these cancer model studies is that MMP inhibitors have been applied early when the tumor diameter is still only a few millimeters. The effect has been to reduce the rate of tumor growth and spread, and thereby extend survival. Treatment of larger tumors is difficult given the limitations of size that must apply in experiments in mice. When treatment has been delayed the effectiveness of the inhibitor has been reduced (Conway *et al.*, 1996) and it is worth noting that tumor regression has not been observed. However, although cancer models provide an indication of what might be expected in the clinic, they lack one important element of human malignancy, namely, stromal tissue, which is known to be the source of most MMP activity. It is possible, therefore, that effects in the clinic may be more pronounced than the activity seen in animals.

V. CLINICAL TRIALS

A. First-Generation Inhibitors

The first-generation MMP inhibitors to be tested in patients were potent compounds that were characterized by poor oral bioavailability. Consequently, alternative routes of administration were explored. Ilomastat was administered in eye drops to normal volunteers and subsequently to patients with corneal ulcers (Galardy et al., 1994). Batimastat was administered directly into the peritoneum or pleural space of patients with malignant effusions. A phase I study of intraperitoneal batimastat was conducted in patients with symptomatic malignant ascites. Patients with any form of malignancy who required paracentesis for symptomatic relief were eligible for the study. Patients received a single intraperitoneal dose of batimastat in 500 mL 5% dextrose (150–1350 mg/m^2) after paracentesis. In this study batimastat was generally well tolerated and there were early signs of efficacy with several patients requiring no further paracentesis for more than 3 months. The intraperitoneal administration of batimastat also gave rise to unexpectedly high and sustained plasma concentrations of the drug with 100–200 ng/mL batimastat still detectable 28 days after a single administration (Beattie et al., 1994). Intraperitoneal batimastat was also administered to cancer patients without acsites in order to study the tolerability of a depot formulation (Wojtowicz-Praga et al., 1996). Further development of batimastat for these indications was hindered by peritoneal irritation and poor tolerability.

A second phase I study of similar design was conducted in patients with malignant pleural effusion. Patients with symptomatic malignant pleural effusion received lower doses of batimastat (15–135 mg/m^2), given intrapleurally in 50 mL 5% dextrose after aspiration of the effusion. Batimastat was well tolerated and again there were early signs that the drug might be effective in palliation of this condition. In patients receiving batimastat at 60–135 mg/m^2, the reduction in the number of aspirations was significant (Macaulay et al., 1995).

Another series of early clinical trials involved the use of tetracycline derivatives with anti collagenase activity. In a placebo-controlled study lymecycline in combination with conventional nonsteroidal anti-inflammatory treatment resulted in a decrease in the severity and duration of reactive arthritis triggered by Clamydia trachomatis (Lauhio et al., 1991). Low-dose doxycycline (20 mg twice daily) was also reported to decrease urinary pyridinoline collagen cross-link excretion in patients with rheumatoid arthritis (Greenwald et al., 1994). In a randomized study of patients with periodontal disease, tetracycline and minocycline gels were shown to reduce significantly the concentration of stromelysin in gingival crevicular fluid. This effect was not observed with metronidazole antimicrobial treatment (Pourtaghi et al., 1996). The relative importance of the antimicrobial and antimetalloproteinase properties in mediating these different effects will become clearer with the development and testing of non-antimicrobial tetracycline derivatives.

B. Second-Generation Inhibitors

The development of low-molecular-weight MMP inhibitors with good oral bioavailability represents an important milestone in this area of research. The great majority of clinical settings for the potential use of these inhibitors involve chronic conditions requiring long-term treatments over months and years. As in many cases of drug design, the early phase of engineering site recognition and potency must give way to the need to overcome a purely practical problem: the ability to deliver the drug. Poor absorption, susceptibility to enzymatic degradation, and high first-pass metabolism may all have contributed to the disappointing plasma concentrations achieved with the first compounds (Beckett et al., 1996). Replacement of specific groups with the intention of shielding the peptide backbone and of increasing hydrogen bonding has led to the development of inhibitors with improved oral bioavailability, without loss of inhibitory potency.

Currently, six MMP inhibitors are believed to be in clinical trials in patients as oral treatments: AG3340 (Agouron), CGS-27023A (Novartis), BAY-9566 (Bayer), and marimastat (British Biotech) in cancer patients and D5410 (Chiroscience) and RO32-3555 (Roche) in patients with rheumatoid arthritis. With the exception of RO32-3555 the inhibitory profile of these compounds is broadly similar. Preliminary results have been presented for marimastat, RO32-3555, and AG3340.

Marimastat displays limited oral bioavailability in rodents and systemic delivery by minipump is required for therapeutic blood concentrations of the drug to be maintained in these animals. However, preliminary results from healthy volunteer studies showed high blood concentrations of marimastat following oral administration with pharmacokinetic parameters indicating once or twice daily administration (Beckett et al., 1996). Results from a phase I study of AG3340 in healthy volunteers indicate that this compound shows a similar pharmacokinetic profile to marimastat, although the terminal elimination half-life appears shorter (Collier et al., 1997). Both compounds were well tolerated in these phase I settings. RO32-3555 has also been studied in healthy volunteers and good oral bioavailability reported (Wood et al., 1996).

Marimastat is currently being tested in a series of trials in cancer patients. Results from the first of these trials were presented at a recent

European Society for Medical Oncology meeting and provide the first indications that the therapeutic potential seen in animal cancer models may be realized in the clinic. A series of studies in patients with advanced malignancy examined the effect of different doses of marimastat on the serum cancer antigens CA125, CEA, PSA, and CA19-9. Serum concentrations of these antigens are followed clinically as surrogate markers of disease progression (Goldenberg *et al.*, 1981; Steinberg *et al.*, 1986; Rubin *et al.*, 1989; Stamey *et al.*, 1989). There was a dose-related reduction in the rate of rise of these markers, with a proportion of patients showing a fall in the absolute cancer antigen serum concentration over the 28-day study period (Millar and Brown, 1996; Poole *et al.*, 1996; Primrose *et al.*, 1996).

In a separate study in patients with advanced gastric cancer, treatment with marimastat was associated with changes in the macroscopic and histological appearance of the tumors consistent with an increase in the quantity of fibrotic stromal tissue. The changes were very similar to those seen in various cancer models and several of the patients appeared to benefit from these alterations in tumor : stroma ratio (Parsons *et al.*, 1996). Marimastat is also being studied in combination with carboplatin in patients with advanced ovarian cancer (Thomas and Adams, 1997). This follows from preclinical studies suggesting increased therapeutic activity for the combination of MMP inhibitors and cytotoxic agents (Anderson *et al.*,1996).

Preliminary indications are that marimastat is generally well tolerated when given for periods of 3 months. The nature of trials in patients with advanced malignancy complicates the analysis of potential side effects and a clearer picture must await randomized placebo-controlled trials. Musculoskeletal pain has emerged as the principal treatment-related side effect with marimastat. The severity and rate of onset of symptoms were found to be dose related and the effects were considered manageable at the dose range selected for future studies. The condition generally resolved rapidly on discontinuation of marimastat and several patients restarted treatment after an interruption of 2–4 weeks (Millar and Brown, 1996; Parsons *et al.*, 1996; Poole *et al.*, 1996; Primrose *et al.*, 1996). The mechanism responsible for the musculoskeletal pain has not been established but it seems likely that it is related to inhibition of metalloproteinase activity in the normal physiologic remodeling of the connective tissue of tendons and joints.

VI. Proteinase Inhibition as Therapy

Proteinase inhibitors are a relatively new class of therapeutic agent. Current examples of inhibitors in clinical use include the HIV aspartic protease inhibitors (Noble and Faulds, 1996); the antifibrinolytic agents, aprotinin and tranexamic acid (Menichetti *et al.*, 1996); and inhibitors of angiotension-converting enzyme, a dipeptidyl carboxypeptidase, captopril and enalapril (Leonetti and Cuspidi, 1996). In the next few years it will become clear whether the MMP inhibitors will also realize their therapeutic potential. The increasing range of diseases in which matrix metalloproteinases are believed to play a role and the intense interest in this field by the pharmaceutical industry suggests that some important therapies will emerge from this research.

In the particular case of cancer, the use of MMP inhibitors represents a fundamentally different approach from conventional treatments. A malignant tumor shows a remarkable ability to adapt to the various cytotoxic agents directed toward it, driven both by its genetic instability and high proliferative rate. Matrix metalloproteinase inhibitors, and other antiangiogenic compounds, differ from cytoreductive treatments in that they are essentially targeting the other component of malignancy, the stroma. The tumor stroma, including the vasculature, plays a vital role in the growth, invasion, and spread of malignant disease. Perhaps this target will be less able than the tumor to evade treatment. In blocking the ability of the tumor to utilize the adjacent tissue for its own purposes, these new treatments may reveal an "Achilles heel" in the malignant phenotype—namely, its reliance on the "collaboration" of nonmalignant tissue.

If significant inhibition of tumor growth and spread can be achieved it will alter the way both surgeons and oncologists view their respective means of intervention. Resection of more widespread disease might become worthwhile if it is known that the residual tumor can be held in check. Equally, the use of radiotherapy and chemotherapy in patients where the responses are short lived should become more worthwhile if the time to relapse can be significantly extended.

ACKNOWLEDGMENTS

I would like to acknowledge the hard work and dedication of the scientists of the British Biotech MMP Inhibitor Programme.

REFERENCES

An, Z., Wang, X., Willmott, N., Chander, S. K., Tickle, S., Docherty, A. J. P., Mountain, A., Millican, A. T., Morphy, R., Porter, J. R., Epemolu, R. O., Kubota, T., Moossa, A. R., and Hoffman, R. M. (1997). Conversion of a highly malignant colon cancer from an aggressive to a controlled disease by oral administration of a metalloproteinase inhibitor. *Clin. Exp. Met.* **15,** 184–195.

Anderson, I. C., Shipp, M. A., Docherty, A. J. P., and Teicher, B. A. (1996). Combination therapy including a gelatinase inhibitor and cytotoxic agent reduces local invasion and metastasis of murine Lewis lung carcinoma. *Cancer Res.* **56,** 715–710.

Basset, P., Wolf, C., Rouyer, N., Bellocq, J. P., Rio, M. C., and Chambon, P. (1994). Stromelysin-3 in stromal tissue as a control factor in breast cancer behavior. *Cancer* **74,** 1045–1049.

Beattie, G. J., Young, H. A., and Smyth, J. F. (1994). Phase I study of intra-peritoneal metalloproteinase inhibitor BB-94 in patients with malignant ascites. Abstract presented at the 8th NCI-EORTC Symposium on New Drug Development, Amsterdam, March 1994.

Beckett, R. P. (1996). Recent advances in the field of matrix metalloproteinase inhibitors (patent update). *Exp. Opin. Ther. Patents* **6,** 1305–1315.

Beckett, R. P., Davidson, A. H., Drummond, A. H., Huxley, P., and Whittaker, M. (1996). Recent advances in matrix metalloproteinase inhibitor research. *Drug Dev. Today* **1,** 16–26.

Brown, P. D., and Giavazzi, R. (1995). Matrix metalloproteinase inhibition: A review of antitumour activity. *Ann. Oncol.* **6,** 967–974.

Butler, T. A., Zhu, C., Mueller, R. A., Fuller, G. C., Lemaire, W. J., and Woessner, J. F. (1991). Inhibition of ovulation in the perfused rat ovary by synthetic collagenase inhibitor SC44463. *Biol. Reprod.* **44,** 1183–1188.

Cawston, T. E. (1996). Metalloproteinase inhibitors and the prevention of tissue breakdown. *Pharm. Ther.* **70,** 163–182.

Chandler, S., Coates, R., Gearing, A., Lury, J., Wells, G., and Bone, E. (1995). Matrix metalloproteinases degrade myelin basic protein. *Neurosci. Lett.* **201,** 223–226.

Chenard, M-P., O'Siorain, L., Shering, S., Rouyer, N., Lutz, Y., Wolf, C., Basset, P., Bellocq, J-P., and Duffy, M. J. (1996). High levels of stromelysin-3 correlate with poor prognosis in patients with breast carcinoma. *Int. J. Cancer.* **69,** 448–451.

Chirivi, R. S., Garofalo, A., Crimmin, M. J., Bawden, L. J., Brown, P. D., and Giavazzi, R. G. (1994). Inhibition of the metastatic spread and growth of B16-BL6 murine melanoma by a synthetic matrix metalloproteinase inhibitor. *Int. J. Cancer* **58,** 460–464.

Collier, M. A., Yuen, G. J., Bansal, S. K., Kolis, S., Chew, T. G., Appelt, K., and Clendeninn, N. J. (1997). A phase I study of the matrix metalloproteinase (MMP) inhibitor AG3340 given in single doses to healthy volunteers. *Proc. Am. Assoc. Cancer Res.* **38,** 13.

Conway, J. G., Trexler, S. J., Wakefield, J. A., Marron, B. E., Emerson, D. L., Bickett, D. M., Deaton, D. N., Garrison, D., Elder, M., McElroy, A., Willmott, N., Docherty, A. J. P., and McGeehan, G. M. (1996). Effect of matrix metalloproteinase inhibitors on tumour growth and spontaneous metastasis. *Clin. Exp. Metastasis* **14,** 115–124.

Davies, B., Brown, P. D., East, N., Crimmin, M. J., and Balkwill, F. R. (1993). A synthetic matrix metalloproteinase inhibitor decreases tumour burden and prolongs survival of mice bearing human ovarian carcinoma xenograft. *Cancer Res.* **53,** 2087–2091.

Dresden, M. H., Heilman, S. A., and Schmidt, J. D. (1972). Collagenolytic enzymes in human neoplasms. *Cancer Res.* **32,** 993–996.

Eccles, S. A., Box, G. M., Court, W. J., Bone, E. A., Thomas, W., and Brown, P. D. (1996). Control of lymphatic and hematogenous metastases of a rat mammary carcinoma by the matrix metalloproteinase inhibitor batimastat (BB-94). *Cancer Res.* **56,** 2815–2822.

Folkman, J. (1996). New perspectives in clinical oncology from angiogenesis research. *Eur. J. Cancer* **32A,** 2535–2539.

Fong, K. M., Kida, Y., Zimmerman, P. V., and Smith, P. J. (1996). TIMP1 and adverse prognosis in non-small cell lung cancer. *Clin. Cancer Res.* **2,** 1369–1372.

Gack, S., Vallon, R., Schmidt, J., Grigoriadis, A., Tuckermann, J., Schenkel, J., Weiher, H., Wagner, E. F., and Angel, P. (1995). Expression of interstitial collagenase during skeletal development of the mouse is restricted to osteoblast-like cells and hypertrophic chondrocytes. *Cell Growth Diff.* **6,** 759–767.

Galardy, R. E., Grobelny, D., Foellmer, H. G., and Fernandez, L. A. (1994). Inhibition of angiogenesis by the matrix metalloproteinase inhibitor N-[2R-2-(hydroxyamido-carbonymethyl)-4-methylpentanoyl)]-L-tryptophan methylamide. *Cancer Res.* **54,** 4715–4718.

Gallegos, N. C., Smales, C., Savage, F. G., Hembry, R. M., and Boulos, P. B. (1995). The distribution of matrix metalloproteinases and tissue inhibitor of metalloproteinases in colorectal cancer. *Surgical Oncol.* **4,** 21–29.

Gersch, I., and Catchpole, H. R. (1949). The organization of ground substance and base-ment membrane and its significance in tissue injury, disease and growth. *Am. J. Anat.* **85,** 457–507.

Gijbels, K., Masure, S., Carton, H., and Opdenakker, G. (1992). Gelatinase in cerebrospi-nal fluid of patients with multiple sclerosis and other inflammatory neurological disor-ders. *J. Neuroimmunol.* **41,** 29–34.

Goldenberg, D. M., Neville, A., and Carter, A. (1981). CEA (carcinoembryonic antigen): Its role as a marker in the management of cancer. *J. Cancer Res. Clin. Oncol.* **101,** 239–242.

Golub, L. M., McNamara, T. F., D'Angelo, G., Greenwald, R. A., and Ramamurthy, N. S. (1987). A non-antibacterial chemically-modified tetracycline inhibits mammalian collagenase activity. *J. Dental Res.* **66,** 1310–1314.

Graham, C. H., and Lala, P. K. (1991). Mechanism of control of trophoblast invasion *in situ. J. Cell. Physiol.* **148,** 228–234.

Grams, F., Crimmin, M., Hinnes, L., Huxley, P., Pieper, M., Tschesche, H., and Bode, W. (1995). Structure determination and analysis of human neutrophil collagenase complexed with a hydoxamate inhibitor. *Biochemistry* **34,** 14012–14020.

Greenwald, R. A., Moak, S. A., and Golub, L. M. (1994). Low dose doxycycline inhibits pyridinoline excretion in selected patients with rheumatoid arthritis. *Ann. N.Y. Acad. Sci.* **732,** 419–421.

Grignon, D. J., Sakr, W., Toth, M., Ravery, V., Angulo, J., Shamsa, F., Pontes, J. E., Crissman, J. C., and Fridman, R. (1996). High levels of tissue inhibitor of metalloprotei-nase-2 (TIMP-2) expression are associated with poor outcome in invasive bladder cancer. *Cancer Res.* **56,** 1654–1659.

Heppner, K. J., Matrisian, L. M., Jensen, R. A., and Rodgers, W. H. (1996). Expression of most matrix metalloproteinase family members in breast cancer respresents a tumor-induced host response. *Am. J. Pathol.* **149,** 273–282.

Hewitt, R. E., Leach, I. H., Powe, D. G., Clark, I. M., Cawston, T. E., and Turner, D. R. (1991). Distribution of collagenase and tissue inhibitor metalloproteinases (TIMP) in colorectal tumours. *Int. J. Cancer.* **49,** 666–672.

Jain, R. K. (1994). Barriers to drug delivery in solid tumours. *Sci. Am.* **271,** 58–65.

Karelina, T. V., Goldberg, G. I, and Eisen, A. Z. (1994). Matrilysin (PUMP) correlates with dermal invasion during appendageal development and cutaneous neoplasia. *J. Invest. Derm.* **103,** 482–487.

Kivirikko, K. I. (1970). Urinary excretion of hydroxyproline in health and disease. *Int. Rev. Connect. Tiss. Res.* **5,** 93–163.

Kleiner, D. E., and Stetler-Stevenson, W. G. (1993). Structural biochemistry and activa-tion of matrix metalloproteinases. *Curr. Opin. Cell Biol.* **5,** 891–897.

Lauhio, A., Leirisalo-Repo, M., Lahdevirta, J., Saikku, P., and Repo, H. (1991). Double-blind, placebo-controlled study of the three month treatment with lymecycline in reactive arthritis, with special reference to *Chlamydia* arthritis. *Arthritis Rheum.* **24,** 6–14.

Leek, R. D., Lewis, C. E., Whitehouse, R., Greenall, M., Clarke, J., and Harris, A. L. (1996). Association of macrophage infiltration with angiogenesis and prognosis in invasive breast carcinoma. *Cancer Res.* **56,** 4625–4629.

Leonetti, G., and Cuspidi, C. (1996). Choosing the right ACE inhibitor—a guide to selection. *Drugs* **49,** 516–535.

Liotta, L. A., and Stetler-Stevenson, W. G. (1990). Metalloproteinases and cancer invasion. *Sem. Cancer Biol.* **1,** 99–106.

Liotta, L. A., Tryggvason, K., Garbisa, S., Robey, P. G., and Abe, S. (1981). Partial purification and characterisation of a neutral protease which cleaves type IV collagen. *Biochemistry* **20,** 100–104.

Lund, L. R., Romer, J., Thomasset, N., Solberg, H., Pyke, C., Bissell, M. J., Dano, K., and Werb, Z. (1996). Two distinct phases of apoptosis in mammary gland involution: proteinase-independent and -dependent pathways. *Development* **122,** 181–193.

Macaulay, V. M., O'Byrne, K. J., Saunders, M. P., Salisbury, A., Long, L., Gleeson, F., Ganesan, T. S., Harris, A. L., and Talbot, D. C. (1995). Phase I study of matrix metalloproteinase (MMP) inhibitor batimastat (BB-94) in patients with malignant pleural effusions. *Br. J. Cancer* **71,** 11.

Menichetti, A., Tritapepe, L., Ruvolo, G., Speziale, G., Cogliati, A., Digiovanni, C., Pacilli, M., and Criniti, A. (1996). Changes in coagulation patterns, blood loss and blood use after cardiopulmonary bypass—aprotinin vs. tranexamic acid vs. epsilon aminocaproic acid. *J. Cardiovasc. Surg.* **37,** 401–407.

Millar, A., and Brown, P. (1996). 360 patient meta-analysis of studies of marimastat, a novel matrix metalloproteinase inhibitor. *Ann. Oncol.* **7,** 123.

Morphy, J. R., Beeley, N. R. A., and Boyce, B. A. (1994). Potent and selective inhibitors of gelatinase A. 2. Carboxylic acids and phosphonic acid derivatives. *Bioorg. Med. Chem. Lett.* **4,** 2747–2752.

Murray, G. I., Duncan, M. E., O'Neil, P., Melvin, W. T., and Fothergill, J. E. (1996). Matrix metalloproteinase-1 is associated with poor prognosis in colorectal cancer. *Nature Med.* **2,** 461–462.

Naito, K., Nakajima, S., Kanbayashi, N., Okuyama, A., and Goto, M. (1993). Inhibition of metalloproteinase activity of rheumatoid arthritis synovial cells by a new inhibitor [BE16627B; L-N-(N-hydroxy-2-isobutylsuccinamoyl)-seryl-L-valine]. *Agents Actions* **39,** 182–186.

Noble, S., and Faulds, D. (1996). Saquinavir—a review of its pharmacology and clinical potential in the management of HIV infection. *Drugs* **52,** 93–112.

O'Byrne, E. M., Parker, D. T., Roberts, E. D., Goldberg, R. L., MacPherson, L. J., Blancuzzi, V., Wilson, D., Singh, H. N., Ludewig, R., and Ganu, V. S. (1995). Oral administration of a matrix metalloproteinase inhibitor, CGS 27023A, protects the cartilage proteoglycan matrix in a partial meniscectomy model of osteoarthritis in rabbits. *Inflam. Res.* **44,** S117–118.

Parsons, S. L., Watson, S. A., Griffin, N. R., and Steele, R. J. C. (1996). An open phase I/II study of the oral matrix metalloproteinase inhibitor marimastat in patients with inoperable gastric cancer. *Ann. Oncol.* **7,** 47.

Pei, D., and Weiss, S. J. (1995). Furin-dependent intracellular activation of the human stromelysin-3 zymogen. *Nature (London)* **375,** 244–247.

Pei, D., Majmudar, G., and Weiss, S. J. (1994). Hydrolytic inactivation of a breast carcinoma cell-derived serpin by human stromelysin-3. *J. Biol. Chem.* **269,** 25849–25855.

Poole, C., Adams, M., Barley, V., Graham, J., Kerr, D., Louviaux, I., Perren, T., Piccart, M., and Thomas, H. (1996). A dose-finding study of marimastat, an oral matrix metalloproteinase inhibitor, in patients with advanced ovarian cancer. *Ann. Oncol.* **7,** 68.

Pourtaghi, N., Radvar, M., Mooney, J., and Kinane, D. F. (1996). The effect of subgingival antimicrobial therapy on the levels of stromelysin and tissue inhibitor of metalloproteinases in gingival crevicular fluid. J. Periodont. **67**, 866–870.

Primrose, J., Bleiberg, H., Daniel, F., Johnson, P., Mansi, J., Neoptolemos, J., Seymour, M., and Van Belle, S. (1996). A dose-finding study of marimastat, an oral matrix metalloproteinase inhibitor, in patients with advanced colorectal cancer. Ann. Oncol. **7**, 35.

Reich, R., Thompson, E. W., Iwamoto, Y., Martin, G. R., Deason, J. R., Fuller, G. C., and Miskin, R. (1988). Effects of inhibitors of plasminogen activator, serine proteinases, and collagenase IV on the invasion of basement membranes by metastatic cells. Cancer Res. **48**, 3307–3312.

Rosenberg, G. A. (1995). Matrix metalloproteinases in brain injury. J. Neurotrauma **12**, 833–842.

Rubin, S. C., Hoskins, W. J., and Hakes, T. B. (1989). CA 125 levels and surgical findings in patients undergoing secondary operations for epithelial ovarian cancer. Am. J. Obstet. Gynecol. **160**, 667–671.

Saarialho-Kere, U. K., Vaalamo, M., Puolakkainen, P., Airola, K., Parks, W. C., and Karjalainen-Lindsberg, M. L. (1996). Enhanced expression of matrilysin, collagenase and stromelysin-1 in gastrointestinal ulcers. Am. J. Pathol. **148**, 519–526.

Seir, C. F. M., Kubben, F. J. G. M., Ganesh, S., Heerding, M. M., Griffioen, G., Hanemaaijer, R., Vankreiken, J. H. J. M., Lamers, C. B. H. W., and Verspaget, H. W. (1996). Tissue levels of matrix metalloproteinases MMP-2 and MMP-9 are related to the overall survival of patients with gastric carcinoma. Br. J. Cancer **74**, 413–417.

Sledge, G. W., Qulali, M., Goulet, R., Bone, E. A., and Fife, R. (1995). Effect of matrix metalloproteinase inhibitor batimastat on breast cancer regrowth and metastasis in athymic mice. J. Natl. Cancer Res. **87**, 1546–1550.

Stamey, T. A., Kabalin, J. W., and McNeal, J. E. (1989). Prostate-specific antigen in the diagnosis and treatment of adenocarcinoma of the prostate II radical prostatectomy treated patients. J. Urol. **141**, 1076–1083.

Steinberg, W., Gelfand, R., and Anderson, K. (1986). Comparison of the sensitivity and specificity of the CA 19-9 and CEA assays in detecting cancer of the pancreas. Gastroenterology **90**, 343–349.

Stetler-Stevenson, W. G., Krutzsch, H. C., and Liotta, L. A. (1989). Tissue inhibitor of metalloproteinase (TIMP-2). A new member of the metalloproteinase inhibitor family. J. Biol. Chem. **264**, 17374–17378.

Strauch, L. (1972). The role of collagenases in tumour invasion. In "Tissue Interactions in Carcinogenesis" (D. Tarin, ed.), pp 399–434, Academic Press, New York.

Strauss, B. H., Robinson, R., Batchelor, W. B., Chisholm, R. J., Natarajan, M. K., Logan, R. A., Mehta, S. R., Levy, D. E., Ezrin, A. M., and Keeley, F. W. (1996). In vivo collagen turnover following experimental balloon angioplasty injury and the role of matrix metalloproteinases. Circ. Res. **79**, 541–550.

Tamaki, K., Tanzawa, K., Kurihara, S., Oikawa, T., Monma, S., Shimada, K., and Sugimura, Y. (1995). Synthesis and structure-activity relationships of gelatinase inhibitors derived from matlystatins. Chem. Pharm. Bull. **43**, 1883–1893.

Taraboletti, G., Garofalo, A., Belotti, D., Drudis, T., Borsotti, P., Scanziani, E., Brown, P., and Giavazzi, R. (1995). Inhibition of angiogenesis and murine hemangioma growth by batimastat, a synthetic inhibitor of matrix metalloproteinases. J. Natl. Cancer Inst. **87**, 293–298.

Taylor, A. C., Levy, B. M., and Simpson, J. W. (1970). Collagenolytic activity of sarcoma tissues in culture. Nature (London) **228**, 366–367.

Thomas, H., and Adams, M. (1997). A phase I clinical trial of the matrix metalloproteinase inhibitor, marimastat, administered concurrently with carboplatin to patients with relapsed ovarian cancer. Eur. J. Cancer **33**, S246.

Thompson, R. W., and Parks, W. C. (1996). Role of matrix metalloproteinases in abdominal aortic aneurysms. *Ann. N.Y. Acad. Sci.* **800**, 157–174.

Wang, X., Fu, X., Brown, P. D., Crimmin, M. J., and Hoffman, R. M. (1994). Matrix metalloproteinase inhibitor BB-94 (batimastat) inhibits human colon tumour growth and spread in a patient-like orthotopic model in nude mice. *Cancer Res.* **54**, 4726–4728.

Watson, S. A., Morris, T. M., Robinson, G., Crimmin, M., Brown, P. D., and Hardcastle, J. D. (1995). Inhibition of organ invasion by metalloproteinase inhibitor, BB-94 (batimastat) in two human colon metastasis models. *Cancer Res.* **55**, 3629–3633.

Watson, S. A., Morris, T. M., Parsons, S. L., Steele, R. J. C., and Brown, P. D. (1996). Therapeutic effect of the matrix metalloproteinase inhibitor, batimastat, in a human colorectal cancer ascites model. *Br. J. Cancer* **74**, 1354–1358.

Wojtowicz-Praga, S., Low, J., Marshall, J., Ness, E., Dickson, R., Barter, J., Sale, M., McCann, P., Moore, J., Cole, A., and Hawkins, M. J. (1996). Phase I trial of a novel matrix metalloproteinase inhibitor batimastat (BB-94) in patients with advanced cancer. *Invest. New Drugs* **14**, 193–202.

Wood, N. D., Aitken, M., Harris, S., Kitchener, S., Mcclelland, G. R., and Sharp, S. (1996). The tolerability and pharmacokinetics of the cartilage protective agent (RO32-3555) in healthy male volunteers. *Br. J. Clin. Pharmacol.* **42**, 676–677.

Zeng, Z., Cohen, A. M., Zhang, Z., Stetler-Stevenson, W. G., and Guillem, J. G. (1995). Elevated tissue inhibitor of metalloproteinase 1 RNA in colorectal cancer stroma correlates with lymph node and distant metastases. *Clin. Cancer Res.* **1**, 899–906.

Zeng, Z. S., Huang, Y., Cohen, A. M., and Guillem, J. G. (1996). Prediction of colorectal cancer relapse and survival via tissue RNA levels of matrix metalloproteinase-9. *J. Clin. Oncol.* **14**, 3133–3140.

Matrix Metalloproteinases in Tissue Repair

William C. Parks,* Barry D. Sudbeck,*
Glenn R. Doyle,* and Ulpu K. Saariahlo-Kere†

*Departments of Medicine and Cell Biology and Physiology, Washington University School of Medicine,
St. Louis, Missouri 63110, and †Department of Dermatology, University of Helsinki, Helsinki, Finland

I. INTRODUCTION: STUDYING THE REGULATION AND FUNCTION OF METALLOPROTEINASES

Matrix metalloproteinases (MMPs) are not expressed in normal, healthy, resting tissues; at least their production and activity, with notable exceptions, are maintained at nearly undetectable levels. In contrast, in any diseased or inflamed tissue and in essentially any cell type grown in culture, some level of metalloproteinase expression is seen. Though the qualitative pattern and quantitative levels of MMPs vary among diseases, tumor types, inflammatory conditions, and cell lines, a reasonably safe generalization is that activated cells express MMPs. This is neither a controversial nor surprising conclusion. After all, matrix is remodeled in situations associated with disease or tissue activation, such as inflammation, cancer, and normal resorption, and metalloproteinases participate in matrix turnover, or at least we think they do.

Before we can understand the regulation and function of metalloproteinases, we need to know where, when, and by which cells these proteinases are produced. By knowing the temporal and spatial patterns of MMP expression, we can formulate testable hypotheses to address how these enzymes are controlled and activated and to determine the function they serve. Although both stimulators and repressors of MMP production have been identified, the actual physiological action of these modulators, if any, in controlling MMP expression *in vivo* is not known. Deciphering the actual role of hormones, cytokines, and other messengers, such as interactions with matrix and cells, in regulating proteolysis is difficult because the study of one or two modulators in culture does not duplicate the *in vivo* environment where multiple factors and conditions can simultaneously influence gene production and proteinase activity. Cell culture models, however, do provide a defined environment for identifying factors and conditions that mediate changes in gene-specific protein production and for characterizing the regulatory mechanisms involved in these processes. It is likely, given that most tissues and cells express MMPs, that multiple factors control proteolytic cascades and that specific regulatory mechanisms operate in different tissues and in different events.

A common theme among many MMP biologists is to understand the mechanisms controlling expression of these proteinases, and much work has been done to identify factors that induce, stimulate, or repress MMP production. We will not summarize everything that is know about MMP regulation. Not only are there too many observations to assimilate and summarize in a cogent, readable, and reasonably succinct review, but much of what is known about the regulation of individual enzymes is included in the separate articles in this book. Instead, we will discuss collagenase-1 (MMP-1) as a paradigm of how cells control production and release of a specific MMP, focusing on the events and factors regulating expression of human collagenase-1 (MMP-1) and how this enzymes serves a distinct function in repair by remodeling matrix. We will emphasize *in vivo* observations that have been followed up with more defined mechanistic studies on isolated cells.

In addition to an unclear picture of the regulation of MMP expression, the precise and actual function of metalloproteinases in the tissue environment is more presumed than proven. These enzymes, at least the matrixin family (see the first article), are called *matrix* metalloproteinases because they *can* degrade matrix molecules. This designation arose largely from numerous studies demonstrating that purified matrix proteins are degraded by purified, activated MMPs when combined under optimal conditions, but few examples exist demonstrating that matrix degradation in a tissue environment is caused *directly* by the catalytic activity of a metalloproteinase. Most studies provide correlative or associative data that, although often quite convincing of a role in matrix turnover, stop short of providing a firm and direct causative link between MMP activity and protein degradation. For example, overexpression of stromelysin-1 in breast epithelium leads to selective degradation of entactin/nidogen in the basement membrane, which is rescued by expression of tissue inhibitor of metalloprotenaise 1 (TIMP-1) (Alexander *et al.*, 1996), and metalloelastase (MMP-12; see the article by Shapiro and Senior) is required for macrophage-mediated elastolysis and tissue damage (Shipley *et al.*, 1996; Hautamaki *et al.*, 1997). Although these elegant studies relied on defined genetic approaches and represent state-of-the-art work on metalloproteinase biology, the findings do not prove that the metalloproteinase in question acts directly on the targeted matrix protein or if matrix degradation is mediated via another enzyme whose activity is dependent on the metalloproteinase. Similarly, modulation of TIMP-1 expression influences the tumorigenic potential of transformed cells (Khokha *et al.*, 1989), but it has not been established if these effects are due to TIMP's ability to inhibit metalloproteinase or to another activity. The concerns raised here are not meant to diminish the findings contributed by any group, but rather underscore the difficulty of determining the precise mechanistic function of a metalloproteinase in a complex environment.

In contrast, because of the restricted substrate specificity of collagenase-1 and collagenase-2 (neutrophil collagenase, MMP-8), the presence of these enzymes likely does indicate that fibrillar collagen is being cleaved. Indeed, excess collagen is evident in tissues of mice that produce collagenase-resistant type I collagen (Liu *et al.*, 1995), and a neoepitope revealed by collagenase cleavage of type II collagen is evident in osteoarthritic cartilage (Billinghurst *et al.*, 1997). These studies, in addition to others, demonstrate that metalloproteinases, at least collagenases, do act on matrix substrates *in vivo*. However, the mere presence of a metalloproteinase in a tissue sample, either normal, inflamed, or diseased, has led investigators to assume that the proteinases identified are involved in remodeling the local extracellular matrix (ECM). Because most MMPs can degrade a wide range of structurally diverse molecules, including adhesive glycoproteins, core proteins of proteoglycans, hydrophobic, insoluble proteins, such as elastin and collagens, and soluble proteins, such as α_1-antitrypsin, it is likely that this class of proteinases can act on many proteins. Thus, as we consider the potential functions and consequences of metalloproteinase activity in a biological process, we would should not limit our thinking to matrix degradation. That stated, in this article, we uphold dogma and discuss the role of collagenase-mediated cleavage of type I collagen in reepithelialization.

II. COLLAGENASE-1 IN TISSUE REPAIR

A. *Collagenase Expression in Keratinocytes*

Shortly after Jerome Gross and Charles Lapiere isolated and characterized collagenase from the regressing anuran tadpole tail (1962), Eisen and Gross (1965), by carefully dissecting the epidermis from the underlying mesenchyme, determined in which tissue compartment this activity was concentrated. Using an *ex vivo* collagen lysis assay, they found that essentially all collagenolytic activity was recovered from the epidermis. Although some, if not most, of the mesenchyme-derived or -associated activity was not recovered, likely due to high-affinity interaction with substrate, these findings indicate that the epidermis is an important source of collagenase. The restriction of enzyme activity to the epidermis suggests that collagen resorption during metamorphosis is mediated by the epithelium and not by cells within the interstitial matrix compartment.

Cellular responses to wounding have often been suggested to recapitulate developmental processes. Although many of the same genes are expressed, such as matrix proteins and cytokines, the response to injury, especially in adult tissues with inflammatory and immune processes, is quite distinct from normal tissue growth (Martin, 1997). Still, development and wound repair share many common features, such as cell migration, new blood vessel formation, and matrix deposition and remodeling, and hence, the proteolytic demands may be similar for these processes. Indeed, following the studies on tadpole metamorphosis, Grillo and Gross (1967) reported that collagenolytic activity is markedly increased in wounded guinea pig skin, and analogously to that seen in the regressing tadpole tail, most collagenolytic activity was detected in the epidermis; interestingly, specifically at the wound margin.

Although both the tadpole and mammalian studies from Gross's laboratory strongly indicated that collagenolytic activity, at least in the skin, is conferred by epithelial cells, dermal skin fibroblasts became one of the more accepted models of collagenase biology. Indeed, many investigators still call collagenase-1 *fibroblast collagenase.* However, *in vivo* observations in a wide variety of normal and disease-associated tissue remodeling events have convincingly demonstrated that collagenase-1 is expressed by numerous cell types, including epidermal cells, smooth muscle cells, endothelial cells, chondrocytes, and macrophages, as well as fibroblasts (Saarialho-Kere *et al.,* 1992, 1993b; Stricklin *et al.,* 1993; Wolfe *et al.,* 1993; Galis *et al.,* 1994; Nikkari *et al.,* 1994). Thus, the term *fibroblast collagenase* is a bit misleading. In skin, one of the most collagen-rich tissues in the body, and thus a tissue with a strong need for collagenase, basal keratinocytes are the primary source of collagenase-1, not fibroblasts.

B. Collagenase Expression in Cutaneous Wounds

Some years after the work from the Gross laboratory in the 1960s, others studies showed that collagenase protein or activity is present in the wound environment (Buckley-Sturrock *et al.,* 1989; Ågren *et al.,* 1992), and it has often been thought that this enzyme is produced by fibroblasts, macrophages, and other cells within the granulation tissue. However, using *in situ* hybridization and immunohistochemistry in a thorough examination of the role of metalloproteinases in cutaneous wounds, chronic ulcers, necrobiotic lesions, blisters, and carcinomas, expression of collagenase-1 by cells in the dermis (i.e., by fibroblasts or macrophages) was seen in less than half of the samples examined (Saarialho-Kere *et al.,* 1992, 1993a, 1995b; Ståhle-Bäckdahl *et al.,* 1994). When detected in the dermis, the expression of collagenase-1 is typically low and confined to a few cells. Furthermore, collagenase mRNA or protein is not detected in any cell in intact healthy skin. However, collagenase-1 is prominently and invariably expressed by basal keratinocytes at the migratory front in all forms of cutaneous wounds, including normally healing wounds and burns in humans and animals and various forms of blisters and chronic ulcers, with injury that breached the basement membrane (Saarialho-Kere *et al.,* 1992, 1993b, 1995b; Stricklin *et al.,* 1993, 1994; Stricklin and Nanney, 1994; Vaalamo *et al.,* 1996, 1997; Okada *et al.,* 1997). Collagenase-1 is not expressed by the hyperproliferative cells just behind the wound front and residing on basement membrane or by suprabasal cells in intact or wounded skin (Fig. 1). Signal for collagenase-1 is always confined to the basal layer of the epidermis and diminishes progressively away from the wound edge. This same restricted pattern of collagenase-1 expression is seen in wounds created in human skin grafted in SCID mice, in human skin wounded *ex vivo,* and in wounded organotypic

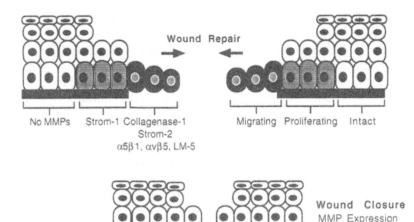

FIG. 1. Spatial patterns of MMP and integrin expression in the epidermis during wound healing. Collagenase-1 is prominently and invariably expressed by migrating basal keratinocytes in all wounds, whether acute or chronic, characterized by disruption of the basement membrane (*black line*). Stromelysin-2, the integrins $\alpha_5\beta_1$ and $\alpha_v\beta_5$, and the basement membrane protein laminin-5 (*LM-5*) are also induced in the migrating cells. Stromelysin-1 is also expressed in the epidermis but by a functionally distinct population of basal keratinocytes. At the completion of reepithelialization, metalloproteinase expression is turned off.

keratinocyte cultures (Inoue *et al.*, 1995; Garlick *et al.*, 1996; Pilcher *et al.*, 1997a) indicating that induction of this MMP is a common response to injury of mammalian skin and suggesting that the activity of collagenase-1 serves a beneficial role in reepithelialization.

C. Collagen Induces Collagenase Expression

Because collagenase-1 was not detected in nonulcerated samples or in intact epidermis, disruption of the basement membrane and subsequent exposure of keratinocytes to the underlying dermal stroma may be a critical determinant for the induction of epidermal collagenolytic activity. Basal keratinocytes normally rest on a basement membrane composed of laminins, entactin/nidogen, proteoglycans, and type IV collagen (Burgeson and Christiano, 1997). During wound healing, keratinocytes migrate from the edge of the wound under a provisional matrix of fibrin and fibronectin (Clark *et al.*, 1982) and over or through the viable dermis, which includes structural macromolecules, such as type I collagen, microfibrils, and elastin, distinct from those in the basement membrane. Thus, loss of contact with the basement membrane and establishment of new cell–matrix interactions with components of the dermal and provisional matrices may be a critical determinant that alters keratinocyte phenotype and induces collagenase-1 production.

Three key *in vivo* observations support the idea that altered cell–matrix interactions regulate the spatially precise pattern of collagenase-1 production in migrating keratinocytes during wound repair. One, immunostaining for type IV collagen or laminin-1, showed that collagenase-1-positive keratinocytes are not in contact with a basement membrane (Saarialho-Kere *et al.*, 1993b, 1995b; Inoue *et al.*, 1995) (Fig. 1) but rather move over the dermal matrix in close apposition to dermal collagen fibers (Pilcher *et al.*, 1997a). Two, collagenase-1 is not expressed by basal keratinocytes in nonulcerative lesions or in blisters, such as bullous pemphigoid, that separate above basement membrane, but this MMP is expressed by migrating keratinocytes in blisters that form below the basement membrane, such as recessive dystrophic epidermolysis bullosa (Saarialho-Kere *et al.*, 1993b, 1995b). Three, in acute wounds (Saarialho-Kere *et al.*, 1993b; Inoue *et al.*, 1995) and in cultured skin equivalent models (Garlick *et al.*, 1996), collagenase-1 expression ceases once reepithelialization is completed and a basement membrane has reformed (Fig. 1). Together, these findings support the idea that keratinocytes acquire a collagenolytic phenotype upon contact with the dermal matrix.

The *in vivo* expression of collagenase-1 by migrating keratinocytes is recapitulated by primary human keratinocytes in culture. When grown in high-calcium medium, keratinocytes form foci of proliferating and differentiated cells surrounded by migrating cells (Pilcher *et al.*, 1997a; Sudbeck *et al.*, 1997a). Reflecting the phenotype of basal cells involved in reepithelialization *in vivo*, collagenase-1 mRNA is expressed only in cells migrating from the colonies of proliferating and differentiated keratinocytes and only if the cells are plated on native type I collagen (Fig. 2). Other molecules that keratinocytes may interact with, such as laminin-1, laminin-5, elastin, fibrinogen/fibrin, fibronectin, and type III collagen, as well as EHS tumor matrix (Matrigel), did not induce collagenase-1 production (Sudbeck *et al.*, 1997b). Cells plated on gelatin generated by heat denaturation or on type I collagen digested with highly purified bacterial collagenase do not express collagenase-1. Furthermore, collagen-mediated induction of collagenase-1 in keratinocytes is detectable by 2 h after plating, similar to kinetics of gene induction seen after wounding normal human skin (Inoue *et al.*, 1995). These findings support the idea that contact with *native* type I collagen is an important, if not the principal, determinant regulating the invariant expression of collagenase-1 by migrating keratinocytes during wound repair. Furthermore, these results imply that interaction with a matrix substrate influences the expression and release of a specific metalloproteinase. In turn, the product of proteolysis, gelatin in the model being discussed here, may repress or no longer sustain MMP expression, thereby providing a feedback signal to control accurately the site and duration of proteinase release. This paradigm implies that the pericellular environment, and the matrix in particular, regulates the pattern and level of MMP expression.

D. Type IV Collagen and Vitronectin

In addition to native type I collagen, contact with vitronectin and type IV collagen also induces collagenase-1 in keratinocytes (Petersen *et al.*, 1990; Sudbeck *et al.*, 1997b). The wound bed is rich in vitronectin (Yamada and Clark, 1996), and contact with this serum glycoprotein may provide an additional matrix-derived signal that can regulate keratinocyte behavior during wound healing. However, the epidermal vitronectin receptor, $\alpha_v\beta_5$, is not present in intact skin but is expressed in migrating basal cells in response to injury (Juhasz *et al.*, 1993). Thus, this matrix protein is likely not involved in rapid induction of collagenase-1 after wounding but may play a role in maintaining the collagenolytic phenotype of migrating keratinocytes. The wound environment is also rich in fibrin and fibronectin, but primary keratinocytes do not attach to these substrates (Guo *et al.*, 1991; Sudbeck *et al.*, 1997b), and keratinocytes in intact skin do not express $\alpha_5\beta_1$ or the α_v subunit (Cavani *et al.*, 1993; Juhasz *et al.*, 1993), the integrins that bind fibronectin and fibrin, respectively (Yamada *et al.*, 1996). As for $\alpha_5\beta_5$, $\alpha_5\beta_1$ is induced in response to wounding and is expressed on the basal surface of the same population of keratinocytes producing collagenase-1 (Fig. 1) (Saarialho-Kere *et al.*, 1993b), and expression of both integrins (Gailit *et al.*, 1994), as well as collagenase-1 (see later discussion), is stimulated by transforming growth factor β type 1 (TGF-β_1). The lack of $\alpha_5\beta_1$ in intact skin supports the conclusion that binding to fibronectin does not induce collagenase-1 expression in keratinocytes. In contrast, fibronectin fragments binding $\alpha_5\beta_1$ stimulate collagenase-1 expression in fibroblasts (Werb *et al.*, 1989) indicating that fundamentally distinct mechanisms control metalloproteinase expression in different cell types.

Although type IV collagen can also mediate induction of collagenase-1 by keratinocytes (Petersen *et al.*, 1990; Sudbeck *et al.*, 1997b), this event is likely an *in vitro* artifact. Because they constitutively express β_1 integrins, keratinocytes in intact skin can interact with type IV collagen in the basement membrane. However, as determined by sensitive techniques, namely, reverse transcription–polymerase chain reaction (RT-PCR) and *in situ* hybridization, keratinocytes in intact skin do not express collagenase-1. In addition, primary keratinocytes grown on EHS tumor matrix, which contains type IV collagen, do not make collagenase-1, presumably because laminin-1 overwhelms any potential inductive effect of type IV collagen. Because type IV collagen in the context of a basement membrane does not induce collagenase-1 expression in keratinocytes, the relevance of its ability to do so in a pure form is questionable. After all, other than that in anchoring plaques, keratinocytes *in vivo* would likely not interact with "pure" type IV collagen.

E. Inhibition of Collagen-Mediated Induction of Collagenase-1

1. Cell–Cell Interactions

In vivo observations by Inoue *et al.* (1995) and Saarialho-Kere *et al.* (1993b) demonstrate that collagenase-1 expression in wounded skin is down-regulated upon completion of reepithelialization. At this stage, cell–cell contacts and keratinocyte interactions with the basement membrane are restored; thus, both or either of these events may regulate the turning off of collagenase-1. Indeed, whereas a strong signal for collagenase-1 mRNA is seen in keratinocytes plated at a subconfluent density, only a weak signal in few cells is seen in confluent cultures (Sudbeck *et al.*, 1997b). Thus, barring cells from migrating (by plating at a high density), and possibly the formation of polarized cell–cell contacts, blocks collagen-mediated induction of collagenase-1. Similarly, collagenolytic activity (presumably, collagenase-3) released by rat mucosal keratinocytes is markedly reduced in confluent cultures compared to the high levels of activity detected in low-density cultures (Lin *et al.*, 1987). Although the mechanism mediating repression of collagenase in high-density culture is not known, cell surface molecules involved in cell–cell interactions are obvious candidates. Analogously, N-CAM-mediated contacts repress metalloproteinase expression in cultured neurons (Edvardsen *et al.*, 1993).

Even though confluent cells are in contact with type I collagen, they do not express collagenase-1. In these experiments, collagenase-1 expression was assessed soon after plating; thus, inhibition of enzyme production was not due to accumulation of keratinocyte-derived basement membrane proteins. These data indicate that contact with type I collagen is not sufficient to induce collagenase-1 expression in keratinocytes; the cells must also have the ability to migrate. Conversely, keratinocyte migration in and of itself does not induce collagenase-1 expression. For example, collagenase-1 is not expressed by keratinocytes involved in reepithelialization of lesions with an intact basement membrane, such as suction blisters or early bullous pemphigoid blisters (Saarialho-Kere *et al.*, 1993b, 1995b). In addition, collagenase-1 is not expressed by cultured keratinocytes on denatured collagen (gelatin) (Sudbeck *et al.*, 1997b), even though these cells can migrate on this substrate (Pilcher *et al.*, 1997a). Thus, induction of collagenase-1 by keratinocytes requires both contact with the appropriate matrix stimulus (native collagen) and the ability of the cells to migrate.

2. BASEMENT MEMBRANE PROTEINS

If contact with dermal matrix mediates the induction of collagenase production by the migrating keratinocyte, then the phenotype of the resting epidermis must be maintained by distinct cell–matrix interactions. In wounded or intact skin, keratinocytes on basement membrane do not express collagenase-1 (Fig. 1). Similarly, primary keratinocytes on a complex mix of basement proteins (EHS tumor matrix) do not express this MMP. Because collagenase-1 expression is induced, albeit artifactually, by *pure* type IV collagen but not by laminin-1 (Sudbeck *et al.*, 1997b), contact with laminin-1 within the newly formed basement membrane may participate in down-regulating collagenase-1 production. Migrating keratinocytes produce and deposit laminin-5 (Fig. 1) (Fini *et al.*, 1996), but this component of the epidermal basement membrane does not inhibit or augment collagen-mediated induction of collagenase-1.

Laminin-1 is deposited in the newly formed basement membrane just behind the migrating front of epidermis (Juhasz *et al.*, 1993) and inhibits keratinocyte migration (Woodley *et al.*, 1988). As reepithelialization progresses, the mass of laminin-1 deposited under the previously migrating keratinocytes would accumulate, providing a site-specific mechanism to down-regulate collagenase-1 expression. Consistent with this idea, relatively small concentrations of laminin-1 can block the inductive effect of native type I collagen (Fig. 2) (Sudbeck *et al.*, 1997b). Assuming one binding event per molecule, one laminin-1 molecule effectively blocked the induction of collagenase-1 mediated by about 3000 collagen molecules. Because type I collagen is so abundant in the dermis, it is not surprising that the inhibitory effect of laminin-1 is so potent.

F. *Integrin-Mediated Induction of Collagenase-1*

In addition to encountering different ECM proteins, migrating keratinocytes also express a distinct pattern of matrix-binding integrins (Fig. 3). Basal keratinocytes constitutively express collagen-binding

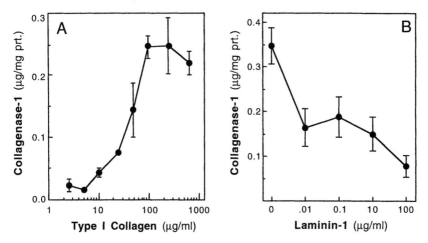

FIG. 2. Collagenase-1 is induced by native type I collagen and repressed by laminin-1. Primary human keratinocytes were grown on dishes precoated with the indicated concentrations of native type I collagen (A) or with 100 μg/mL type I collagen containing the indicated concentrations of laminin-1 (B). Medium was collected 48 h later, and collagenase-1 levels were determined by ELISA and normalized to total protein. Data represent the mean ± SD of triplicate samples (From Sudbeck, B. D., Pilcher, B. K., Welgus, H. G., and Parks (1997). Induction and repression of collagenase-1 by keratinocytes is controlled by distinct components of different extracellular matrix compartments. *J. Biol. Chem.* **272,** 22103–22110, with permission.)

integrins, such as $\alpha_1\beta_1$, $\alpha_2\beta_1$, and $\alpha_3\beta_1$ (Hertle *et al.*, 1992; Pellegrini *et al.*, 1992; Juhasz *et al.*, 1993). The location of $\alpha_2\beta_1$ on the basolateral surface of keratinocytes led to the idea that this integrin has a function in cell–cell interactions (Carter *et al.*, 1990; Symington *et al.*, 1993), but no epidermal ligand has been found for this integrin (Tenchini *et al.*, 1993). Because the epidermis is not normally in contact with type I collagen, it is tempting to speculate that the basal production of collagen-binding integrins, besides being involved in potential cell–cell and basement membrane interactions in the intact skin, keeps keratinocytes primed and ready to respond to injury. The skin is the principal barrier to the outside environment and provides protection from invasion by pathogenic organisms. Hence, it is reasonable that the epidermis is equipped to respond rapidly to injury and begin wound closure rather than relying on cues from later events, such as release of factors from inflammatory cells. Furthermore, a mechanistic connection between collagenase expression and the recognition of type I collagen by the producing cell seems intuitive. After all, cells likely do not release proteases indiscriminately, especially an enzyme-like collagenase-1 with such a defined substrate specificity, but rather rely on precise cell–matrix interactions to remodel adjacent connective tissue accurately. Indeed, collagen-mediated induction of collagenase-1 mRNA and keratinocyte migration are potently inhibited by α_2-blocking antibody but not by antibodies against the α_1 or α_3 subunits (Pilcher *et al.*, 1997a). These data do not indicate that $\alpha_2\beta_1$ binding to collagen is directly required for keratinocyte migration but rather that this cell–matrix interaction mediates induction of collagenase-1, which, in turn, is essential for cell movement (see later discussion).

To facilitate movement, migrating keratinocytes disrupt or loosen cell–cell and cell–matrix contacts, the latter being typified by the diffuse location of the hemidesmosomal integrin, $\alpha_6\beta_4$, at the wound front (Fig. 3). In addition, keratinocytes selectively express other integrins, such as $\alpha_5\beta_1$ and $\alpha_v\beta_5$, and these receptors are present on the same keratinocytes that express collagenase-1 (Saarialho-Kere *et al.*, 1993b).

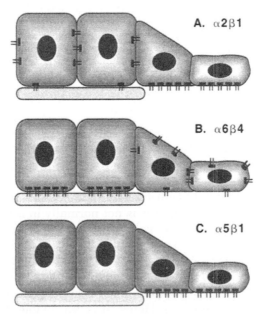

FIG. 3. Distinct patterns of integrins in intact and wounded epidermis. (A) Some constitutively expressed intrigrins, such as $\alpha_2\beta_1$, have a basolateral localization in intact skin. At the wound edge, however, this receptor accumulates at the basal surface of the migratory front. (B) In contrast, $\alpha_6\beta_4$, a component of hemidesmosomes, is located on the basal surface in intact skin and is dispersed along the surface of cells at the wound edge. (C) Other integrins, such as $\alpha_5\beta_1$ and $\alpha_v\beta_3$ are induced in basal keratinocytes at the wound edge and are located on the basal surface.

Although induction of collagenase-1 by synovial fibroblasts grown on fibronectin fragments is mediated by $\alpha_5\beta_1$ (Werb *et al.*, 1989; Huhtala *et al.*, 1995), because, as stated, the $\alpha_5\beta_1$ receptor is not present in intact skin, it is doubtful that this integrin transduces the rapid induction of collagenase-1 expression by keratinocytes following injury.

G. TIMP-1 in Wounds

Although dermal collagen induces collagenase-1 expression, other factors in the wound environment may affect collagenase-1 expression and activity. Keratinocytes are capable of secreting TIMP-1 *in vitro*, but TIMP-1 mRNA seldom colocalizes with collagenase-1 mRNA in migrating keratinocytes in wounds (Saarialho-Kere *et al.*, 1992; Stricklin *et al.*, 1993, 1994; Stricklin and Nanney, 1994). The distinct localization of enzyme and inhibitor suggests that keratinocyte-derived collagenase-1 acts without impedance from TIMP-1, and this actually makes sense. As for most biological processes, matrix degradation is a precise event. Proteinases are produced and released on demand from cells activated to degrade matrix proteins. By means of specific cell surface receptors, the cell recognizes a particular matrix molecule and is instructed to produce the appropriate metalloproteinase, which is then released into the pericellular space where it degrades its specific substrate. It is possible that sites of matrix degradation are isolated by reorganization of cell membrane, analogous to the ruffled border of osteoclasts. As the cell moves beyond the site of matrix degradation, excess or spent proteinase would spill into the open extracellular space. Thus, TIMPs may act in the tissue environment to neutralize "spent" proteinases thereby preventing excessive and unwanted degradation away from the sites of metalloproteinase production. The presence of TIMP-1 beneath the epidermis may provide a mechanism to contain collagenolytic activity to the epidermal front.

H. Cytokines and Collagenase-1 Expression by Keratinocytes

Expression of collagenase-1 is also regulated by many soluble factors (see the following article), and these may have a role in modulating expression in normal wounds and overexpression of proteinase in chronic wounds (see later discussion). Indeed, many cytokines are present in the wound environment (Mauviel and Uitto, 1993), and the epidermis is the source of many soluble mediators (Kupper, 1990; McKay and Leigh, 1991). Whereas contact with dermal collagen is necessary to induce gene expression, the pattern and quantity of cytokines may regulate the net output of collagenase-1 by keratinocytes. Collagenase-1 expression is up-regulated by hepatocyte growth factor

(HGF), but interestingly, this stimulation was seen only in keratino-cytes plated on collagen, not in cells on gelatin or EHS tumor matrix (Dunsmore *et al.*, 1996). These data further support the contention that cell–matrix interactions are required for collagenase-1 expression in the epidermis.

In addition, the regulation of collagenase-1 by soluble factors is differ-ent in keratinocytes than in other cell types, and these cell-specific mechanisms may be contribute to controlling the site and level of MMP expression in skin. For example, modulation of intracellular calcium levels blocks secretion of collagenase-1 from keratinocytes without af-fecting mRNA levels (Sudbeck *et al.*, 1997a), whereas these same condi-tions stimulate active synthesis in fibroblasts (Unemori and Werb, 1988; Lohi *et al.*, 1994; Sudbeck *et al.*, 1997a). Furthermore, fibroblast growth factor 2 (FGF-2) and keratinocyte growth factor (FGF-9/KGF) repress collagenase-1 production in keratinocytes but stimulate en-zyme production in mesenchymal cells (Pilcher *et al.*, 1997b). Interest-ingly, FGF-2 and KGF are not seen in migrating keratinocytes, where collagenase-1 is exclusively expressed, but rather bind and apparently activate the hyperproliferative population (Kurita *et al.*, 1992; Werner *et al.*, 1992) whose daughters become, in part, the migratory/collagenase-1-expressing keratinocytes. Thus, these related cytokines, which bind to the same receptor, may down-regulate collagenase-1 expression in cultured cells by promoting differentiation to a prolifera-tive phenotype. *In vivo*, blocking KGF signaling by a dominant-negative receptor strategy impairs reepithelialization by blocking an important stimulus of keratinocyte proliferation (Werner *et al.*, 1993, 1994). It is not clear, however, if hindered proliferation would also lead to a reduc-tion in the steady-state number of keratinocytes with a migratory and collagenolytic phenotype. Contact with dermal collagen may be suffi-cient to mediate this change in keratinocyte behavior.

TGF-β_1 is considered to be a fibrogenic factor because it stimulates production of structural matrix proteins and TIMP-1, while repress-ing expression of collagenase-1 in fibroblasts (Mauviel and Uitto, 1993; Zhang *et al.*, 1995). In keratinocytes, however, TGF-β_1 stimulates collagenase-1 production (Mauviel *et al.*, 1996). Similarly, interferon-γ up-regulates collagenase-1 production in keratinocytes (Tamai *et al.*, 1995) but inhibits expression in dermal fibroblast (Varga *et al.*, 1995). Collectively, these findings indicate that the regulation of collagenase-1 in keratinocytes is unique compared to that in other cell types. Be-cause TGF-β_1 receptors are identical in keratinocytes and fibroblasts, cell-specific intracellular signaling pathways likely dictate whether collagenase-1 is stimulated or repressed. Indeed, Mauviel *et al.* (1996) demonstrated that composition of Jun–Fos heterodimers interacting with the collagenase-1 promoter in response to TGF-β_1 differs between keratinocytes and fibroblasts leading to opposite responses in gene transcription and, consequently, enzyme production.

The distinct regulation of collagenase-1 in keratinocytes compared to that in mesenchymal cells suggests that this MMP serves a fundamentally different role in the wounded epidermis than it does in other tissue compartments. In addition, the stage at which collagenase-1 activity is needed in the epidermis may differ from the temporal demands in the interstitial compartment. Indeed, when collagenase-1 is expressed in the migrating epidermis, its activity would potentially be detrimental in the dermal wound bed where deposition of new collagen is required for repair and tissue strength. As discussed later, collagenase-1 released by keratinocytes probably has little do to with gross matrix remodeling but rather modifies the underlying matrix to facilitate cell movement.

III.   The Role of Collagenase-1 in Reepithelialization

A.   Collagenase-1 and Cell Migration

The invariable expression of collagenase-1 by basal keratinocytes in all forms of wounds and the confinement of its expression to periods of active reepithelialization suggest that this enzyme facilitates cell migration during wound repair. Beyond directly remodeling structural proteins, such as during morphogenesis and tissue resorption, MMPs are thought to break down ECM barriers that impede cell migration. Clearly, this is a reasonable role for these proteinases in facilitating cell movement through a three-dimensional matrix, as is seen during blastocyst invasion (Librach et al., 1991), angiogenesis (Fisher et al., 1994), and extravasation and infiltration of inflammatory cells (Shipley et al., 1996). During normal reepithelialization, however, keratinocytes migrate along a path of least resistance, dissecting underneath the scab while remaining superficial to the underlying viable dermis and wound bed (Stenn and Malhotra, 1992). Thus, epidermal repair involves cell migration in a two-dimensional plane rather than through a three-dimensional matrix-rich environment. Recent findings from our group indicate that keratinocytes use collagenase-1 to cleave collagen to gelatin, thereby providing a substrate that is more conducive to migration. Without collagenolytic activity, keratinocytes do not migrate on a collagen-containing matrix (Fig. 4).

The specificity of MMP-1 in promoting keratinocyte movement was demonstrated in distinct experiments using two different migration assays. For one assay, human keratinocytes were plated at a confluent density on type I collagen within a cloning chamber, and cell migration

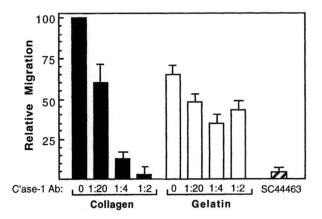

FIG. 4. Keratinocyte migration on type I collagen is collagenase-1 dependent. Primary human keratinocytes were plated on culture slides coated with colloidal gold particles and type I collagen or gelatin and the indicted titers of affinity-purified collagenase-1 antibody or 25-μM SC44463. Keratinocyte migration was quantified 20 h later by quantifying the linear tracks created in the gold substrate. The data shown are the means ± SEM of duplicate samples from four experiments. (Reproduced from *The Journal of Cell Biology,* Pilcher *et al.,* 1997, **137,** 1445–1147 by copyright permission of The Rockefeller University Press.)

was monitored for 24–96 h after removal of the chamber. For the other assay, keratinocytes were plated at low density on chamber slides coated with a colloidal gold–type I collagen mixture, and the migration of single cells, which leave a path of phagokinetic tracks devoid of gold particles, is quantified. In these assays, keratinocyte migration on native collagen was completely blocked by treatment of broad-acting peptide hydroxymate inhibitor, SC44463 (Fig. 4). This compound, similar to other synthetic MMP inhibitor (see article by Brown), is a substrate-based inhibitor containing a hydroxamic acid moiety, which chelates the active site zinc cation and renders MMPs catalytically inactive and has a K_i of about 1 nM for pure collagenase-1 in solution (Moore and Spilburg, 1986). Another hydroxymate, SC44201, is a stereoisomer of SC44463 differing only in the plane of a single hydrogen atom and is a weak inhibitor of MMP activity, with a relative K_i for all MMPs over 1000-fold greater than that of SC44463 (Moore and Spilburg, 1986). Consistent with the idea that keratinocyte migration on type I collagen is MMP dependent, SC44201 did not affect cell movement (Pilcher *et al.,* 1997a).

In another experiment, HaCaTs, a spontaneously transformed line of human keratinocytes (Boukamp *et al.,* 1988) that does not express collagenase-1 in response to collagen (Sudbeck *et al.,* 1997a), did not migrate on a type I collagen matrix but moved efficiently on denatured type I collagen (gelatin) (Pilcher *et al.*, 1997a). Epidermal growth factor, which induces MMP-1 production by HaCaT cells, resulted in the ability of these cells to migrate across a type I collagen matrix, and this effect was completely inhibited by the peptide hydroxymate compound. Because SC44463 inhibits a broad spectrum of metalloproteinases and because keratinocytes have the potential to express other MMPs, namely, 92-kDa gelatinase (Sudbeck *et al.*, 1997a), stromelysin-1, and stromelysin-2 (Windsor *et al.*, 1993; Saarialho-Kere *et al.*, 1994), these experiments in and of themselves do not demonstrate that the activity of collagenase-1 is required for keratinocyte migration on collagen.

The requirement of collagenolytic activity for cell motility was tested by assessing keratinocyte migration on a collagenase-resistant mutant type I collagen. Human collagenases cleave fibrillar type I collagen at Gly_{775}–Ile_{776} in the $\alpha_1(I)$-chains and at Gly_{775}–Leu_{776} in the $\alpha_2(I)$-chain (see the article by Jeffrey), and these sites are conserved among mammalian type I collagens. This cleavage renders the molecule thermally unstable and susceptible to further degradation by other proteinases. Steven Krane and co-workers made a double substitution of Pro for Glu_{774} and Ala_{777} and Met for Ile_{776} in the region of the collagenase cleavage site of the murine $\alpha_1(I)$-chain, rendering the molecule resistant to proteolysis by collagenases (Wu *et al.*, 1990; Liu *et al.*, 1995). Keratinocytes plated on collagenase-resistant mutant collagen do not migrate, yet the cells express collagenase-1 and adhere equally to those on wild type collagen (Pilcher *et al.*, 1997a) indicating that the cell-recognition site is distinct from the enzyme cleavage site.

Because all three human collagenases can make the same cleavage in type I collagen, the experiment with the mutant substrate did not demonstrate that collagenase-1 is specifically required for keratinocyte migration on a collagen matrix. Although RT-PCR analysis indicated that normal human keratinocytes express only collagenase-1 and do not produce collagenase-2 (MMP-8) or collagenase-3 (MMP-13) (Pilcher *et al.*, 1997a), functional data are needed to demonstrate a dependence on collagenase-1. Reagents that selectively block the activity of collagenase-1 verified that this MMP alone is required for keratinocyte on a collagen-containing matrix. Keratinocyte migration is completely inhibited by affinity-purified anti-MMP-1 antiserum (Fig. 4), and the concentrations of collagenase-1 antibody that inhibit cell migration are the same as those that block activity of pure enzyme in solution (Pilcher *et al.*, 1997a). Interestingly, collagenase-1 antibody does not affect keratinocyte migration on gelatin (Fig. 4), supporting the idea that keratinocytes need the catalytic activity of collagenase-1 to denature native collagen to migrate on this matrix. Cell migration on gelatin is slightly reduced compared to that on collagen, probably because the cells do not adhere as efficiently to the heat-denatured substrate. Furthermore, treatment with a new generation hydoxymate compound that inhibits all MMPs *except collagenase-1* (collagenase sparing) does not affect keratinocyte migration (J. A. Dumin and W. C. Parks, unpublished observations). Importantly, these data demonstrate that the proteolytic activity of collagenase-1, and not that of any other MMP acting on cleaved collagen or another substrate, is required for keratinocyte migration on native type I collagen.

B. *Collagenase-1 Facilitates Cell Directionality*

Based on the above preceding observations, we propose that collagenase-1 acting on its principal substrate in the dermis, type I collagen, provides migrating keratinocytes with a mechanism to maintain their course and directionality in the wound environment during reepithelialization (Fig. 5). As discussed, basal keratinocytes constitutively express the type I collagen-binding integrin $\alpha_2\beta_1$ along their basolateral surfaces. In wounded epidermis, migrating keratinocytes continue to express their collagen-binding receptors (Hertle *et al.*, 1992; Cavani *et al.*, 1993; Juhasz *et al.*, 1993), but $\alpha_2\beta_1$ becomes redistributed and concentrated at the frontobasal end of the cells (Guo *et al.*, 1991). This redistribution places $\alpha_2\beta_1$, in particular, where it would likely come into intimate contact with dermal type I collagen (Fig. 5). Although basal keratinocytes also express $\alpha_1\beta_1$ and $\alpha_3\beta_1$, keratinocytes preferentially use $\alpha_2\beta_1$ to bind to type I collagen (Lange *et al.*, 1994), and as stated, blocking the ability of this integrin to bind type I collagen completely inhibits collagenase-1 expression and, consequently, keratinocyte migration (Pilcher *et al.*, 1997a).

The $\alpha_2\beta_1$ binds native collagen with high affinity (Staatz *et al.*, 1989); thus, clustering of this integrin at the forward edge of keratinocytes may actually tether the cells to the matrix, rendering them unable to migrate. Therefore, the proteolytic activity of collagenase-1 may aid in dissociating keratinocytes from these high-affinity attachments to a collagen matrix. As stated, collagenase-1 does not degrade fibrillar type I collagen but rather makes a single, site-specific cleavage through the triple helix about 3/4 the length from the N terminus. The resultant TC^A and TC^B fragments are thermally unstable at body temperature and can spontaneously denature into gelatin (see article by Jeffrey). Besides being highly susceptible to complete degradation by different proteinases, gelatin binds $\alpha_2\beta_1$ with a much lower affinity than does native collagen (Staatz *et al.*, 1989). Thus, by cleaving type I collagen, which then denatures into gelatin, collagenase-1 effectively mediates the loosening of the tight contacts keratinocytes might establish with the dermal matrix. Similarly, treatment of collagen substrata with *Clostridial* collagenase, which is enzymatically distinct from human collagenase-1 and which is used therapeutically for wound debridement (Sorodd and Sasvary, 1994), enhances keratinocyte migration (Herman, 1996). This function is distinct from the often suggested idea that migrating cells use metalloproteinases to remove matrix barriers that may physically impede movement.

Although collagenase-1 facilitates keratinocyte migration by affecting the conformation of type I collagen and, consequently, the avidity with which cells interact with it, one can argue that this is an inherently inefficient mechanism. If activated keratinocytes migrate *over* the viable dermis by interacting with provisional matrix proteins, rather than *through* matrix, then why do they need to cleave type I collagen? Why would they adhere to the dermis with high affinity if their objective is to close the wound as quickly as possible? The answer, we believe, is twofold. First, contact with dermal collagen provides keratinocytes with an unambiguous spatial signal that the basement membrane is breached and that a wound has occurred. Second, the process of interacting with and then cleaving type I collagen provides keratinocytes with a mechanism to determine and maintain their directionality during reepithelialization. In support of this idea, cells plated on a native collagen matrix organize the substrate into tense fibrils between foci of keratinocytes (J. A. Dumin and W. C. Parks, unpublished observations). Individual cells can be seen moving along these organized fibrils, suggesting that stressed collagen provides a directional path for keratinocytes during reepithelialization. Blocking antibodies to $\alpha_2\beta_1$ inhibit matrix organization and keratinocyte migration, and these data provide a mechanistic and functional link among the matrix (collagen), the cell surface ($\alpha_2\beta_1$), and collagenase-1 expression and activity.

An important observation relevant to the directionality hypothesis is that collagenase-1 production is induced in keratinocytes by native type I collagen but not by denatured forms of the molecule (Sudbeck *et al.*, 1994, 1997a). Thus, collagenase-1 acting on collagen creates a mediator that does not support or maintain its own production. The conversion of collagen to gelatin would replace the inductive stimulus with a neutral substrate (gelatin) and, in stationary cells, collagenase-1 expression would decline. Indeed, collagenase-1 expression is rapidly turned off at the completion of reepithelialization (Inoue *et al.*, 1995). Although cell–cell contacts may be involved in this process, the initial expression of collagenase-1 may mediate cleavage of the collagen substrate, thereby neutralizing the inductive effect of the underlying matrix. If keratinocytes continue to interact with type I collagen, presumably by migrating, then they would continue to express collagenase-1, which they do throughout reepithelialization.

During wound healing *in vivo,* collagenase-1 cleavage of collagen would leave a trail of denatured collagen (gelatin), which would not attract keratinocyte attachment. Using high-affinity interactions with native type I collagen as a "molecular compass," keratinocytes could then bind to components of the provisional matrix to support motility. Because gelatin does not induce or maintain collagenase-1 expression, if keratinocytes were to begin to stray, they would not have the proteinase needed to invade the dermis. The activity of collagenase-1 would allow keratinocytes to break away from collagen while they continually extend and interact with new native collagen molecules present in the superficial plane of the viable dermis. In a stratified epithelium, cell migration is thought to proceed in a leap-frog fashion whereby the cell at the front extends along and attaches to the matrix and is replaced by cells coming from behind and above (Stenn and Malhotra, 1992). Thus, these interrelated collagenolytic and migratory processes likely occur within a limited microenvironment. To reiterate, this process is distinct from the proteolytic activities associated with the spread and metastasis of epidermal cancer. For example, cells derived from squamous cell carcinomas express collagenase-3 (MMP-13), a distinct MMP that has gelatinolytic activity (see article by Jeffrey) and is not expressed in normal epidermis (Johansson *et al.,* 1997; Vaalamo *et al.,* 1997). Thus, the qualitative pattern of MMP expression may significantly influence how and to what extent cells move through or on tissue.

Once the $\alpha_2\beta_1$–collagen contacts are loosened, keratinocytes would then be able to use other integrins to migrate on other connective tissue proteins (Fig. 5). In response to injury, various matrix proteins, particularly fibronectin and vitronectin, are expressed and deposited in high concentrations into the provisional wound bed (Yamada and Clark, 1996). Furthermore, additional integrins, namely, fibronectin-binding $\alpha_5\beta_1$ and vitronectin-binding $\alpha_v\beta_5$, are induced in the same migrating keratinocytes that express collagenase-1 (Hertle *et al.,* 1992; Cavani *et al.,* 1993; Juhasz *et al.,* 1993; Saarialho-Kere *et al.,* 1993b). The $\alpha_5\beta_1$ integrin is expressed by keratinocytes migrating out of skin explants, and blocking the function of this integrin inhibits keratinocyte migration on fibronectin matrices (Takashima and Grinnell, 1984, 1985). Similarly, keratinocytes migrate on vitronectin, and antibodies against the $\alpha_v\beta_5$ integrin inhibit this response (Kim *et al.,* 1994). Thus, fibronectin and vitronectin are among the wound matrix proteins that can provide a substrate for keratinocyte migration during reepithelialization (Fig. 5).

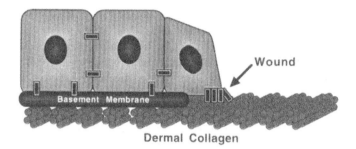

Dermal Collagen

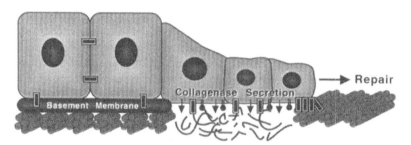

FIG. 5. Collagenase-1-dependent keratinocyte migration during epidermal wound repair. After wounding, basal keratinocytes dislodge from the basement membrane and contact dermal type I collagen. Collagen-binding integrins (□), such as $\alpha_2\beta_1$, are constitutively expressed on basal keratinocytes and are seen on the basal and lateral surfaces of cells in intact skin, but in wounded skin, these receptors accumulate on the frontobasal surface of migrating keratinocytes. The $\alpha_2\beta_1$ binds dermal collagen, and this high-affinity interaction may induce collagenase-1 expression as well as that of new integrins, such as $\alpha_5\beta_1$ and $\alpha_v\beta_5$ (⬇, ↓). High-affinity binding to collagen, however, may hinder cell motility. Cleavage of type I collagen by collagenase-1, and its subsequent conversion to gelatin, would reduce the affinity of $\alpha_2\beta_1$'s interaction with the matrix, thereby allowing the keratinocytes to use the newly expressed integrins to migrate on their ligands, such as fibronectin and vitronectin, which are abundant in the wound bed matrix. The high-affinity interaction of $\alpha_2\beta_1$ with dermal collagen, but not with gelatin, provides the migrating cells with a mechanism to control their direction and to remain superficial during reepithelialization.

IV. Expression of Other Metalloproteinases in Wound Repair

A. Stromelysins

Other MMPs are expressed by human keratinocytes during wound repair. For example, stromelysin-2 (MMP-10) is produced by the same basal keratinocytes that express collagenase-1, whereas stromelysin-1 (MMP-3) is seen in a spatially distinct population of basal keratinocytes adjacent to, but removed from, the wound edge (Fig. 1) (Saarialho-Kere *et al.,* 1994; Madlener *et al.,* 1996; Vaalamo *et al.,* 1996). Stromelysin-2 is closely related to stromelysin-1, both by structure and substrate specificity (see the article by Nagase), but until it was seen in wound healing, stromelysin-2 was only found in a few cancer cells (Sirum and Brinckerhoff, 1989; Murphy *et al.,* 1991). Unlike collagenase-1 and stromelysin-1, expression of stromelysin-2 is confined strictly to the epidermis. Because stromelysins can activate procollagenase (again, see Nagase article), stromelysin-2 may have a role in regulating collagenolytic activity at the migratory front of the epidermis. Similar to collagenase-1, stromelysin-2 may facilitate keratinocyte migration by degrading noncollagenous matrix molecules or by removing damaged basement membrane.

Stromelysin-1 is also expressed in the basal epidermis, but the keratinocytes expressing this metalloproteinase are removed from the migrating front and are in contact with an intact basement membrane (Fig. 1). Because it is produced by proliferating keratinocytes, stromelysin-1 is probably not involved in reepithelialization per se but rather is needed for restructuring the newly formed basement membrane. In further contrast to stromelysin-2, stromelysin-1 is abundantly expressed by dermal fibroblasts in the granulation tissue associated with wounds (Saarialho-Kere *et al.,* 1994; Vaalamo *et al.,* 1996). Because of its broad substrate specificity, stromelysin-1 may be an important enzyme in remodeling the dermal matrix during wound repair and may function with collagenase-3, which is expressed in the same location at the same stage of repair (Vaalamo *et al.,* 1997). Similar patterns of MMP expression are seen in wounded rodent skin (Okada *et al.,* 1997). Stromelysin-1 is also produced following injury of cultured airway epithelial cells (Buisson *et al.,* 1996) suggesting that this metalloproteinase may serve a common role in the repair of diverse epithelial structures. As suggested in the introduction of this article, the catalytic activity of stromelysin-1 (and of stromelysin-2 as well) may have functions beyond or unrelated to matrix remodeling. It is hoped that the precise role of these metalloproteinase in tissue repair will be revealed in studies with genetically defined animals.

Because stromelysin-2 and collagenase-1 are expressed by the same cells (Fig. 1), contact of migrating keratinocytes with the dermal matrix may influence production of both these enzymes, and indeed, stromelysin-2 is expressed in keratinocytes cultured on a type I collagen substratum (Saarialho-Kere *et al.,* 1993b). Whatever does control stromelysin-2 production in keratinocytes is likely to be different from what regulates stromelysin-1 expression, especially in light of the distinct localization of these enzymes in the wound environment. Although both enzymes share a high degree of amino acid sequence homology and similar proteolytic properties (see Nagase article), the promoter regions of these two genes are quite disparate (Sirum and Brinckerhoff, 1989). Thus, whereas stromelysin-1 is synthesized by many cell types and is stimulated by a variety of cytokines, stromelysin-2 production is seemingly limited to epithelial cells (Windsor *et al.*, 1993). Furthermore, although the expression of both enzymes is induced by phorbol ester, tumor necrosis factor α (TNF-α), epidermal growth factor, KGF, and TGF-β_1, stromelysin-2 production is not influenced by interleukin-1, interleukin-6, or platelet-derived growth factor (Brinckerhoff *et al.*, 1992; Windsor *et al.*, 1993; Madlener *et al.*, 1996; Pilcher *et al.*, 1997b).

In contrast to collagenase-1 and stromelysin-2, stromelysin-1-producing keratinocytes are in contact with an underlying basement membrane, just as are resting keratinocytes in intact skin, which do not produce stromelysin-1. Thus, the primary stimulus for stromelysin-1 expression in wound healing is probably not cell interaction with a matrix molecule but rather exposure to a soluble factor. Although the identity of this inducing factor is not known, HGF stimulates stromelysin-1 expression in keratinocytes (Dunsmore *et al.*, 1996). Because this factor also up-regulates collagenase-1 production, HGF may be a common regulator of metalloproteinase expression in the epidermis. Other cytokines, such as TNF-α (MacNaul *et al.*, 1990), interleukin-1 (Frisch and Ruley, 1987), epidermal growth factor (Kerr *et al.*, 1988), and platelet-derived growth factor (Matrisian, 1992), stimulate stromelysin-1 expression in cultured fibroblasts and may also be active on keratinocytes. However, and similar to collagenase-1, the response of stromelysin-1 to some cytokines differs between keratinocytes and mesenchymal cells. For example, TGF-β_1 and interferon γ down-regulate stromelysin-1 expression in interstitial cells, both in cell culture (Edwards *et al.*, 1987; Varga *et al.*, 1995) and *in vivo* (Quaglino *et al.*, 1991), but augment enzyme production by keratinocytes (Windsor *et al.*, 1993; Tamai *et al.*, 1995). However, because stromelysin-1 is expressed in both keratinocytes and dermal fibroblasts at the same stage of repair, the production of these cytokines would have to be precisely localized and controlled to induce biosynthesis of stromelysin-1 by keratinocytes without inhibiting its production in the dermis.

B. Gelatinases

Other matrix metalloproteinases with broad catalytic activity, such as the 72- and 92-kDa gelatinases (MMP-2/Gel A and MMP-9/Gel B, respectively), may be important in releasing keratinocytes from the basement membrane prior to lateral movement at the beginning of epithelial wound healing, and both gelatinases are transiently seen in epidermal cells shortly after wounding (Salo *et al.,* 1994; Okada *et al.,* 1997). Gel B is produced by human keratinocytes plated on plastic (Salo *et al.,* 1991), and its expression is stimulated by growth on collagen and by TGF-β_1 (Woodley *et al.,* 1986; Sarret *et al.,* 1992). In addition, up-regulation of 92-kDa gelatinase is seen in the epithelial layer of healing rabbit corneal wounds (Fini *et al.,* 1992), and hence, the healing response in the cornea may be distinct from that in the skin. Because neutrophils store 92-kDa gelatinase but do not actively make it (Ståhle-Bäckdahl and Parks, 1993), this enzyme may be involved in a wound healing response when these cells are present.

C. Matrilysin

Interestingly, collagenase-1 and stromelysin-1 are not expressed in damaged epithelia of other tissues (Saarialho-Kere *et al.,* 1996; Parks *et al.,* 1997). Instead, matrilysin (MMP-7), which is not produced in the epidermis but is produced by the epithelium of most other tissues (see Wilson and Matrisian article), is expressed in injured lung, intestine, and inflamed exocrine epithelial (Saarialho-Kere *et al.,* 1995a, 1996; Parks *et al.,* 1997). Thus, induction of metalloproteinases is a common and potentially required response of epithelial cells involved in repair processes, and matrilysin is apparently the predominant MMP used by injured lung, intestine, and possibly other tissues. Similar to the induction of collagenase-1 in skin, changes in cell–matrix interactions could modulate matrilysin expression. In addition to the programmed responses of differentiated cells, the collection of constitutive matrix-binding receptors on an epithelial cell and the composition of the underlying interstitial matrix may dictate the pattern of metalloproteinase expression among epithelia. Thus, whereas collagenase-1 facilitates migration of keratinocytes over the collagen-rich matrix of dermis, matrilysin would be a more appropriate proteinase to remodel lung and intestinal interstitial matrix components, which include elastin, adhesive glycoproteins, and proteoglycans.

V. Regulation of Collagenase-1 in Wound Fibroblasts

In addition to the epidermis, collagenase-1 is expressed by dermal fibroblasts in normally healing wounds and chronic ulcers (Saarialho-Kere *et al.,* 1992, 1993b, 1995b; Girard *et al.,* 1993; Vaalamo *et al.,* 1996, 1997; Okada *et al.,* 1997). Typically, however, only a few scattered positive cells are seen, and the signal per cell is much less than that detected in keratinocytes. Dermal expression of collagenase-1 may be limited to certain stages of wound repair, such as resolution of granulation tissue. In contrast to the idea that cell–matrix interactions are required for collagenase-1 induction in keratinocytes, enzyme expression in stromal cells may be more dependent on cytokines. Stimulation of collagenase-1 gene expression by fibroblasts is mediated by an interleukin-1 autocrine loop (West-Mays *et al.*, 1995, 1997) and is augmented by other factors, such as TNF-α epidermal growth factor, and platelet-derived growth factor (Edwards *et al.*, 1987; Unemori *et al.*, 1991). Supportive of the requirement of inflammatory mediators for expression of collagenase-1 by stromal cells, no expression of collagenase-1 is seen in samples of fibrotic ulcers lacking any inflammation or in acute wounds made *ex vivo*, which are also devoid of an inflammatory infiltrate (Saarialho-Kere *et al.*, 1993b; Inoue *et al.*, 1995).

Cell–matrix interactions may also influence metalloproteinase production in dermal fibroblasts. Collagenase-1 expression is markedly increased in dermal fibroblasts grown in collagen (Mauch *et al.*, 1989), and as for keratinocytes, this induction is mediated by $\alpha_2\beta_1$ (Langholz *et al.*, 1995; Riikonen *et al.*, 1995). However, because fibroblasts, unlike keratinocytes, normally reside in a collagen-rich environment, the cells must possess a mechanism to repress collagenase-1 expression in resting tissue where they likely contact collagen. Indeed, many studies have indicated that changes in cell shape are required for expression of collagenase-1 by fibroblasts in collagen gels (Mauch *et al.*, 1989; Werb *et al.*, 1989; Lambert *et al.*, 1992; Xu and Clark, 1997); thus, factors that control fibroblast movement, shape, and contractility may also regulate the expression of MMP-1 in activated cells. In addition, matrix fragments that may be generated and encountered in the wound environment can induce collagenase-1 expression in fibroblasts. Collagenase-1 is not expressed by synovial fibroblasts plated on intact fibronectin, but elevated levels of the enzyme are produced by cells grown on a 120-kDa, RGD-containing fragment of fibronectin (Werb *et al.*, 1989). The inductive effect of this fibronectin fragment is mediated by cooperative signaling through $\alpha_5\beta_1$ and $\alpha_4\beta_1$ integrins (Huhtala *et al.*, 1995). Interestingly, this effect is markedly enhanced by exposure to tenascin (Tremble *et al.*, 1994), an abundant and early deposited component of the wound bed matrix (Latijnhouwers *et al.*, 1996).

Cell–cell interactions may also regulate collagenase-1 expression in fibroblasts. Production of a number of MMPs, including collagenase-1 and stromelysin-1, is stimulated by EMMPRIN (extracellular matrix metalloproteinase inducer), a transmembrane glycoprotein that is attached to the surface of many types of malignant human tumor cells (Guo *et al.*, 1997). EMMPRIN is also expressed by keratinocytes (De-Castro *et al.*, 1996), and this observation led to the idea that migrating keratinocytes would induce collagenase-1 in dermal fibroblasts. However, *in vivo* observations do not support this idea. In human wounds, fibroblasts just under the migratory front of the epidermis— and hence the dermal cells with the greatest potential of interacting with EMMPRIN on keratinocytes—do not express collagenase-1 (Saarialho-Kere *et al.*, 1993b, 1995b; Vaalamo *et al.*, 1996, 1997). These dermal cells do, however, express stromelysin-1 (Saarialho-Kere *et al.*, 1994; Vaalamo *et al.*, 1996, 1997), and epidermal EMMPRIN may regulate expression of this and other MMPs. Thus, as in keratinocytes, the regulation of collagenase-1 production in fibroblasts is seemingly complex, being affected by multiple and diverse stimuli. Importantly, the degradative activity of collagenase-1 in the epidermis and dermis may be involved in distinct processes related to healing that are accomplished by the different cellular compartments. Whereas keratinocytes may cleave dermal collagen at the surface to aid migration and promote reepithelialization, stromal collagenolytic activity may be needed for tissue remodeling associated with granulation and scar formation.

VI. METALLOPROTEINASES IN CHRONIC WOUNDS

The failure of wounds to heal represents a major health care problem, and excess proteinases are thought to be a detriment to proper healing and wound closure. Many studies have shown that metalloproteinases are elevated in various forms of chronic ulcers. In our examination of many samples of pyogenic granuloma, pyoderma gangrenosum, and decubitus and stasis ulcers, the levels of collagenase-1 mRNA were markedly higher and the transcripts were seen over much longer distances of the basal epidermis than the levels and pattern seen in normally healing wounds (Saarialho-Kere *et al.*, 1993b). In addition, the levels of gelatinases, mostly Gel B released from neutrophils, in chronic wound fluid are elevated well above the levels seen in fluid collected from normal wounds (Wysocki *et al.*, 1993; Ågren, 1994; Bullen *et al.*, 1995). These and similar observations have led to a commonly accepted idea that excess metalloproteinases impair wound healing by indiscriminately degrading matrix, cytokines, and other components of the wound environment that are needed for repair. Although this is seemingly a reasonable pathogenic mechanism, and one that holds for other conditions, such as osteoarthritis, no experiments have been done to show that inhibition of metalloproteinase activity or expression in chronic wounds facilitates healing. Because metalloproteinases are expressed in normal wounds and likely serve a needed and beneficial role in repair, treatment of chronic ulcers with MMP inhibitors may have little or no effect—but the experiment still needs to be done.

In other studies, one of us (U. S.-K.) reported that the expression of collagenase-1 and stromelysins was not appreciably different between chronic ulcers and acute wounds (Vaalamo et al., 1996). In previous reviews (Parks, 1995; Mignatti et al., 1996; Parks and Sires, 1996), we suggested that stromelysin-1 and stromelysin-2 were expressed only in chronic wounds and that activity of these enzymes contributed to an inability to heal. However, it has now been shown that the stromelysins are produced by the same populations of basal keratinocytes in both normal and chronic wounds (Saarialho-Kere et al., 1994; Madlener et al., 1996; Vaalamo et al., 1996; Okada et al., 1997). Thus, aberrant production of these enzymes is not a common feature of poorly healing wounds. Interestingly, TIMP-1 levels are markedly reduced in chronic ulcers (Bullen et al., 1995; Vaalamo et al., 1996) suggesting that a proteinase–antiproteinase imbalance contributes to excess damage and impaired healing in many lesions. Whereas overexpression of collagenase-1, coupled with decreased production of TIMP-1, may impair healing, insufficient proteinase activity may lead to the accumulation of wound-associated tissue and delays in wound closure (Ågren and Mertz, 1994). Thus, the properly regulated, site-specific expression of collagenase-1, as well as other MMPs, may be needed for promoting efficient wound repair.

VII. SUMMARY

In this article, we discussed cutaneous wound healing as a model for understanding the role and regulation of metalloproteinases, with an emphasis on collagenase-1, in the tissue environment. Our findings indicate that collagenase-1 expressed as keratinocytes contacts dermal collagen after injury and that its activity is required for reepithelialization over the collagen-rich dermis. Importantly, we do not believe that keratinocyte-derived collagenase-1 functions to remodel matrix grossly; rather that it alters how cells interact with collagen. As stated at the beginning of this article, determining the function that a particular metalloproteinase serves in remodeling is complicated by the potential ability of these enzymes to do more than degrade components of the extracellular matrix. As we enter an era with more use and creation of genetically defined animals, we should have the reagents to determine the processes and factors that regulate metalloproteinase production and activation and to assess and identify the function of these enzymes.

ACKNOWLEDGMENTS

We thank our colleagues, Jo Ann Dumin, Sarah Dunsmore, Catherine Fliszar, Alice Pentland, Brian K. Pilcher, Jill Roby, and Howard G. Welgus, for their assistance with the work discussed in this article. In addition, we thank Dr. Dan Getman and his associates at Monsanto-Searle for the synthetic metalloproteinase inhibitors. Our work on metalloproteinases is supported by grants from the National Institutes of Health, the Monsanto-Searle/Washington University Research agreement, and the Council for Tobacco Research.

REFERENCES

Ågren, M. S. (1994). Gelatinase activity during wound healing. *Br. J. Dermatol.* **131,** 634–640.
Ågren, M. S., and Mertz, P. M. (1994). Are excessive granulation tissue formation and retarded wound contraction due to decreased collagenase activity in wounds in tight-skin mice? *Br. J. Dermatol.* **131,** 337–340.
Ågren, M. S., Taplin, C. J., Woessner, J. F., Eaglstein, W. H., and Mertz, P. M. (1992). Collagenase in wound healing: Effect of wound age and type. *J. Invest. Dermatol.* **99,** 709–714.
Alexander, C. M., Howard, E. W., Bissell, M. J., and Werb, Z. (1996). Rescue of mammary epithelial cell apoptosis and entactin degradation by a tissue inhibitor of metalloproteinases-1 transgene. *J. Cell Biol.* **135,** 1669–1677.
Billinghurst, R. C., Dahlberg, L., Ionescu, M., Reiner, A., Bourne, R., Rorabeck, C., Mitchell, P., Hambor, J., Diekmann, O., Tschesche, H., Chen, J., Van Wart, H., and Poole, A. R. (1997). Enhanced cleavage of type II collagen by collagenases in osteoarthritic articular cartilage. *J. Clin. Invest.* **99,** 1534–1545.
Boukamp, P., Petrussevska, R. T., Breitkreutz, D., Hornung, J., Markham, A., and Fusenig, N. E. (1988). Normal keratinization in a spontaneously immortalized aneuploid human keratinocyte cell line. *J. Cell Biol.* **106,** 761–771.
Brinckerhoff, C. E., Sirum-Connolly, K. L., Karmilowicz, M. J., and Auble, D. (1992). Expression of stromelysin and stromelysin-2 in rabbit and human fibroblasts. *Matrix* Suppl. 1, 165–175.
Buckley-Sturrock, A., Woodward, S. C., Senior, R. M., Griffin, G. L., Klagsbrun, M., and Davidson, J. M. (1989). Differential stimulation of collagenase and chemotatic activity in fibroblasts derived from rat wound repair tissue and human skin by growth factors. *J. Cell. Physiol.* **138,** 70–78.
Buisson, A. C., Gilles, C., Polette, M., Zahm, J. M., Birembaut, P., and Tournier, J. M. (1996). Wound repair-induced expression of a stromelysins is associated with the acquisition of a mesenchymal phenotype in human respiratory epithelial cells. *Lab. Invest.* **74,** 658–669.
Bullen, E. C., Longaker, M. T., Updike, D. L., Benton, R., Ladin, D., Hou, Z., and Howard, E. W. (1995). Tissue inhibitor of metalloproteinases-1 is decreased and activated gelatinases are increased in chronic wounds. *J. Invest. Dermatol.* **104,** 236–240.
Burgeson, R. E., and Christiano, A. M. (1997). The dermal-epidermal junction. *Cur. Opin. Cell Biol.* **9,** 651–658.
Carter, W. G., Wayner, E. A., Bouchard, T. S., and Kaur, P. (1990). The role of integrins alpha 2 beta 1 and alpha 3 beta 1 in cell–cell and cell–substrate adhesion of human epidermal cells. *J. Cell Biol.* **110,** 1387–1404.
Cavani, A., Zambruno, G., Marconi, A., Manca, V., Marchetti, M., and Giannetti, A. (1993). Distinctive integrin expression in the newly forming epidermis during wound repair. *J. Invest. Dermatol.* **101,** 600–604.
Clark, R. A. F., Lanigan, J. M., DellaPelle, P., Manseau, E., Dvorak, H. F., and Colvin, R. B. (1982). Fibronectin and fibrin provide a provisional matrix for epidermal cell migration during wound re-epithelization. *J. Invest. Dermatol.* **79,** 264–269.

DeCastro, R., Zhang, Y., Guo, H., Kataoka, H., Gordon, M. K., Toole, B., and Biswas, G. (1996). Human keratinocytes express EMMPRIN, an extracellular matrix metalloproteinase inducer. *J. Invest. Dermatol.* **106,** 1260–1265.

Dunsmore, S. E., Rubin, J. S., Kovacs, S. O., Parks, W. C., and Welgus, H. G. (1996). Mechanisms of hepatocyte growth factor stimulation of keratinocyte metalloproteinase production. *J. Biol. Chem.* **271,** 24576–24582.

Edvardsen, K., Chen, W., Rucklidge, G., Walsh, F. S., Öbrink, B., and Bock, E. (1993). Transmembrane neural cell-adhesion molecule (NCAM), but not glycosyl-phosphatidylinositol-anchored NCAM, down-regulates secretion of matrix metalloproteinases. *Proc. Natl. Acad. Sci. USA* **90,** 11463–11467.

Edwards, D. R., Murphy, G., Reynolds, J. J., Whitham, S. E., Docherty, A. J. P., Angel, P., and Heath, J. K. (1987). Transforming growth factor modulates the expression of collagenase and metalloproteinase inhibitor. *EMBO J.* **6,** 1899–1904.

Eisen, A. Z., and Gross, J. (1965). The role of epithelium and mesenchyme in the production of a collagenolytic enzyme and a hyaluronidase in the anuran tadpole. *Devel. Biol.* **12,** 408–418.

Fini, M. E., Parks, W. C., Rinehart, W. B., Matsubara, M., Girard, M., Cook, J. R., Jeffrey, J. J., Burgeson, R. E., Raizman, M., Kreuger, R., and Zieske, J. (1996). Role of matrix metalloproteinases in failure to re-epithelialize following corneal injury. *Am. J. Pathol.* **149,** 1287–1302.

Fini, M. E., Girard, M. T., and Matsubara, M. (1992). Collagenolytic/gelatinolytic enzymes in corneal wound healing. *Acta Ophthalmol.* **70,** 26–33.

Fisher, C., Gilbertson-Beadling, S., Powers, E. A., Petzold, G., Poorman, R., and Mitchell, M. A. (1994). Interstitial collagenase is required for angiogenesis *in vitro*. *Dev. Biol.* **162,** 499–510.

Frisch, S. M., and Ruley, H. E. (1987). Transcription from the stromelysin promoter is induced by interleukin-1 and repressed by dexamathasone. *J. Biol. Chem.* **262,** 16300–16304.

Gailit, J., Welch, M. P., and Clark, R. A. (1994). TGF-β1 stimulates expression of keratinocyte integrins during re-epithelialization of cutaneous wounds. *J. Invest. Dermatol.* **103,** 221–227.

Galis, Z. S., Sukhova, G. K., Lark, M. W., and Libby, P. (1994). Increased expression of matrix metalloproteinases and matrix degrading activity in vulnerable regions of human atherosclerotic plaques. *J. Clin. Invest.* **94,** 2493–2503.

Garlick, J. A., Parks, W. C., Welgus, H. G., and Taichman, L. B. (1996). Re-epithelialization of human oral keratinocytes *in vitro*. *J. Dent. Res.* **75,** 912–918.

Girard, M. T., Matsubara, M., Kublin, C., Tessier, M. J., Cintron, C., and Fini, M. E. (1993). Stromal fibroblasts synthesize collagenase and stromelysin during long-term tissue remodeling. *J. Cell Sci.* **104,** 1001–1011.

Grillo, H. C., and Gross, J. (1967). Collagenolytic activity during mammalian wound repair. *Devel. Biol.* **15,** 300–317.

Gross, J., and Lapiere, C. M. (1962). Collagenolytic activity in amphibian tissues: A tissue culture assay. *Proc. Natl. Acad. Sci. USA* **48,** 1014–1022.

Guo, H., Zucker, S., Gordon, M. K., Toole, B. P., and Biswas, C. (1997). Stimulation of matrix metalloproteinase production by recombinant extracellular matrix metalloproteinase inducer from transfected Chinese hamster ovary cells. *J. Biol. Chem.* **272,** 24–27.

Guo, M., Kim, L. T., Akiyama, S. K., Gralnick, H. R., Yamada, K. M., and Grinnell, F. (1991). Altered processing of integrin receptors during keratinocyte activation. *Exp. Cell. Res.* **195,** 315–322.

Hautamaki, R. D., Kobayashi, D. K., Senior, R. M., and Shapiro, S. D. (1997). Requirement for macrophage elastase for cigarette smoke-induced emphysema. *Science* **277,** 2002–2004.

Herman, I. M. (1996). Stimulation of human keratinocyte migration and proliferation *in vitro*—insights into the cellular responses to injury and wound healing. *Wounds— Compend. Clin. Res. Pract.* **8,** 33–41.

Hertle, M. D., Kubler, M.-D., Leigh, I. M., and Watt, F. M. (1992). Aberrant integrin expression during epidermal wound healing and in psoriatic epidermis. *J. Clin. Invest.* **89,** 1892–1901.

Huhtala, P., Humphries, M. J., McCarthy, J. B., Tremble, P. M., Werb, Z., and Damsky, C. H. (1995). Cooperative signaling by $\alpha_5\beta_1$ and $\alpha_4\beta_1$ integrins regulates metalloproteinase gene expression in fibroblasts adhering to fibronectin. *J. Cell Biol.* **129,** 867–879.

Inoue, M., Kratz, G., Haegerstrand, A., and Ståhle-Bäckdahl, M. (1995). Collagenase expression is rapidly induced in wound-edge keratinocytes after acute injury in human skin, persists during healing, and stops at re-epithelialization. *J. Invest. Dermatol.* **104,** 479–483.

Johansson, N., Westermarck, J., Leppa, S., Hakkinen, L., Koivisto, L., Lopez-Otin, C., Peltonen, J., Heino, J., and Kähäri, V. M. (1997). Collagenase 3 (matrix metalloproteinase 13) gene expression by HaCaT keratinocytes is enhanced by tumor necrosis factor alpha and transforming growth factor beta. *Cell Growth Differ.* **8,** 243–250.

Juhasz, I., Murphey, G. F., Yan, H.-C., Herlyn, M., and Albelda, S. M. (1993). Regulation of extracellullar matrix proteins and integrin cell substratum adhesion receptors on epithelium during cutaneous wound healing *in vivo. Am. J. Pathol.* **143,** 1458–1469.

Kerr, L. D., Holt, J. T., and Matrisian, L. M. (1988). Growth factors regulate transin gene expression by c-*fos*-dependent and c-*fos*-independent pathways. *Science* **240,** 1424–1427.

Khokha, R., Waterhouse, P., Yagel, S., Lala, P. K., Overall, C. M., Norton, G., and Denhardt, D. T. (1989). Antisense RNA-induced reduction in murine TIMP levels confers oncogenicity on Swiss 3T3 cells. *Science* **243,** 947–950.

Kim, J. P., Zhang, K., Chen, J. D., Kramer, R. H., and Woodley, D. T. (1994). Vitronectin-driven human keratinocyte locomotion is mediated by the alpha v beta 5 integrin receptor. *J. Biol. Chem.* **269,** 26926–26932.

Kupper, T. (1990). The activated keratinocyte: A model for inducible cytokine production by non-bone marrow-derived cells in cutaneous inflammatory and immune responses. *J. Invest. Dermatol.* 94 (Suppl. 6), 146S–149S.

Kurita, Y., Tsuboi, R., Ueki, R., Rifkin, D. B., and Ogawa, H. (1992). Immunohistochemical localization of basic fibroblast growth factor in wound healing sites of mouse skin. *Arch. Dermatol. Res.* **284,** 193–197.

Lambert, C. A., Soudant, E. P., Nusgens, B. V., and Lapiere, C. M. (1992). Pretranslational regulation of extracellular matrix macromolecules and collagenase expression in fibroblasts by mechanical forces. *Lab. Invest.* **66,** 444–51.

Lange, T. S., Bielinsky, A. K., Kirchberg, K., Bank, I., Herrmann, K., Krieg, T., and Scharffetter-Kochanek, K. (1994). Mg^{2+} and Ca^{2+} differentially regulate beta 1 integrin-mediated adhesion of dermal fibroblasts and keratinocytes to various extracellular matrix proteins. *Exp. Cell Res.* **214,** 381–388.

Langholz, O., Rockel, D., Mauch, C., Kozlowska, E., Bank, I., Krieg, T., and Eckes, B. (1995). Collagen and collagenase gene expression in three-dimensional collagen lattices are differentially regulated by alpha-1-beta-1 and alpha-2-beta-1 integrins. *J. Cell Biol.* **131,** 1903–1915.

Latijnhouwers, M. A., Bergers, M., Van Bergen, B. H., Spruijt, K. I., Andriessen, M. P., and Schalkwijk, J. (1996). Tenascin expression during wound healing in human skin. *J. Pathol.* **178,** 30–35.

Librach, C. L., Werb, Z., Fitzgerald, M. L., Chiu, K., Corwin, N. M., Esteves, R. A., Grobelny, D., Galardy, R., Damsky, C. H., and Fisher, S. J. (1991). 92-kD type IV collagenase mediates invasion of human cytotrophoblasts. *J. Cell Biol.* **113,** 437–449.

Lin, H.-Y., Wells, B. R., Taylor, R. E., and Birkedal-Hansen, H. (1987). Degradation of type I collagen by rat mucosal keratinocytes. Evidence for secretion of a specific epithelial collagenase. *J. Biol. Chem.* **262,** 6823–6831.

Liu, X., Wu, H., Byrne, M., Jeffrey, J. J., Krane, S., and Jaenisch, R. (1995). A targeted mutation at the known collagenase cleavage site in mouse type I collagen impairs tissue remodeling. *J. Cell. Biol.* **130,** 227–237.

Lohi, J., Kähäri, V. M., and Keski-Oja, J. (1994). Cyclosporin A enhances cytokine and phorbol ester-induced fibroblast collagenase expression. *J. Invest. Dermatol.* **102,** 938–944.

MacNaul, K. L., Chartrain, N., Lark, M., Tocci, M. J., and Hutchinson, N. I. (1990). Discoordinate expression of stromelysin, collagenase, and tissue inhibitor of metalloproteinases-1 in rheumatoid human synovial fibroblasts. *J. Biol. Chem.* **265,** 17238–17245.

Madlener, M., Mauch, C., Conca, W., Brauchle, M., Parks, W. C., and Werner, S. (1996). Growth factor regulation of stromelysin-2 expression by keratinocytes: Implications for normal and imparied wound healing. *Biochemistry J.* **320,** 659–664.

Martin, P. (1997). Wound healing—aiming for perfect skin regeneration. *Science* **276,** 75–81.

Matrisian, L. M. (1992). The matrix-degrading metalloproteinases. *BioEssays* **14,** 455–463.

Mauch, C., Adelmann, G. B., Hatamochi, A., and Krieg, T. (1989). Collagenase gene expression in fibroblasts is regulated by a three-dimensional contact with collagen. *FEBS Lett.* **250,** 301–305.

Mauviel, A., and Uitto, J. (1993). The extracellular matrix in wound healing: role of the cytokine network. *Wounds—Compend. Clin. Res. Pract.* **5,** 137–152.

Mauviel, A., Chung, K. Y., Agarwal, A., Tamai, K., and Uitto, J. (1996). Cell-specific induction of distinct oncogenes of the jun family is responsible for differential regulation of collagenase gene expression by transforming growth factor-beta in fibroblasts and keratinocytes. *J. Biol. Chem.* **271,** 10917–10923.

McKay, I. A., and Leigh, I. M. (1991). Epidermal cytokines and their roles in cutaneous wound healing. *Br. J. Dermatol.* **124,** 513–518.

Mignatti, P., Rifkin, D. B., Welgus, H. G., and Parks, W. C. (1996). Proteinases and tissue remodeling. *In* "The Molecular and Cellular Biology of Wound Repair, second edition" (R. A. F. Clark, ed.), pp. 427–474. Plenum Press, New York.

Moore, W. M., and Spilburg, C. A. (1986). Purification of human collagenases with a hydroxamic acid affinity column. *Biochemistry* **25,** 5189–5195.

Murphy, G., Cockett, M. I., Ward, R. V., and Docherty, A. J. P. (1991). Matrix metalloproteinase degradation of elastin, type IV collagen and proteoglycan. A quantitative comparison of the activities of 95 kDa and 75 kDa gelatinases, stromelysins-1 and -2 and punctuated metalloproteinase (PUMP). *Biochem. J.* **277,** 277–279.

Nikkari, S. T., Jarvelainen, H. T., Wight, T. N., Ferguson, M., and Clowes, A. W. (1994). Smooth muscle cell expression of extracellular matrix genes after arterial injury. *Am. J. Pathol.* **144,** 1348–1356.

Okada, A., Tomasetto, C., Lutz, Y., Bellocq, J.-P., Rio, M.-C., and Bassett, P. (1997). Expression of matrix metalloproteinases during rat skin wound healing: Evidence that membrane type-1 matrix metalloproteinase is a stromal activator of pro-gelatinase A. *J. Cell Biol.* **137,** 67–78.

Parks, W. C. (1995). The production, role, and regulation of matrix metalloproteinases in the healing epidermis. *Wounds—Compend. Clin. Res. Pract.* **7,** 23A–37A.

Parks, W. C., and Sires, U. I. (1996). Matrix metalloproteinases and skin biology. *Curr. Opin. Dermatol.* **3,** 240–247.

Parks, W. C., Dunsmore, S. E., Saarialho-Kere, U. K., Roby, J. D., and Welgus, H. G. (1997). Matrilysin is expressed by conducting airway epithelium and alveolar type II pneumocytes in cystic fibrotic lung. Submitted for publication.

Pellegrini, G., De Luca, M., Orecchia, G., Balzac, F., Cremona, O., Savoia, P., Cancedda, R., and Marchisio, P. C. (1992). Expression, topography, and function of integrin receptors are severely altered in keratinocytes from involved and uninvolved psoriatic skin. *J. Clin. Invest.* **89,** 1783–1795.

Petersen, M. J., Woodley, D. T., Stricklin, G. P., and O'Keefe, E. J. (1990). Enhanced synthesis of collagenase by human keratinocytes cultured on type I or type IV collagen. *J. Invest. Dermatol.* **94,** 341–346.

Pilcher, B. K., Dumin, J. A., Sudbeck, B. D., Krane, S. M., Welgus, H. G., and Parks, W. C. (1997a). The activity of collagenase-1 is required for keratinocyte migration on a type I collagen matrix. *J. Cell Biol.* **137,** 1445–1457.

Pilcher, B. K., Gaither-Ganim, J., Parks, W. C., and Welgus, H. G. (1997b). Cell type-specific inhibition of keratinocyte collagenase-1 expression by FGF-2 and FGF-7: A common receptor pathway. *J. Biol. Chem.* **272,** 18147–18154.

Quaglino, D., Nanney, L. B., Ditesheim, J. A., and Davidson, J. M. (1991). Transforming growth factor-beta stimulates wound healing and modulates extracellular matrix gene expression in pig skin: Incisional wound model. *J. Invest. Dermatol.* **97,** 34–42.

Riikonen, T., Westermarck, J., Koivisto, L., Broberg, A., Kähäri, V.-M., and Heino, J. (1995). Integrin alpha 2 beta 1 is a positive regulator of collagenase (MMP-1) and collagen alpha 1(I) gene expression. *J. Biol. Chem.* **270,** 13548–13552.

Saarialho-Kere, U. K., Vaalamo, M., Karjalainen-Lindsberg, M.-L., Airola, K., Parks, W. C., and Puolakkainen, P. (1996). Enhanced expression of matrilysin, collagenase, and stromelysin-1 in gastrointestinal ulcers. *Am. J. Pathol.* **148,** 519–526.

Saarialho-Kere, U. K., Crouch, E. C., and Parks, W. C. (1995a). Matrix metalloproteinase matrilysin is constitutively expressed in human exocrine epithelium. *J. Invest. Dermatol.* **105,** 190–196.

Saarialho-Kere, U. K., Vaalamo, M., Airola, K., Niemi, K.-M., Oikarinen, A. I., and Parks, W. C. (1995b). Interstial collagenase is expressed by keratinocytes which are actively involved in re-epithelialization in blistering skin diseases. *J. Invest. Dermatol.* **104,** 982–988.

Saarialho-Kere, U. K., Kovacs, S. O., Pentland, A. P., Parks, W. C., and Welgus, H. G. (1994). Distinct populations of keratinocytes express stromelysin-1 and -2 in chronic wounds. *J. Clin. Invest.* **94,** 79–88.

Saarialho-Kere, U. K., Chang, E. S., Welgus, H. G., and Parks, W. C. (1993a). Expression of interstitial collagenase, 92 kDa gelatinase, and TIMP-1 in granuloma annulare and necrobiosis lipoidica diabeticorum. *J. Invest. Dermatol.* **100,** 335–342.

Saarialho-Kere, U. K., Kovacs, S. O., Pentland, A. P., Olerud, J., Welgus, H. G., and Parks, W. C. (1993b). Cell–matrix interactions modulate interstitial collagenase expression by human keratinocytes actively involved in wound healing. *J. Clin. Invest.* **92,** 2858–2866.

Saarialho-Kere, U. K., Chang, E. S., Welgus, H. G., and Parks, W. C. (1992). Distinct localization of collagenase and TIMP expression in wound healing associated with ulcerative pyogenic granuloma. *J. Clin. Invest.* **90,** 1952–1957.

Salo, T., Mäkela, M., Kylmäniemi, M., Autio-Harmainen, H., and Larjava, H. (1994). Expression of matrix metalloproteinase-2 and -9 during early human wound healing. *Lab. Invest.* **70,** 176–182.

Salo, T., Lyons, J. G., Rahemtulla, F., Birkedal-Hansen, H., and Larjava, H. (1991). Transforming growth factor-β1 up-regulates type IV collagenase expression in cultured human keratinocytes. *J. Biol. Chem.* **266,** 11436–11441.

Sarret, Y., Woodley, D. T., Goldberg, G. I., Kroneberger, A., and Wynn, K. C. (1992). Constitutive synthesis of a 92-kDa keratinocyte-derived type IV collagenase is enhanced by type I collagen and decreased by type IV collagen matrices. *J. Invest. Dermatol.* **99,** 836–841.

Shipley, J. M., Wesselschmidt, R. L., Kobayashi, D. K., Ley, T. J., and Shapiro, S. D. (1996). Metalloelastase is required for macrophage-mediated proteolysis and matrix invasion in mice. *Proc. Natl. Acad. Sci. USA* **93,** 3942–3946.

Sirum, K. L., and Brinckerhoff, C. E. (1989). Cloning of the genes for human stromelysin and stromelysin 2: Differential expression in rheumatoid synovial fibroblasts. *Biochemistry* **28,** 8691–8698.

Sorodd, H. S., and Sasvary, D. H. (1994). Collagenase ointment and polymyxin B sulfate/bacitracin spray versus silver sulfadiazine cream in partial-thickness burns: A pilot study. *J. Burn Care Rehabil.* **15,** 13–17.

Staatz, W. D., Rajpara, S. M., Wayner, E. A., Carter, W. G., and Santoro, S. A. (1989). The membrane glycoprotein Ia-IIa (VLA-2) complex mediates the Mg^{+2}-dependent adhesion of platelets to collagen. *J. Cell Biol.* **108,** 1917–1924.

Ståhle-Bäckdahl, M., and Parks, W. C. (1993). 92 kDa gelatinase is actively expressed by eosinophils and secreted by neutrophils in invasive squamous cell carcinoma. *Am. J. Pathol.* **142,** 995–1000.

Ståhle-Bäckdahl, M., Inoue, M., Giudice, G. J., and Parks, W. C. (1994). 92-kD gelatinase is produced by eosinophils at the site of blister formation in bullous pemphigoid and cleaves the extracellular domain of the 180 kDa bullous pemphigoid autoantigen. *J. Clin. Invest.* **93,** 2202–2230.

Stenn, K. S., and Malhotra, R. (1992). Epithelialization. *In* "Wound Healing: Biochemical and Clinical Aspects" (I. K. Cohen, R. F. Diegelmann, and W. J. Linblad, eds.), pp. 115–127. W. B. Saunders Co., Philadelphia.

Stricklin, G. P., and Nanney, L. B. (1994). Immunolocalization of collagenase and TIMP in healing human burn wounds. *J. Invest. Dermatol.* **103,** 488–492.

Stricklin, G. P., Li, L., and Nanney, L. B. (1994). Localization of mRNAs representing interstitial collagenase, 72-kDa gelatinase, and TIMP in healing porcine burn wounds. *J. Invest. Dermatol.* **103,** 352–358.

Stricklin, G. P., Li, L., Jancic, V., Wenczak, B. A., and Nanney, L. B. (1993). Localization of mRNAs representing collagenase and TIMP in sections of healing human burn wounds. *Am. J. Pathol.* **143,** 1657–1666.

Sudbeck, B. D., Pilcher, B. K., Pentland, A. P., and Parks, W. C. (1997a). Modulation of intracellular calcium levels inhibits secretion of collagenase-1 by migrating keratinocytes. *Mol. Biol. Cell* **8,** 811–824.

Sudbeck, B. D., Pilcher, B. K., Welgus, H. G., and Parks, W. C. (1997b). Induction and repression of collagenase-1 by keratinocytes is controlled by distinct components of different extracellular matrix compartments. *J. Biol. Chem.* **272,** 22103–22110.

Sudbeck, B. D., Parks, W. C., Welgus, H. G., and Pentland, A. P. (1994). Collagen-mediated induction of keratinocyte collagenase is mediated by tyrosine kinase and protein kinase C activities. *J. Biol. Chem.* **269,** 30022–30029.

Symington, B. E., Takada, Y., and Carter, W. G. (1993). Interaction of integrins $\alpha 3\beta 1$ and $\alpha 2\beta 1$: Potential role in keratinocyte intercellular adhesion. *J. Cell Biol.* **120,** 523–535.

Takashima, A., and Grinnell, F. (1985). Fibronectin-mediated keratinocyte migration and initiation of fibronectin receptor function *in vitro*. *J. Invest. Dermatol.* **85,** 304–308.

Taskashima, A., and Grinnell, F. (1984). Human keratinocyte adhesion and phagocytosis promoted by fibronectin. *J. Invest. Dermatol.* **83,** 352–358.

Tamai, K., Ishikawa, H., Mauviel, A., and Uitto, J. (1995). Interferon-gamma coordinately upregulates matrix metalloprotease (MMP)-1 and MMP-3, but not tissue inhibitor of metalloproteases (TIMP), expression in cultured keratinocytes. *J. Invest. Dermatol.* **104,** 384–390.

Tenchini, M. L., Adams, J. C., Gilberty, C., Steel, J., Hudson, D. L., Malcovati, M., and Watt, F. M. (1993). Evidence against a major role for integrins in calcium-dependent intercellular adhesion of epidermal keratinocytes. *Cell Adhes. Commun.* **1,** 55–66.

Tremble, P., Chiquet-Ehrismann, R., and Werb, Z. (1994). The extracellular matrix ligands fibronectin and tenascin collaborate in regulating collagenase gene expression in fibroblasts. *Mol. Biol. Cell* **5,** 439–453.

Unemori, E. N., and Werb, Z. (1988). Collagenase expression and endogenous activation in rabbit synovial fibroblasts stimulated by the calcium ionophore A23187. *J. Biol. Chem.* **263,** 16252–16259.

Unemori, E. N., Hibbs, M. S., and Amento, E. P. (1991). Constitutive expression of 92-kD gelatinase (type IV collagenase) by rheumatoid synovial fibroblasts and its induction in normal human fibroblasts by inflammatory cytokines. *J. Clin. Invest.* **88,** 1656–1662.

Vaalamo, M., Mattila, L., Johansson, N., Kariniemi, A. L., Karjalainen-Lindsberg, M. L., Kähäri, V. M., and Saarialho-Kere, U. (1997). Distinct populations of stromal cells express collagenase-3 (MMP-13) and collagenase-1 (MMP-1) in chronic ulcers but not in normally healing wounds. *J. Invest. Dermatol.* **109,** 96–101.

Vaalamo, M., Weckroth, M., Puolakkainen, P., Kere, J., Saarinen, P., Lauharanta, J., and Saarialho-Kere, U. K. (1996). Patterns of matrix metalloproteinase and TIMP-1 expression in chronic and normally healing cutaneous wounds. *Br. J. Dermatol.* **135,** 52–59.

Varga, J., Yufit, T., and Brown, R. E. (1995). Inhibition of collagenase and stromelysin gene expression by interferon-γ in human dermal fibroblasts in mediated in part via induction of tryptophan degradation. *J. Clin. Invest.* **96,** 475–481.

Werb, Z., Tremble, P. M., Behrendtsen, O., Crowley, E., and Damsky, C. H. (1989). Signal transduction through the fibronectin receptor induces collagenase and stromelysin gene expression. *J. Cell. Biol.* **109,** 877–889.

Werner, S., Smola, H., Liao, X., Longaker, M. T., Krieg, T., Hofschneider, P. H., and Williams, L. T. (1994). The function of KGF in morphogenesis of epithelium and reepithelialization of wounds. *Science* **266,** 819–822.

Werner, S., Weinberg, W., Liao, X., Blessing, M., Peters, K. G., Yuspa, S., Weiner, R. I., and Williams, L. T. (1993). Expression of a dominant-negative FGF receptor mutant in the epidermis of transgenic mice reveals a role of FGF in keratinocyte organization and differentiation. *EMBO J.* **12,** 2635–2643.

Werner, S., Peters, K. G., Longaker, M. T., Fuller-Pace, F., Banda, M., and Williams, L. T. (1992). Large induction of keratinocyte growth factor in the dermis during wound healing. *Proc. Natl. Acad. Sci. USA* **89,** 6896–6900.

West-Mays, J. A., Sadow, P. M., Tobin, T. W., Strissel, K. J., Cintron, C., and Fini, M. E. (1997). Repair phenotype in corneal fibroblasts is controlled by an interleukin-1 alpha autocrine feedback loop. *Invest. Ophthalmol. Vis. Sci.* **38,** 1367–1379.

West-Mays, J. A., Strissel, K. J., Sadow, P. M., and Fini, M. E. (1995). Competence for collagenase gene expression by tissue fibroblasts requires activation of an interleukin 1 alpha autocrine loop. *Proc. Natl. Acad. Sci. USA.* **92,** 6768–6772.

Windsor, L. J., Grenett, H., Birkedal-Hansen, B., Bodden, M. K., Engler, J. A., and Birkedal-Hansen, H. (1993). Cell type-specific regulation of SL-1 and SL-2 genes. Induction of the SL-2 gene but not the SL-1 gene by human keratinocytes in response to cytokines and phorbolesters. *J. Biol. Chem.* **268,** 17341–17347.

Wolfe, G. C., MacNaul, K. L., Beuchel, F. F., McDonnell, J., Hoerrner, L. A., Lark, M. W., Moore, V. L., and Hutchinson, N. I. (1993). Differential *in vivo* expression of collagenase messenger RNA in synovium and cartilage. *Arthritis Rheum.* **36,** 1540–1457.

Woodley, D. T., Bachmann, P. M., and O'Keefe, E. J. (1988). Laminin inhibits human keratinocyte migration. *J. Cell. Physiol.* **136,** 140–146.

Woodley, D. T., Kalebec, T., Baines, A. J., Link, W., Prunieras, M., and Liotta, L. (1986). Adult human keratinocytes migrating over nonviable dermal collagen produce collagenolytic enzymes that degrade type I and type IV collagen. *J. Invest. Dermatol.* **4,** 418–423.

Wu, H., Byrne, M. H., Stacey, A., Goldring, M. B., Birkhead, J. R., Jaenisch, R., and Krane, S. M. (1990). Generation of collagenase-resistant collagen by site-directed mutagenesis of murine pro-a1(I) collagen gene. *Proc. Natl. Acad. Sci. USA* **87,** 5888–5892.

Wysocki, A. B., Staiano-Coico, L., and Grinnell, F. (1993). Wound fluid from chronic leg ulcers contains elevated levels of metalloproteinases MMP-2 and MMP-9. *J. Invest. Dermatol.* **101,** 64–68.

Xu, J., and Clark, R. A. (1997). A three-dimensional collagen lattice induces protein kinase C-zeta activity: Role in alpha2 integrin and collagenase mRNA expression. *J. Cell Biol.* **136,** 473–483.

Yamada, K. M., and Clark, R. A. F. (1996). Provisional matrix. *In* "The Molecular and Cellular Biology of Wound Repair, 2nd ed.," (R. A. F. Clark, ed.), pp. 51–94, Plenum Press, New York.

Yamada, K. M., Gailit, J., and Clark, R. A. F. (1996). Integrins in wound repair. *In* "The Molecular and Cellular Biology of Wound Repair, 2nd ed.," (R. A. F. Clark, ed.), pp. 311–338, Plenum Press, New York.

Zhang, M. C., Giro, M., Quaglino, D., Jr., and Davidson, J. M. (1995). Transforming growth factor-beta reverses a posttranscriptional defect in elastin synthesis in a cutis laxa skin fibroblast strain. *J. Clin. Invest.* **95,** 986–994.

Regulation of
Matrix Metalloproteinase
Gene Expression

M. Elizabeth Fini, Jeffery R. Cook,
Royce Mohan, and Constance E. Brinckerhoff*

*Vision Research Laboratories of the New England Medical Center and Departments of Ophthalmology and Anatomy and Cellular Biology, Tufts University School of Medicine, Boston, Massachusetts 02111; and the *Departments of Medicine and Biochemistry, Dartmouth Medical School, Hanover, New Hampshire 03755*

Matrix Metalloproteinases

I. Introduction

Matrix metalloproteinases (MMPs) are expressed by the resident cells of tissues or invading inflammatory cells during tissue remodeling events *in vivo*. As discussed in other articles in this volume, MMP activity in tissues is regulated at multiple levels including conversion of proenzyme to the activated form, and complexing with specific inhibitors. However, these controls seem to be important primarily for fine tuning the rate of collagenolysis since (at least for the secreted MMPs) the major level of regulation is at synthesis. Thus, resident tissue cells, as a rule, synthesize collagenase only upon demand for remodeling, and the presence of collagenase in a tissue indicates that remodeling is occurring. Gelatinase A may be an exception; this enzyme can be isolated in large amounts from apparently quiescent tissues (Fini and Girard, 1990a; Matsubara *et al.*, 1991a, 1991b; Fini *et al.*, 1992), and control at the level of proenzyme activation may be the key to regulation in this case (Sato *et al.*, 1994; Okada *et al.*, 1997; Stetler-Stevenson, 1996). Gelatinase A aside, the primary nature of regulation at the level of gene expression makes this an important area in which to gain further understanding if we are to learn how to control extracellular matrix (ECM) degradation.

Studies on the regulation of MMP gene expression began with the first known MMP, collagenase, and studies were in progress long before the gene or cDNA for collagenase was cloned. Work during the precloning period took advantage of the latent collagenase activity assay as a measure of changes in gene expression (for example, Johnson-Muller and Gross, 1978). In general, this assay is an accurate reflection of gene expression, however, it can be subject to a certain level of misinterpretation in that the capacity for procollagenase activation is influenced by the relative cosynthesis of activators and inhibitors. Assays that probed directly for levels of protein using a monospecific antibody to the collagenase protein were more accurate (Nagase *et al.*, 1983). Use of the *in vitro* translation technique coupled with the antibody allowed deduction of changes in mRNA levels following various cell treatments (Nagase *et al.*, 1981; Brinckerhoff *et al.*, 1982). Such indirect methods are cumbersome, however; more rapid advances were made possible with molecular cloning. The first MMP cDNA clone was a short, ~500-bp fragment representing only the 3' untranslated region of the rabbit collagenase message. Nevertheless, it allowed the investigation of factors regulating mRNA steady-state levels (Gross *et al.*, 1984; Fini *et al.*, 1986a). CDNA cloning methodologies were improving rapidly at that time and it was only a few more years before a full-length clone for collagenase was developed and sequenced from the human species

(Goldberg *et al.*, 1986; Whitham *et al.*, 1986; Brinckerhoff *et al.*, 1987) and from the rabbit species (Fini *et al.*, 1987b). At about the same time, development and sequencing of a stromelysin full-length cDNA clone from the human species (Whitham *et al.*, 1986; Wilhelm *et al.*, 1987; Saus *et al.*, 1988) and from the rabbit species (Fini *et al.*, 1987b) were also reported. Using the collagenase cDNA clones as probes, the rabbit (Fini *et al.*, 1987a) and human (Collier *et al.*, 1988a) collagenase gene and promoter were also isolated. In the meantime, investigators performing differential cloning of genes induced in fibroblasts by oncogenic transformation (Matrisian *et al.*, 1985, 1986) and UV light (Angel *et al.*, 1986; Angel *et al.*, 1987a, 1987b) discovered, by sequence comparison, that they had isolated cDNAs and genes for stromelysin and collagenase, respectively. Convergence of work from all of these groups resulted in delineation of what is now called the matrix metalloproteinase gene family (Murphy *et al.*, 1988) and provided the first studies on regulation of MMP gene expression at the transcriptional level. This early promoter work was also groundbreaking in a way that extended beyond studies on tissue remodeling; the concept of activation of gene expression through enhancer binding proteins was new at that time and the collagenase gene (Angel *et al.*, 1987a, 1987b; Chiu *et al.*, 1988), along with the metallothionein gene (Lee *et al.*, 1987), offered an important model for the first studies on regulation by transcription factor AP-1.

This article begins from the studies initiated when transcriptional regulation studies on the MMP genes were just beginning and there were only a handful of known MMPs. Now, 10 years later, the MMP gene family is much larger. These enzymes have overlapping substrate specificities and differential expression patterns; therefore, it is generally assumed that individual members of the MMP family play different biological roles. There is little direct evidence at this point in time to support this idea; studies currently ongoing with specific blocking reagents and in transgenic knock-out mice should be most revealing in this respect. Nevertheless, the wealth of correlative evidence provides strong support for the importance of investigations into differential regulatory mechanisms for MMP expression.

For the purposes of this article, we have chosen to focus on the first MMPs for which genes were cloned and about which the most knowledge has accumulated: collagenase, the stromelysins, gelatinase B, and gelatinase A. Expression of each of these MMPs is controlled in unique ways and this group comprehensively illustrates the range of different mechanisms for controlling MMP gene expression. For each gene, we begin by describing the *in vivo* expression patterns since, not only do these expression patterns provide clues to function, but also to important aspects of regulation. We have begun with the first MMP, collagenase, and we cover the details of this gene in greatest detail. We subsequently relate the expression pattern and gene activation mechanisms for the other MMP genes to this prototype.

II. COLLAGENASE

In rabbits, two collagenase genes have been identified that are very closely related (Fini *et al.*, 1986a, 1986b, 1987a), designated as interstitial collagenase or MMP-1. Only one of these genes has been studied in detail, although the second gene is probably only a duplicate version of the first. In humans, three very different collagenase genes have been identified. Interstitial collagenase or collagenase-1 (MMP-1) is positioned at human cytogenetic locus 11q22.2-q22.3 (Formstone *et al.*, 1993) and neutrophil collagenase or collagenase-2 (MMP-8) is also found at this site. The newly described collagenase-3 (MMP-13) is found at cytogenetic locus 14q11-q12 (Pendas *et al.*, 1995). The majority of the functional work on the human promoter has been done on the MMP-1 gene. The collagenase from rat (Quinn *et al.*, 1990), mouse (Henriet *et al.*, 1992), and *Xenopus* (Brown *et al.*, 1996; Fini *et al.*, 1996b) is most closely related by sequence, as well as by its unique enzymatic activity features, to human collagenase-3 (Freije *et al.*, 1994; Krane *et al.*, 1996). However, the gene structure, at least in the mouse (Schorpp *et al.*, 1995) and rat (Rajakumar and Quinn, 1996), is much like that of human collagenase-1 and rabbit collagenase. There are discrepancies in the literature, therefore, as to the proper name for this gene; is it the homologue of MMP-1 or MMP-13, or is it a separate gene altogether? In support of the latter hypothesis, the mouse collagenase gene is located on chromosome 9 at cytogenetic locus A1-A2 in proximity to the curly whiskers (cw) locus (Schorpp *et al.*, 1995); this site does not appear to be syntenic with that of the human MMP-13 gene. Recently, a new *Xenopus* collagenase was cloned that is quite different in sequence from any of the known collagenases (Stolow *et al.*, 1996), so the question of how the collagenase genes in the different species are related still remains quite open.

For the purposes of this article, we refer to the rodent/*Xenopus* collagenase as MMP-13 and we will use the word *collagenase* to signify both the MMP-1 and MMP-13 genes. Our discussion will be confined to expression studies on these collagenase genes only.

A. *Expression of Collagenase In Vivo*

Collagenase expression occurs at a number of sites during normal and abnormal tissue remodeling, suggesting a role in a diverse range of processes from uterus involution and wound healing, to cancer and arthritis. There is a very large historical literature on collagenase expression in tissues; however, because of the general lack of specific molecular probes at the time these studies were done, the results have led to conflicting opinion as to the cell source of collagenase. The first study to report the existence of a collagenase localized this activity to the skin epidermis in resorbing tadpole tails (Gross and Lapiere, 1962) and subsequent studies also identified activity in epidermis in other species, as well as at other epithelial sites. However, with the advent of cell culture technology, it became clear that latent collagenase could be synthesized at high levels by mesenchymal cell types such as skin fibroblasts, synovial cells, and chondrocytes. This suggested that mesenchymal cells might be the major source of collagenase during such *in vivo* events as metamorphosis, but that the enzyme is activated by factors produced by overlying epithelial cells; hence, the localization of enzymatic activity to the interface between the two tissues. In support of this concept, *latent* enzymatic activity has been found in association with the repair fibroblasts in wounds (Girard *et al.*, 1993) and several labs have presented tissue culture models for collagenase activation by factors from the epidermis (for example, He *et al.*, 1989). For these reasons, attention became focused away from epithelial cell types as a source of collagenase, and the few reports of its production by cultured cells were often regarded as the result of culture conditions. Opinion underwent a second revision with the clear demonstration by *in situ* localization methods that MMP-1 collagenase mRNA and protein are expressed in the basal cells of the migrating repair epidermis (Saarialho-Kere *et al.*, 1993a, 1993b). Neovascularization is a component of repair and collagenase expression occurs in vascular cell types as well as fibroblasts in wounds. Furthermore, collagenase is expressed by macrophages that invade a tissue during the inflammatory stage of repair. These and other data have made it clear that collagenase expression is not specific for mesenchymal cells and, in fact, almost any cell type can express collagenase given the appropriate conditions.

Expression of the MMP-13 collagenase of mice begins at embryonic day 14.5 (E14.5) during development (Gack *et al.*, 1995). The major sites of synthesis are areas of endochondral and intramembranous bone formation, such as the mandible, maxilla, clavicle, scapula, vertebrae, and dorsal portion of the ribs. The highest level of collagenase expression occurs in the metaphyses and diaphyses of the long bones in hypertrophic chondrocytes and osteoblastic cells localized along the newly formed bone trabeculae, but not in osteoclasts. These data suggest that the major role for collagenase during development (at least in rodents) is in skeletal morphogenesis. In support of this idea, knock-out mice for c-fos, an AP-1 transcription factor involved in collagenase gene expression (discussed in detail in later sections), show developmental bone abnormalities, and this corrrelates with reduced expression of collagenase at these sites (Gack *et al.,* 1994). Further support is provided by the skeletal system abnormalities that develop with aging in a transgenic mouse in which the collagenase cleavage site in type I collagen has been mutated (X. Liu *et al.,* 1995). The developmental role of collagenase seems to be the same in *Xenopus;* expression of the MMP-13 collagenase is not seen until the time of tadpole metamorphosis when it is induced to high levels in the resorbing tail (Wang and Brown, 1993), localized to the mesenchymally derived skeletal rudiments in this area (D. D. Brown, personal communication).

In species that express MMP-1, a different developmental localization pattern is seen. In humans, an important site of MMP-1 collagenase expression during development is also the epidermis as well as other skin tissues. Thus, at 8 weeks of gestation, collagenase-1 expression is localized to the periderm, the basal cell epidermal keratinocytes, the dermal fibroblasts, and the surrounding matrix (McGowan *et al.,* 1994). At 12 weeks, expression is also seen in dermal and subcutaneous blood vessels, around developing nerves, and in the developing hair follicle. A collagenase from the frog, *Rana pipiens,* was recently cloned and found to be most related to MMP-1; the *in situ* localization studies with this probe have demonstrated synthesis in the epidermis of metamorphosing tadpole tails (Oofusa and Yomori, 1994) as opposed to the mesenchymal location of expression exhibited by the MMP-13 gene of *Xenopus.*

Further studies must be done to learn whether the localization differences seen for MMP-13 and MMP-1 reflect true differences in specific gene expression pattern or simply species differences. Interestingly, although collagenase expression in mouse skin has not been observed by *in situ* localization studies, mice with a mutation at the collagenase cleavage site in type I collagen (preventing cleavage by MMP-1 collagenase) develop skin defects related to scleroderma (X. Liu *et al.,* 1995). This suggests that some sort of collagenolytic enzyme is also expressed in the skin of mice, although the exact nature of the enzyme and its specific tissue site of synthesis is currently unknown.

Remodeling-induced expression patterns of MMP-1 and MMP-13 demonstrate the same sort of differential regulation seen in embryos. An excellent example is skin and corneal wounds. Induced expression of human MMP-1 collagenase occurs in the migrating epidermis of skin wounds and both MMP-1 collagenase and MMP-13 collagenase are expressed by the proliferating fibroblasts within the repairing dermis (Saarialho-Kere *et al.,* 1993a, 1993b; Okada *et al.,* 1997; Vaalamo *et al.,* 1997). Expression of the rat MMP-13 collagenase is also localized to the fibroblasts of skin wounds (Porras-Reyes *et al.,* 1991). In the repairing cornea, induced expression of collagenase occurs in the repair fibroblasts of rabbit (MMP-1) and human (MMP-1), but not in rat (MMP-13) (Girard *et al.,* 1993; Fini *et al.,* 1996a). In these examples, the overall pattern of collagenase expression appears to reflect the repertoire of collagenase genes found in the species under examination. However, this cannot explain all examples of differential expression. For example, MMP-1 collagenase expression occurs in epithelial cells migrating to resurface corneal wounds of human (MMP-1), but no collagenase expression has been detected at this location in either rat (MMP-13) or rabbit (MMP-1) (Fini *et al.,* 1996a). This variability of expression pattern may reflect species differences in regulatory mechanisms that should be kept in mind when these mechanisms are under investigation. It also suggests that different species use different strategies to control tissue remodeling.

B. *Regulation of Collagenase Gene Expression in Cultured Cells*

The large number of agents that induce collagenase expression have been extensively defined by studies with cultured fibroblasts, and have also been studied in other cell types including skin keratinocytes, osteoblasts, chondrocytes, and myometrial smooth muscle cells. Most of these studies have been on the MMP-1 gene; however, a few involve MMP-13. We will make note of these special cases as we go along in the following paragraphs.

Historically, there was much interest in tissue interactions as a means of controlling collagenase expression in fibroblasts or monocytes (for example, Wahl *et al.,* 1975; Newsome and Gross, 1979; Dayer *et al.,* 1977; Johnson-Muller and Gross, 1978). Since that time, many diffusible extracellular ligands for cell surface receptors have been identified that could act as paracrine mediators of collagenase expression *in vivo.* The first of these were the inflammatory cytokines, interleukin 1α (IL-1α) and interleukin-1β (IL-1β), which mediate fibroblast collagenase expression stimulated by leukocytes or epithelial cells (Mizel *et al.,* 1981; Postlethwaite *et al.,* 1983; Saklatvala *et al.,* 1984; Girard *et al.,* 1991; Strissel *et al.,* 1997). A number of other cytokines have subsequently been found to induce collagenase expression including tumor necrosis factor α (TNF-α) (Mawatari *et al.,* 1989; Mitchell and Cheung, 1991; Lyons *et al.,* 1993), anti-inflammatory cytokines such as IL-10 (Reitamo *et al.,* 1994), growth-promoting cytokines such as platelet-derived growth factor (PDGF) or epidermal growth factor

(EGF) (Chua *et al.,* 1985; Bauer *et al.,* 1985; Tan *et al.,* 1995), and antiviral cytokines such as interferon β (IFN-β) (Sciavolino *et al.,* 1994) or interferon γ (IFN-γ) (Tamai *et al.,* 1995). The lipid mediator of inflammation, platelet activating factor (PAF) induces collagenase expression in rabbit corneal epithelium (Bazan *et al.,* 1993). Vascular endothelial growth factor (VEGF) induces collagenase expression in human endothelial cells (Unemori *et al.,* 1992). Bacterial lipopolysaccaride (LPS) (Kuter *et al.,* 1989) or the acute phase reactant, serum amyloid A (Brinckerhoff *et al.,* 1989), also induce collagenase expression in fibroblasts. MMP-13 collagenase expression is induced in cultures of rat uterine smooth muscle cells by treatment with serotonin (Wilcox *et al.,* 1994a, 1994b).

In addition to mediating short-range tissue interactions, many of the stimulators just described can also disseminate via the blood to act in a more widespread way. In addition, many bonafide blood-borne hormones are known to induce collagenase expression. MMP-13 collagenase expression is induced in cultures of rat bone cells by vitamin D or parathyroid hormone (Meikle *et al.,* 1992; Rajakumar and Quinn, 1996). Thyroid hormone induces expression of MMP-13 collagenase in *Xenopus* cell lines (Kanamori and Brown, 1993). Prolactin induces MMP-13 collagenase expression in rat fibroblasts and *Xenopus* mesenchymal cells (P. Sadow, M. Mody, D. Mullady, and M. E. Fini, unpublished results).

Historical studies revealed that sometimes direct interaction between cells was necessary for induction of collagenase, suggesting that a cell surface molecule must be involved. For example, the interaction between carcinoma cells and fibroblasts induces expression of collagenase in the fibroblasts. This interaction is now know to be mediated by a stimulatory factor that is expressed on the surface of epithelial cell types known as EMMPRIN, a member of the immunoglobulin superfamily (Biswas *et al.,* 1995). Collagenase expression is also induced in fibroblasts or monocytes by direct cell–cell interaction with T lymphocytes via the cell surface receptor CD3 (Miltenburg *et al.,* 1995) or CD40 (Malik *et al.,* 1996). Cell–matrix interactions can also be inductive for collagenase expression. Thus, expression is induced as fibroblasts spread on the 120-kDa fragment of fibronectin (Werb *et al.,* 1989; Huhtala *et al.,* 1995) or on vitronectin plus decorin (Huttenlocher *et al.,* 1996). In epidermal keratinocyte cultures, attachment and spreading on a collagen substrate induces collagenase expression (Sudbeck *et al.,* 1994).

In addition to induction via ligand–receptor interactions, collagenase expression can also be induced by a number of other conditions and agents. Agents such as crystals of monosodium urate monohydrate

(Gross *et al.*, 1984) and calcium phosphate crystals (McCarthy *et al.*, 1991) induce collagenase expression in cultured fibroblasts. Several cellular stress conditions have also been shown to induce collagenase gene expression including heat shock (Vance *et al.*, 1989; Hitraya *et al.*, 1995), UV light (Stein *et al.*, 1989; Petersen *et al.*, 1992), and "aging" of cell cultures via multiple passaging (Kumar *et al.*, 1992). Collagenase expression is further induced in cultured fibroblasts by agents or conditions that stimulate alteration of the actin cytoskeleton, an event that occurs during many remodeling situations *in vivo* such as wound contraction or cell migration. This class of stimulators includes the act of phagocytosing latex beads, the act of contracting a collagen gel, treatment with trypsin, and treatment with cytochalasin B (Aggeler *et al.*, 1984; Unemori and Werb, 1986). Interaction of fibroblasts with certain ECM molecules that bind cell surface integrin receptors can also lead to remodeling of the actin cytoskeleton, and this secondary effect is probably the true inducer of collagenase expression in these cases. For example, when fibroblasts are treated with an antibody to $\alpha_5\beta_1$ fibronectin receptor or plated on an RGD peptide (Werb *et al.*, 1989), cytoskeletal remodeling occurs and collagenase expression is induced. However, collagenase expression induced as fibroblasts spread on the 120-kDa fragment of fibronectin (Werb *et al.*, 1989; Huhtala *et al.*, 1995) occurs without discernible alterations in the cytoskeleton. Thus, it seems likely that fibroblast–matrix interactions can send two different signals for collagenase expression: one initiated indirectly via stimulation of cytoskeletal remodeling and one initiated by direct ligation of cell surface integrin receptors.

C. Activation of Collagenase Gene Transcription at the Promoter

Transcription, mRNA half-life, and mRNA translation can each determine the rate at which a protein is synthesized. While there is ample evidence for the importance of translational control in regulating expression of ECM proteins, this has not been revealed to be an important regulatory mechanism for MMPs. In contrast, several studies have implicated a major role for mRNA half-life in the control of collagenase expression (Brinckerhoff *et al.*, 1986; Delany and Brinckerhoff, 1992; Vincenti *et al.*, 1994), as well as expression of other MMPs (Overall *et al.*, 1991). Nevertheless, control at the level of transcription still appears to be the major factor determining MMP gene expression, and it is this level of control on which we will concentrate for the rest of this chapter. The reader might find it useful during this and the following sections on MMP transcriptional promoters to refer to Fig. 1, which is a summary of the information presented in the text.

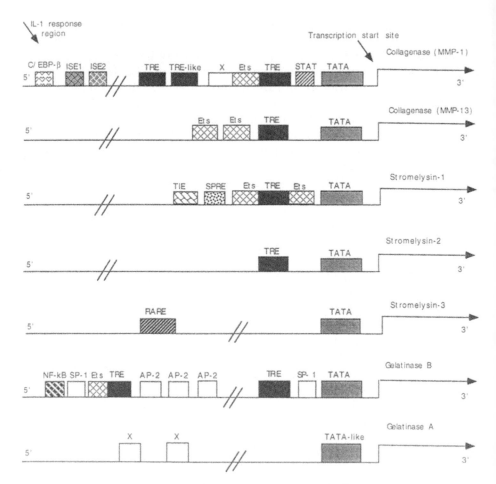

FIG. 1. Schematic structure of MMP gene promoters. The diagram represents a compilation of data on functionally characterized elements in the different transcriptional promoters of the MMP genes, as discussed in the text. The physical linkage of specific elements and their rough spacing are delineated, but spacing is not to scale. A diagonal double line separates promoter elements downstream and upstream of base −200. TATA, TATA box for transcriptional initiation; TATA-like, noncanonical TATA box; TRE, TPA (PMA) response element; TRE-like, noncanonical TRE; Ets, transcription factor Ets response element; ISE1 and ISE2, immortalization-sensitve elements; C/EBP-β, CCAAT/enhancer-binding protein-beta response element; SPRE, stromelysin PDGF response element; TIE, TGF-beta inhibitory element; NF-κB, transcription factor NF-κB response element; SP-1, transcription factor SP-1 response element; AP-2, transcription factor AP-2 response element; RARE, retinoic acid response element; X, response element of unknown binding specificity. The region for IL-1 response is indicated with an arrow; specific elements are not yet mapped.

As mentioned in the historical introduction, the transcription factor, AP-1, is central to collagenase gene transcription (Angel *et al.*, 1987a, 1987b, 1988; Schonthal *et al.*, 1988). AP-1 is a group of different dimeric complexes consisting of members of the jun and fos gene families—part of the larger bZip family of transcription factors. AP-1 complexes regulate expression of a number of genes (including collagenase) involved in cell replication, cell invasion, and response of the cell to potentially damaging stresses. When constitutively activated by genetic mutation, AP-1 family members act as cellular oncogenes causing progression of the cell to a state of uncontrolled growth and invasiveness. AP-1 complexes exist *de novo* in most cell types but often in an inactive state; exogenous stimulators ultimately cause AP-1 complexes to become active. This occurs, for example, when cells are treated with tumor promoters. Because of this connection with tumor formation, several labs interested in oncogenesis have used collagenase as their model gene for investigation into the mechanisms by which tumor-promoting phorbol esters and oncogenes alter transcription. This work combined with work from labs with a primary interest in basic mechanisms of tissue remodeling has resulted in rapid advances in the field.

Early studies showed that a short DNA element whose 5' end is located at base -73 in the human gene promoter is all that is necessary for basal and PMA-stimulated collagenase promoter activity in HeLa cells when linked to a DNA fragment containing a TATA box (required for proper transcriptional initiation) and a start site for transcription (Angel *et al.*, 1987a, 1987b). This element, which has become known as the TRE (TPA or PMA response element), was shown to bind AP-1 transcription complexes. A second DNA element located adjacent to the TRE was subsequently demonstrated to act synergistically with AP-1 to enhance levels of transcriptional activation by PMA and by non-nuclear oncoproteins such as Ha-ras (Gutman and Wasylyk, 1990). This element binds members of the transcription factor ets gene family (also known as PEA3). The close spacing of the TRE and ets-binding element was shown to be important for synergistic activation of transcription.

In contrast to the situation in transformed cell lines, the TRE is not sufficient for activity of the rabbit collagenase promoter in normal rabbit fibroblasts (Auble and Brinckerhoff, 1991); the smallest piece of promoter DNA conferring PMA-inducibility extends to base -110 from the transcription start site. Mutagenesis studies demonstrated a requirement for the TRE, the adjacent ets-binding site, and a third closely-linked DNA element containing the sequence with unknown binding specificity. In addition to conferring response to PMA, this fragment of DNA contains the elements necessary to drive transcription in response to all other stimulators of collagenase expression tested thus far. For example, in fibroblast cultures this promoter drives reporter gene expression in response to exogenously applied IL-1 (Vincenti *et al.*, 1994), TNF-α or IFN-β (Sciavolino *et al.*, 1994), heat shock (Hitraya *et al.*, 1995), and ligation of the $\alpha_5\beta_1$ integrin receptor (Tremble *et al.*, 1995). In monocytes, LPS and zymosan also activate this minimal collagenase promoter (Pierce *et al.*, 1996).

The rat MMP-13 collagenase gene promoter is constructed somewhat differently from the human and rabbit MMP-1 collagenase promoters (see Fig. 1); however, the TRE is conserved and required for parathyroid hormone induction in a bone tumor cell line (Rajakumar and Quinn, 1996). This is also true for the human MMP-13 gene promoter (Pendas *et al.*, 1997). However, in contrast to the MMP-1 gene, no synergistic effect was observed between the TRE and ets-binding elements in the human MMP-13 gene promoter. This could explain some of the observed differences in expression pattern between MMP-1 and MMP-13.

Very recently, a fourth response element located within the confines of the minimal MMP-1 collagenase promoter—a STAT transcription factor binding site—was delineated in the human gene (Korzus *et al.*, 1997). The STAT response element is located just downstream of the TRE, between bases -53 and -45. This location, abutting the proximal TRE, is important for the capacity of the STAT response element to cooperate with the TRE for efficient response to a newly defined collagenase inducer, oncostatin M. It will be interesting to learn whether this element also contributes to response to other stimulators because STAT transcription factors are found in most cell types.

Additional sequences located upstream of base -110 in the rabbit collagenase promoter can further up-regulate promoter activity in response to PMA. Inclusion of distal sequences even further upstream specifically enhance activation by IL-1 (Vincenti *et al.*, 1996). Some of the PMA activating sequences have been characterized. Thus, a second TRE with 5' end at position -186 enhances promoter activation and, interestingly, can compensate for the more proximal TRE (White and Brinckerhoff, 1995). Stimulation of the promoter by PMA also utilizes a third noncanonical TRE element located between bases -182 and -141, which, like the canonical TREs, binds c-Fos, c-Jun, and Jun-D (Chamberlain *et al.*, 1993). Evidence suggests that this element is not completely equivalent to the canonical TREs since it has a somewhat lower affinity for AP-1 binding protein (Chamberlain *et al.*, 1993). In fact, it is becoming clear that different TRE-related elements are distinguished by their ability to bind different sets of AP-1 heterodimers, and dimers composed of other members of the larger b-Zip family of transcription factors (Smith *et al.*, 1993). The different AP-1 complexes have different strengths as transriptional activators, and (as discussed later) some of these complexes act as negative regulators. Therefore, preferential binding of specific factors, which may be up- or down-regulated under certain circumstances, could be very important for fine-tuning the level of collagenase gene expression in a given situation. It will be important to characterize the noncanonical TREs in the colla-genase gene further, in order to understand why they may play a part in modulating the action at the canonical TREs.

Data are beginning to accumulate on the molecular mechanisms by which unique AP-1 complexes differentially affect transcription. These effects appear to be mediated by accessory proteins that bind to specific AP-1 family members. For example, v-Jun is a stronger transcriptional activator than c-Jun of the collagenase promoter in certain cell types. These functional differences between v-Jun and c-Jun result from a deletion in v-Jun (referred to as "delta deletion") that seems to weaken the interaction of Jun with a negative cellular regulator molecule (Havarstein et al., 1992). The nature of the cellular inhibitor is not known but will be important to determine. In fact, the characterization of both positively and negatively acting transcriptional accessory pro-teins that bind AP-1 is likely to be an area of intense research in the next few years.

D. Inhibition of Collagenase Gene Transcription at the Promoter

In comparison to the many known agents that stimulate collagenase expression, there are far fewer inhibitors. Limiting this class even further is the fact that these agents can also act as stimulators under the right conditions. Thus, dexamethasone, which has long been recog-nized as an inhibitor of collagenase expression in fibroblasts (Brincker-hoff and Harris, 1981), acts as a stimulator of collagenase expression in cultures of bone cells (Delany et al., 1995). The steroid hormone, estrogen, has been ascribed stimulatory effects on collagenase expres-sion in cervix (Rajabi et al., 1991). Estrogen is a stimulator of collagen-ase gene expression in rat myometrial cells, but progesterone is an inhibitor (Wilcox et al., 1994b). Retinoic acid is another classic inhibitor of collagenase gene expression in fibroblasts (Brinckerhoff and Harris, 1981), but thyroid hormone, which binds cellular receptors of the same class as retinoids, stimulates collagenase gene expression in a Xenopus cell line (Kanamori and Brown, 1993). The bifunctional cytokine TGF-β, which can affect cell growth in either a positive or negative way, can also act as an inhibitor (Edwards et al., 1987) or a stimulator (Chua et al., 1985; Mauviel et al., 1996) of collagenase expression. In fact, both actions can occur in the same experiment; thus, TGF-β was shown to act as an inhibitor of collagenase gene expression early in the PMA induction time course, but its action becomes stimulatory at later times after its application (Fini *et al.*, 1995). Anti-tumor-promoting phenolic antioxidant compounds also have been described as both positive and negative regulators of collagenase gene expression (Yoshioka *et al.*, 1995; Yokoo and Kitamura, 1996a; Sato *et al.*, 1996). Calcium ionophor induces collagenase gene expression in cultures of rabbit synovial fibroblasts (Unemori and Werb, 1988) but inhibits expression in cultures of human keratinocytes (Sudbeck *et al.*, 1997). An inhibitory action of prostaglandins has also been described in fibroblast cultures that is mediated via increases in cAMP; however, these agents have a stimulatory action on the collagenase gene in human monocyte cultures (Corcoran *et al.*, 1992).

Although most inhibitors of collagenase expression can also act as stimulators, there are still a few inhibitory agents that have not been shown to act bifunctionally to our knowledge. This list includes the insulin-like growth factors, which inhibit collagenase synthesis in bone cell cultures (Canalis *et al.*, 1995). The E1A oncoprotein encoded by the adenoviruses, which can act to immortalize cells, also acts as an inhibitor of collagenase expression (Hagmeyer *et al.*, 1993; van Dam *et al.*, 1993). As a last example, physical association between two rat glioma cells inhibits their capacity to synthesize collagenase. This effect is mediated by transmembrane neural cell adhesion molecule (NCAM) but not glycosyl-phosphatidylinositol-anchored NCAM (Edvardsen *et al.*, 1993). Although there are no reports on a stimulatory function for these inhibitory agents, it seems likely that such reports will be forthcoming in the future. This is because the currently existing studies on inhibitors of collagenase expression have all identified the same mechanism for inhibition: direct interference with transcriptional activity at the TRE. Therefore, inhibitory effects might be easily switched to stimulatory effects by slight changes in the relevent molecular interactions.

To restate the point: All inhibitors of collagenase expression (for which a mechanism is known) operate through AP-1 and the TRE. However, there are distinctive differences in the way that each of the specific inhibitors accomplishes its action on the TRE, and the inhibitors can be grouped into different mechanistic classes based on these differences. One class of inhibitors works by activating a distinct class of AP-1 complexes, which (like Fos-Jun heterodimers) bind to the collagenase gene TRE, but which (unlike Jun-Fos) act to reduce transcriptional activity. For example, the b-Zip family member, ATF-2, is induced in fibroblasts by inhibitory agents that elevate the intracellular levels of cAMP. C-jun/ATF-2 heterodimers bind both canonical and noncanonical TREs. This binding activates transcription from the noncanonical TRE found in the c-Jun gene; however, it inhibits transcription from the collagenase gene TRE (De Cesare et al., 1995). As another example, PMA induces expression of three Jun family members, c-Jun, JunB, and JunD, in cultures of human fibroblasts; however, the TGF-β inhibitor preferentially induces JunB, and only JunB is induced by the cAMP inhibitor (Angel and Karin, 1992). JunB binds the collagenase TRE as effectively as c-Jun or JunD; however, it is a poor inducer of gene activation when compared to c-Jun or JunD. This suggests that JunB may exert its negative action on transcription in a dominant-negative fashion, by competing with positively acting Jun complexes for TRE binding. In contrast to its usual effect on fibroblasts, TGF-β stimulates collagenase gene expression in cultures of skin keratinocytes and this correlates with induced expression of c-Jun; JunB expression is not induced (Mauviel et al., 1996). As a third example, anti-tumor-promoting phenolic antioxidant compounds selectively induce expression of Fra-1, a member of the fos gene family, over c-Fos. Fra-containing AP-1 heterodimers bind to the collagenase TRE, but have low transactivation potential (Yoshioka et al., 1995). The Epstein–Barr virus BZLF1 gene product is a fourth example of this class of inhibitors. BZLF is a transcription factor that is partially homologous to c-Fos and can bind both canonical and noncanonical TREs. However, BZLF1 protein cannot stimulate the collagenase promoter and, in fact, it inhibits promoter activation by c-Jun and c-Fos (Sato et al., 1992).

Retinoids and steroid hormones constitute a second mechanistic family of collagenase promoter inhibitors. These agents typically influence gene expression by binding to their specific cellular receptors, which then bind to unique response elements within the promoters of genes. In contrast, the mechanism by which these agents inhibit collagenase expression is via interaction of the cellular receptor with AP-1 transcription complexes (Jonat et al., 1990; Yang-Yen et al., 1990; Schule et al., 1990; Lopez et al., 1993). Band shift analysis of the retinoic acid receptor suggests that the receptor inhibits AP-1 binding to the TRE (Schule et al., 1990; Pan et al., 1992); however, transfection studies in intact cells indicate that retinoic acid receptors form a complex at the collagenase promoter's TRE, tethered to the DNA by c-Jun (Pan et al., 1995; Schroen and Brinckerhoff, 1996). Similarly, studies on the glucorticoid receptor performed in intact cells demonstrate that the receptor antagonizes AP-1 activity at the TRE without affecting DNA site occupation, and that the hormone receptor does not need to bind to the DNA in order to do this (Konig et al., 1992). Despite the fact that DNA binding by the receptor is not required, analysis of mutant retinoic acid receptors by transfection revealed that the DNA-binding domain is important for the inhibition of AP-1 activity (Schule *et al.*, 1990). While DNA binding and activation of transcription by the glucocorticoid receptor requires dimerization, repression appears to be a function of receptor monomers (Heck *et al.*, 1994). Furthermore, repression of AP-1 activity by the glucocorticoid receptor is ligand independent (W. Liu *et al.*, 1995), unlike the transcription activation function. These and other studies indicate that repression of AP-1 activity and transactivation functions of retinoid and steroid receptors are separable.

As discussed earlier, although retinoids and steroid hormones generally act as inhibitors of collagenase gene expression, they can also act as stimulators under the appropriate conditions. This action also involves the TRE rather than specific receptor response elements. Thus, interaction of the estrogen receptor with the AP-1 complex in the appropriate cells stimulates the transcriptional efficiency of this complex (Webb *et al.*, 1995). Unlike the situation for inhibition, the DNA binding domain of the steroid receptor is not required for AP-1 activation. The oncogenic form of the thyroid hormone receptor, v-ErbA, also has a positive influence on AP-1 activity and acts as a dominant-negative oncoprotein by overcoming the repression of the AP-1 activity induced by retinoic acid receptors or by the cellular form of the thyroid hormone receptor, c-ErbAα (Desbois *et al.*, 1991; Ways *et al.*, 1993).

Like the hormone receptors discussed earlier, the E1A oncoprotein encoded by the adenoviruses also acts to inhibit collagenase promoter activity by binding AP-1 complexes. This binding has a negative effect on transcriptional activation by AP-1 complexes that bind canonical TREs such as the proximal TRE in the collagenase promoter; on the other hand, E1A binding activates many AP-1 complexes that recognize noncanonical TREs such as that found in the c-Jun gene (Hagmeyer *et al.*, 1993; van Dam *et al.*, 1993). The effect of E1A on AP-1 complexes is independent of the retinoblastoma gene product, which is known to control E1A activity in other situations. By use of a series of E1A deletion proteins, the repression function was found to require two E1A sequence elements, one within the nonconserved E1A N terminus, and the second within a portion of conserved region 1. These domains possess binding sites for several cellular transcription accessory proteins, including p300, Dr1, YY1, CBP, and the TBP subunit of TFIID (Dorsman *et al.*, 1995; Song *et al.*, 1995; Lee *et al.*, 1996; Smits *et al.*, 1996). The p300 and CBP proteins, in particular, are required as coactivators for transcription mediated by AP-1 hetero- or homodimers. P300 and CBP are therefore, common mediator proteins through which E1A gains control over AP-1-regulated genes in host cells infected with Adenovirus.

E. Signaling Pathways for Expression of Collagenase

Well before its physiologic relevence was understood, PMA was recognized as a strong activator of collagenase gene expression (Brinckerhoff *et al.,* 1979). This, combined with the fact that it was readily available resulted in the popularity of PMA as the inducer for mechanistic studies on collagenase expression. It is now known that PMA initiates intracellular signal transduction through activation of protein kinase C (PKC), by virtue of its similarity to the second messenger, diacylglycerol. Activation of signaling pathways by PMA stimulates expression of the genes for c-Jun, c-Fos, and other AP-1 transcription factors. PMA also activates endogenous AP-1 by ultimately stimulating (through the signaling pathways that it initiates) phosphorylation of some sites and dephosphorylation of other sites on the protein components (Angel and Karin, 1991). The complexity of PKC activation has increased with the increase in size of the PKC gene family; it has been becoming clear that the different family members vary in their biochemical properties. Several recent publications have provided evidence that the different PKC gene family members are not equivalent in their capacity to control activity of AP-1 responsive promoters such as collagenase. In one study, seven PKC isozymes were cotransfected into cells along with a human collagenase promoter containing the PEA3/AP-1 motif, or a porcine urokinase plasminogen activator (uPA) promoter, which did not contain this motif (Reifel-Miller *et al.,* 1996). The results demonstrated that only PKC delta, epsilon, and eta could transactivate the collagenase promoter while all the isozymes transactivated the uPA promoter. It will be interesting to learn the molecular mechanism behind the selective effects of the different PKC isoforms.

That PKC is essential for PMA-mediated induction of collagenase has been shown by the lack of collagenase gene response when PKC is down-regulated, or when its activity is inhibited with agents such as staurosporine (for example, Sudbeck *et al.,* 1994). However, PKC does not directly phosphorylate AP-1; instead, it phosphorylates some intermediate signaling molecule, which then becomes responsible for transmitting the stimulatory signal. A great deal is now known about the nature of signaling cascades that lead to AP-1 activation; however, which of these pathways is utilized by the various stimulating agents to activate the collagenase promoter is only beginning to be defined. Dominant-negative inhibition of the mitogen-activated protein kinases (MAP kinases) p42mapk and p44mapk inhibits collagenase promoter activation (Pages *et al.,* 1993), suggesting the involvement of these kinases in the signaling cascade. Several studies have also implicated tyrosine protein kinases in activating the collagenase gene promoter.

Thus, cotransfection of cells with a c-src expression plasmid activates the promoter (Schonthal *et al.*, 1988; Vincenti *et al.*, 1996a, 1996b). Also, inhibitors of src-related tyrosine kinases such as herbimycin C or genistein inhibit collagenase gene expression in keratinocytes plated on collagen (Sudbeck *et al.*, 1994), or in fibroblasts treated with PMA or IL-1 (Vincenti *et al.*, 1996a, 1996b). A dominant-negative Raf-1 kinase mutant strongly interferes with both PMA and UV-induced AP-1 activation, indicating the common involvement of this kinase in signal transduction initiated by each stimulus (Radler-Pohl *et al.*, 1993). Collagenase promoter activation by oncostatin M via the combined action of AP-1 and STAT transcription factors is also dependent on the Raf-1 protein kinase (Korzus *et al.*, 1997).

Other signaling molecules involved in collagenase gene activation have also been identified. Okadaic acid, a nonphorbol ester tumor promoter, is capable of strongly activating the collagenase promoter in a process that requires the TRE. This agent is a specific inhibitor of protein phosphatases 1 and 2A, suggesting that these molecules also participate in signaling to the collagenase gene (Kim *et al.*, 1990). As mentioned earlier, the second messenger cAMP is important in collagenase gene response to prostaglandins. Like prostaglandins, treatment of cells with a membrane-permeable form of cAMP can stimulate or inhibit collagenase gene expression at the promoter level depending on the cell type (Grove *et al.*, 1989; Civitelli *et al.*, 1989; Salvatori *et al.*, 1992). Intracellular levels of cAMP are stimulated by activation of protein kinase A. In a bone cell line, parathyroid hormone stimulates collagenase gene expression by activation of protein kinase A (Scott *et al.*, 1992), which ultimately induces the transcription factor, CREB, to bind to the TRE (Rajakumar and Quinn, 1996).

F. Regulation of Collagenase Gene Expression via Autocrine Loops

Early work with PMA (Brinckerhoff *et al.*, 1982), as well as with several other strong and readily available collagenase inducers such as urate crystals (Gross *et al.*, 1984) or cytochalasin B (Frisch *et al.*, 1987), in normal fibroblasts revealed that collagenase expression was induced relatively late after stimulators were applied. In fact, more recent studies with cDNA probes show that PMA stimulates a biphasic response in early passage corneal fibroblasts: A low level of induction occurs over the first 5 h, then a dramatic increase occurs sometime after 5 h (Fini and Girard, 1990b). Induction in response to cytochalasin B appears to be monophasic, but (like the second phase of response to PMA) also begins quite late—some time after 8 h in corneal fibroblasts

(West-Mays *et al.*, 1997). Treatment with agents that inhibit protein synthesis blocked induction of mRNA levels by PMA and cytochalasin B (Frisch *et al.*, 1987). New protein synthesis has also been shown to be required for induction of collagenase gene expression in fibroblasts by UV light exposure (Kramer *et al.*, 1993). Protein synthesis, in contrast, is not required for collagenase gene response to PMA in certain transformed cells lines (Angel *et al.*, 1987a, 1987b). These observations have suggested that efficient expression of collagenase in normal fibroblasts involves more than a simple activation of preexisting transcription factors as it does in transformed cell lines: It also requires the products of early response genes.

The early response gene products required for collagenase expression in normal fibroblasts might simply be transcription factors; perhaps there is not enough of AP-1 or ets factors preexisting in normal cells. However, the increasing evidence that fibroblasts express many cytokines and growth factors has led several groups to explore the hypothesis that these may be the all-important products of early genes required. There are now many cases in which this hypothesis has proven to be true (Table I). Thus, the autocrine cytokine, IL-1α, is absolutely required as an intermediate for induction of collagenase synthesis in response to cytochalasin B in rabbit fibroblast cultures (Fini *et al.*, 1994a; West-Mays *et al.*, 1995), and this autocrine mediates about 90% of the collagenase synthesis induced by PMA. Autocrine IL-1α, along

TABLE I

AGENTS AND CONDITIONS THAT STIMULATE COLLAGENASE EXPRESSION VIA
AUTOCRINE INTERMEDIATES

Agent	Autocrine	Cell type	Reference
Long-term culture	IL-1	Human fibroblasts	Kumer *et al.*, 1992
Ultraviolet light	IL-1α and bFGF	Human fibroblasts	Kramer *et al.*, 1993 Wlaschek *et al.*, 1993
PMA	IL-1α	Rabbit fibroblasts	Fini *et al.*, 1994a
PMA	TNF-α	Human U937 monocyte/ macrophage	Callaghan *et al.*, 1996
Seratonin	IL-1	Rat myometrial cells	Wilcox *et al.*, 1994a
Ligation of integrins by RGD peptide	IL-1α	Rabbit chondrocytes	Arner and Tortorella, 1995
Cytochalasin B	IL-1α	Rabbit fibroblasts	West-Mays *et al.*, 1995
Trypsin	IL-1α	Rabbit fibroblasts	West-Mays *et al.*, 1995
Plating on collagen	EGF	Human skin keratinocytes	Pilcher *et al.*, in preparation

with autocrine bFGF also mediates the UV light-induced stimulation of collagenase synthesis by human dermal fibroblasts (Kramer *et al.*, 1993; Wlaschek *et al.*, 1993). Collagenase induced as a result of cell "aging" after long-term culture (Kumar *et al.*, 1992) also occurs through IL-1α. In early passage cultures of uterine smooth muscle cells, an IL-1 autocrine loop is required for collagenase expression in response to the hormone seratonin (Wilcox *et al.*, 1994a). In chondrocytes, IL-1α mediates expression of another metalloproteinase, stromelysin, in response to ligation of integrin receptors by the RGD peptide (Arner and Tortorella, 1995). In U937 monocyte/macrophage cells, TNF-α acts as the autocrine mediator of collagenase expression stimulated by PMA (Callaghan *et al.*, 1996). Finally, collagenase expression induced by plating keratinocytes on collagen is controlled through an EGF autocrine loop (Pilcher *et al.*, in preparation). Together, these studies have contributed to the emergence of a new paradigm for understanding the control of collagenase expression, which dictates that signals from diverse stimulators for collagenase gene expression are routed outside the cell via a small set of cytokine intermediates; these cytokines then send the signal back into the cell and to the collagenase gene by ligation of their specific receptor on the cell surface. Importantly, the extracellular routing of the stimulatory signal offers a convenient point for therapeutic intervention in cases of pathological collagenase expression.

Appreciation of the autocrine cytokine routing paradigm as a mechanism for regulation of collagenase gene expression helps us to focus our collagenase gene regulatory studies onto those fewer signaling pathways activated by the autocrine intermediates. In addition, it makes it clear that an understanding of mechanisms activating the collagenase gene requires an understanding of mechanisms activating expression of autocrine cytokines such as IL-1α. Surprisingly, while there is a fair amount of work published on IL-1β gene regulation, little has been published on the IL-1α gene other than the sequence of its 5'-flanking DNA; this work has revealed potential binding sites for transcription factors NF-κB, AP-1, and an Ad2MLTF-like binding site (Furutani *et al.*, 1986). Since IL-1α autoregulates its own expression and is also known to activate NF-κB, it might be hypothesized that NF-κB is involved in controlling IL-1α gene expression. In support of this hypothesis, transfection of fibroblasts with a dominant-negative form of NF-κB inhibits the induction of IL-1α gene expression (West-Mays *et al.*, submitted). In addition, treatment of cells with antioxidants, which interfere with the signaling action of reactive oxygen species, inhibits both NF-κB activation and IL-1α gene expression (Cook and Fini, in preparation). These studies implicate both NF-κB and reactive oxygen species in controlling expression of collagenase through their effects on IL-1α gene expression.

Cell culture models are clearly important for revealing basic mechanisms of gene expression, but exactly what the *in vivo* situation is that they reflect is not always certain. In fact, it has long been known that, unlike early passage fibroblast cultures, quiescent fibroblasts that have been freshly isolated from a nonremodeling tissue cannot make collagenase in response to PMA or cytochalasin B (Johnson-Muller and Gross, 1978; Kuter *et al.*, 1989; Fini and Girard, 1990b), calling into question the meaning of early passage results. It seemed important, therefore, to understand the mechanisms leading to acquisition of competency for collagenase expression. An important clue was the finding that, while freshly isolated cells cannot respond to PMA or CB, they are fully competent to express collagenase when treated with IL-1 (Girard *et al.*, 1991). Recently, it was demonstrated that aquisition of competency for collagenase expression in response to PMA or cytochalasin B is the result of the acquired capacity of the IL-1α gene to respond to its own product and generate a positive IL-1α feedback loop (West-Mays *et al.*, 1995). These studies suggest that there are actually two steps for IL-1α gene activation; that is, before the gene can be activated by stimulators, cells must first acquire the competency for activation. Interestingly, this culture "activation" bears much in common with the activation of quiescent corneal fibroblasts at the edge of a wound. It was recently shown that these cells become competent to express IL-1α as they take on the wound healing phenotype and begin to express collagenase (Girard *et al.*, 1993; West-Mays *et al.*, 1997). Therefore, the activation of fibroblast competency for expression of cytokines such as IL-1α may be an important mechanism by which tissue remodeling is controlled during the repair process.

Given the new knowledge regarding the autocrine loop paradigm for collagenase gene regulation, it could be hypothesized that, in addition to direct effects on AP-1 inhibitors of collagenase gene expression might also act by blocking synthesis or activity of IL-1α. In fact, TGF-β and dexamethasone block the IL-1α autocrine loop in fibroblast cutures treated with PMA or CB (West-Mays *et al.*, submitted). In contrast, retinoic acid does not block IL-1α gene expression but acts only by direct action on the collagenase gene. The knowledge of the importance of autocrine loops also has suggested an additional class to be added to the list of inhibitors of collagenase gene expression, that is, agents that block autocrine growth factors loops such as IL-1 receptor antagonist (IL-1ra) (Fini *et al.*, 1994a), antioxidants (West-Mays *et al.*, submitted), and the anticancer drug suramin (Sachsenmaier *et al.*, 1994). Suramin also acts to inhibit activation of the collagenase gene directly in normal diploid fibroblasts, independently of IL-1α (West-Mays *et al.*, submitted). Perhaps this occurs through the inhibition of G protein signaling molecules, another known target of suramin (Sachsenmaier *et al.*, 1994). Finally, as mentioned earlier, antioxidants block expression of IL-1α, and this is likely to be the mechanism by which they inhibit collagenase expression (Cook and Fini, in preparation).

To finish this section, it seems pertinent to consider what cytokines could contribute to activation of the collagenase promoter that PMA, which also stimulates intracellular signaling pathways culminating in activation of transcription factor AP-1, cannot contribute. One study reports that TNF-α stimulates prolonged activation of the gene for c-jun, a component of the AP-1 transcription complex, contrasting with its transient activation by PMA (Brenner *et al.*, 1989). Therefore, more c-jun protein would be available for activation of the collagenase gene in the case of stimulation by TNF-α. Perhaps this gives some explanation for the role of autocrine cytokines induced by PMA in enhancing the action of PMA on the collagenase gene promoter. However, much more work needs to be done to fully understand the molecular nature of this phenomenon.

G. Summary: Mechanisms Controlling Collagenase Gene Expression

To summarize this section, we have reviewed the current state of knowledge regarding the agents, signaling pathways, and transcriptional mechanisms for activation and repression of the collagenase gene, the first member of the MMP family to be studied. Work during the past 10 years has enabled us to group the large number of stimulators into specific mechanistic classes and it has been revealed that all of these classes are linked by their common reliance on the promoter response element known as the TRE. Nevertheless, evidence suggests that other important transcriptional regulators remain to be defined. In addition to the TRE, a new concept of regulation by autocrine cytokine loops has recently emerged that also provides a common link between the many known collagenase stimulators. Much basic knowledge has been obtained on signaling pathways that control TRE binding activity, and the knowledge that many diverse stimulators work through common cytokine intermediates simplifies this task. However, knowlege of the specific pathways required for collagenase gene expression lags behind the state of the art. Further work will be required to delineate the similarities and differences between the signaling pathways utilized by the different classes of regulators.

III. Stromelysins

Two different stromelysin genes have been identified in rats (Matrisian *et al.*, 1986; Breathnach *et al.*, 1987) and humans (Sirum and Brinckerhoff, 1989; Quinones *et al.*, 1989); one of these genes has also been identified in rabbits (Sirum-Connolly and Brinckerhoff, 1991). The proteins encoded by these two genes share high similarity at the amino acid level and have similar activity and substrate preferences. They have been designated stromelysin-1 and stromelysin-2. A third gene, stromelysin-3, has a divergent sequence, and the substrate profile of the enzyme is unique. Its inclusion as a member of the stromelysin subgroup was based on its greater sequence similarity to stromelysins than to other members of the MMP family rather than on its biochemical properties. Both stromelysin-1 and stromelysin-2 are located (along with collagenase-1 and collagenase-2) at human cytogenetic locus 11q22.3-23 (Jung *et al.*, 1990; Formstone *et al.*, 1993). Stromelysin-3 is located at human cytogenetic locus 22q11.2 (Levy *et al.*, 1992).

A. *Signaling Pathways for Expression of Stromelysin-1*

To our knowledge, the embryonic expression pattern for stromelysin-1 has not been determined. In postnatal mice, stromelysin-1 is constitutively expressed in heart and lung. As is true for collagenase, expression of stromelysin occurs under a variety of normal and pathological remodeling conditions, and the expression patterns of these two enzymes often, but not always, overlap. One example of stromelysin-1 expression that has received wide attention occurs in the developing ducts of the murine mammary gland, where evidence with transgenic animals suggests a specific role for this enzyme in branching morphogenesis (Talhouk *et al.*, 1992; Witty *et al.*, 1995).

In normal diploid fibroblast cultures, the stromelysin-1 gene responds to the same stimuli as collagenase and expression typically occurs coordinately with collagenase-1. For example, PMA, CB, and IL-1 each induce coordinate expression of collagenase and stromelysin in early passage cultures of rabbit synovial fibroblasts, and dexamethasone or retinoic acid coordinately inhibit expression (Brinckerhoff *et al.*, 1987; Fini *et al.*, 1987b; Frisch *et al.*, 1987; McCachren *et al.*, 1989). However, there are now a number of examples of noncoordinate expression. Thus, in primary cultures of rheumatoid synovial cells (as opposed to early passage cultures), stromelysin is expressed preferentially over collagenase in response to IL-1 or TNF-α (MacNaul *et al.*, 1990). Also, nerve growth factor (NGF) induces stromelysin expression in PC12 cells as it stimulates them to undergo neuronal differentiation (Machida *et al.*,

1989) and this induction appears to be independent of collagenase expression. These and related studies suggest that the signaling pathways for stromelysin-1 expression and/or response elements in the stromelysin-1 promoter would have unique features not seen in the collagenase promoter.

We know of no differences between signaling pathways for activation of stromelysin-1 and collagenase. As is the case for collagenase, PKC mediates activation of stromelysin-1 in response to PMA and other stimulators, and cAMP and PKA can mediate both stimulation or inhibition, depending on the cell type. For example, IL-1 stimulation of stromelysin-1 transcription in rheumatoid synovial fibroblasts appears to involve an inhibitory prostaglandin-dependent pathway mediated by cAMP and protein kinase A, and a stimulatory PKC-dependent pathway controlled by PMA (Case *et al.*, 1990). Down-regulation of PKC by PMA inhibits subsequent induction of stromelysin-1 gene expression by EGF in rat fibroblasts, suggesting that this kinase is required to mediate EGF signaling to the stromelysin gene (McDonnell *et al.*, 1990). Addition of a membrane-permeable form of cAMP to fibroblasts inhibits EGF-mediated stimulation of stromelysin-1 expression (Kerr *et al.*, 1988). On the other hand, both PKC and PKA are involved in stimulating stromelysin-1 expression in response to nerve growth factor in PC12 cells (Machida *et al.*, 1989).

Other components of signaling pathways to the stromelysin-1 gene have also been described. Stromelysin-1 was originally identified in rat fibroblasts as a gene induced by transformation with polyoma virus (Matrisian *et al.*, 1985); at that time, it was called *transin.* Much like collagenase, transin expression was also shown to be induced by transformation of cells with oncogenes v-src, H-ras, and erb-B (Matrisian *et al.*, 1985; Kerr *et al.*, 1988). Mouse 3T3 cells transfected with a ras expression vector produced higher levels of stromelysin-1 as well (Sistonen *et al.*, 1989). Also, like collagenase, stromelysin-1 gene induction by PDGF is dependent on the protein kinase, Raf-1, in human fibroblasts (Kirstein *et al.*, 1996).

Although not specifically reported in most cases, a careful perusal of the literature indicates that autocrine cytokine loops also mediate stromelysin-1 expression along with collagenase expression in response to the relevant stimulators. One study specifically documents this case: the activation of stromelysin expression in chondrocytes by RGD peptide mediated by IL-1α (Arner and Tortorella, 1995). Examination of total protein profiles published in papers from one of our labs reveal that stromelysin-1 expression is activated coordinately with collagenase in rabbit fibroblasts treated with PMA or CB and induction of both enzymes is inhibited by cotreatment with IL-1ra (Fini *et al.*, 1994a).

Interestingly, in freshly isolated corneal fibroblast cultures that cannot activate the IL-1α autocrine loop, stromelysin appears to be preferentially stimulated over collagenase by treatment with PMA (Girard *et al.*, 1991; West-Mays *et al.*, 1995; Strissel *et al.*, 1997a, 1997b). This suggests that the stromelysin gene might be capable of more direct response to PMA than the collagenase gene, a hypothesis that needs to be tested.

B. Activation and Inhibition of the Stromelysin-1 Transcriptional Promoter

The gene for stromelysin-1 may respond to the same stimulators, utilizing the same signaling pathways as collagenase. However, accumulating data have revealed that the stromelysin-1 promoter includes some unique response aspects. The elements required for induction of the human stromelysin-1 gene by PMA, EGF, and IL-1β were reported by one of our labs to be located on a 307-bp fragment, which includes about 270 bp of 5'-flanking DNA (Sirum and Brinckerhoff, 1989). A second group reported that a 1.3-kb fragment of the human stromelysin promoter was similarly inducible by IL-1β and suppressible by dexamethasone (Quinones *et al.*, 1989). Inducibility decreased as the promoter was shortened but IL-1 inducibility was maintained until the promoter was cut down to base −53. DNA sequence analysis revealed the presence of a TRE at position −70 to −64 in the DNA that was removed, suggesting involvement in promoter activation (Quinones *et al.*, 1989); this response element is conserved in the rabbit gene (Sirum-Connolly and Brinckerhoff, 1991). In fact, the final level of constitutive and IL-1 or PMA-induced transcriptional activation of the promoter was reduced by mutation of this TRE (Buttice *et al.*, 1991; Quinones *et al.*, 1989); however, the fold-induction by IL-1 or TPA was not altered. Further analysis revealed that elements similar to those that bind ets transcription factors, located between bases −86 to −71 and −63 to −54, could respond to IL-1 in the absence of the TRE (Quinones *et al.*, 1994). This region was shown to mediate activation by transcription factors of the ets gene family, including c-ets (Wasylyk *et al.*, 1991). The TRE also mediates activation by c-ets; it does not bind c-ets but is activated indirectly through c-ets stimulation of c-Jun and c-Fos gene expression (Wasylyk *et al.*, 1991). Both ets-responsive elements also mediate activation by the non-nuclear oncoproteins Ha-ras, v-Src, and v-Mos.

Early pioneering experiments indicated that both c-fos and c-jun contribute to activation of the stromelysin-1 gene promoter by the growth factors PDGF or EGF in rat fibroblasts and their action occurs at the TRE (Kerr *et al.*, 1988; McDonnell *et al.*, 1990). Stimulation of stromelysin-1 gene expression by PDGF was blocked by a selective reduction in c-fos synthesis using antisense oligonucleotides; however, this did not completely block induction by EGF (Kerr *et al.*, 1988), suggesting that each growth factor utilizes unique induction mechanisms. Expression of stromelysin-1 and collagenase could not be induced by either PDGF or EGF in a c-fos-deficient embryonic fibroblast line derived from c-fos knock-out mice, nor could these genes be induced by UV light (Schreiber *et al.*, 1995). Transformation with polyoma virus middle T antigen, however, restored inducibility to the stromelysin-1 gene, but, interestingly, this did not restore inducibility for collagenase. Together, these results provide further support for the idea that stromelysin-1 promoter activation can occur independently of transcription factors that bind the TRE, by utilizing mechanisms that are not utilized by the collagenase promoter. Activation of stromelysin-1 in rat PC12 cells by NGF is also independent of the TRE. A *cis*-acting element that lies within a 12-bp region between positions -241 and -229 of the promoter was shown to participate in this response. The element is not apparently related to other known response elements (deSouza *et al.*, 1995).

In addition to the ets-binding sites and the TRE, a third element in the stromelysin promoter, the stromelysin PDGF response element (SPRE), is required for efficient response to PDGF and other mitogens (Sanz *et al.*, 1995; Kirstein *et al.*, 1996). During mitogenic induction, these promoter elements appear to be activated by two separate pathways. Thus expression of a dominant-negative version of Raf-1 interferes with the induction through the ets-binding element and the TRE, but not through SPRE. Expression of a dominant-negative form of lambda/iota PKC affected expression only through SPRE. The protein binding to the SPRE, PBP, was cloned from a fibroblast cDNA library and was found to encode a novel transcription factor with a putative leucine zipper, a nuclear localization signal, and a basic domain with homology to the DNA-binding domains of Fos and Jun. PBP can heterodimerize with c-Jun to transactivate the SPRE site. Cross-talk therefore occurs between different enhancer elements during mitogenic induction of the human stromelysin-1 gene.

As is the case for the collagenase promoter, activation of the stromelysin-1 promoter is inhibited by adenovirus E1A protein (van Dam *et al.*, 1989), the glucocorticoid receptor (Offringa *et al.*, 1988), and the retinoic acid receptor (Nicholson *et al.*, 1990), through the TRE. TGF-beta also inhibits activation of the stromelysin gene (Kerr *et al.*, 1988) and the stromelysin-1 promoter (Kerr *et al.*, 1990), as it does the collagenase gene and promoter. However, this inhibition occurs by utilization of a unique 10-bp element located upstream of the TRE, which binds Fos protein. Perhaps, this element constitutes a noncanonical TRE; further work on this element needs to be performed to investigate this idea.

C. Expression of Stromelysin-2 and Activation of the Promoter

As is true of stromelysin-1, the embryonic expression pattern for stromelysin-2 has not been determined to our knowledge. In postnatal mice, stromelysin-2 is expressed in heart and kidney; these are not the same tissues that express stromelysin-1 (discussed earlier). Only a few studies on remodeling situations that involve expression of stromelysin-2 have been published, however, the available reports generally reveal a very different temporal or spatial expression pattern from stromelysin-1 and the other MMPs. For example, transcripts for stromelysin-2 are detectable in late secretory and menstrual endometrium, but not in proliferative endometrium (Rodgers et al., 1994). In human skin repair collagenase-1 and stromelysin-1 are expressed at the migrating front where the keratinocytes are not in contact with a basement membrane, whereas stromelysin-2 is expressed behind the migrating front (Saarialho-Kere et al., 1994).

There are also only a few reports on expression of stromelysin-2 in cell culture, however, again these reveal great differences with respect to stromelysin-1. Thus, stromelysin-2 was originally cloned from rat genomic DNA and was shown to be expressed along with stromelysin-1 in several transformed rat embryonic fibroblast lines. However, unlike stromelysin-1, stromelysin-2 is not expressed by cultures of normal diploid fibroblasts from rat (Breathnach et al., 1987) and was inducible in human fibroblasts only to very low levels (Sirum and Brickerhoff, 1989). On the other hand, human skin keratinocytes express easily detectable levels of stromelysin-2 along with stromelysin-1 (Windsor et al., 1993).

We are aware of only two reports describing the sequence of the putative stromelysin-2 gene promoter; it appears that no functional studies have been done. It was found that there was little sequence similarity between the 5'-flanking region of the rat stromelysin-2 gene and that for stromelysin-1, except that a sequence related to the stromelysin-1 gene's TRE was also found at the same place in the stromelysin-2 gene (Breathnach et al., 1987). The sequence of about 300 bp of DNA flanking the 5' end of the human stromelysin-1 and stromelysin-2 genes was determined and aligned (Sirum and Brinckerhoff, 1989). About 50% percent overall sequence identity was found with stretches up to 80% identical, including the TRE.

D. Expression of Stromelysin-3 and Activation of the Promoter

The cDNA for the newest member of the stromelysin class of MMPs, stromelysin-3, was cloned from human breast tumors. In these tumors, stromelysin-3 was found to be expressed specifically in the normal stromal cells surrounding the carcinoma rather than in the carcinoma cells themselves (Basset *et al.*, 1990). Most of the literature since that time has been concerned with the role of this gene in oncogenesis; it is expressed in a wide variety of tumors and its expression correlates with greater invasive characteristics of the tumor cells (Rouyer *et al.*, 1994). Stromelysin-3 is also expressed during mouse embryonic development: in trophoblastic cells at the site of embryonic implantation, and in a variety of developing embryonic tissues (Lefebvre *et al.*, 1995). The highest levels of stromelysin-3 expression are observed during development of the external features of limb, tail, and snout, and during bone and spinal cord morphogenesis. In limb, tail, and snout, stromelysin-3 is expressed in mesenchymal cells at the interface between epithelial cells and stroma and adjacent to epithelial cells undergoing apoptosis. It is not known how this expression pattern compares to that for stromelysin-1 or stromelysin-2. In postnatal mice, stromelysin-3 is expressed in the mammary glands during branching morphogenesis, as is stromelysin-1 (Witty *et al.*, 1995). In adult humans, the gene is reported to be expressed in basal epithelial cells of repairing lung in organ culture (Buisson *et al.*, 1996), and in the fibroblasts of repairing skin (Wolf *et al.*, 1992). In menstrual endometrium, stromelysin-3 is expressed in the proliferative endometrium, while stromelysin-2 is expressed in the late secretory and menstrual endometrium. Stromelysin-3 is expressed in the intestine of the metamorphosing *Xenopus* tadpole at sites where apoptosis is occurring (Patterton *et al.*, 1995).

Like collagenase, stromelysin-3 is expressed by normal fibroblasts in culture. In these cells, expression is stimulated by retinoic acid at the same time that collagenase expression is being repressed (Anglard *et al.*, 1995; Guerin *et al.*, 1997). Another study describes stimulation of stromelysin-3 by retinoic acid in cultured fibroblasts (Anderson *et al.*, 1995); this discrepancy is not completely understoood but may potentially be explained by differential response to dose (Guerin *et al.*, 1997). In a *Xenopus* cell line, expression of stromelysin-3 is also induced by treatment with thyroid hormone (which binds to receptors of the same superfamily as those bound by retinoic acid). Interestingly, the gene responds quite rapidly to treatment and this response is not inhibited by cycloheximide. In contrast, induction of collagenase expression by thyroid hormone in these cells occurs with much slower kinetics. These data suggest that the stromelysin-3 gene, unlike any of the other MMPs that we have discussed thus far, is an early response gene and induction may, therefore, proceed without routing through autocrine intermediates. The same is also true for gelatinase B, as is discussed in detail in the next section.

Sequence analysis of a 1.4-kb fragment of DNA located upstream of the start site for transcription (Anglard *et al.*, 1995) revealed a canonical TATA box-like sequence, the 5' end of which is located at base −31. There was no consensus TRE apparent within this region. A consensus retinoic acid response element was identified at base −385. Transient transfection revealed that an upstream fragment of 3.4 kb was functional and activated by retinoic acid receptors in the presence of retinoic acid. Thus, the stromelysin-3 gene appears to be regulated by retinoic acid quite differently from collagenase and stromelysin-1 (and probably stromelysin-2), which utilize the TRE. Considering the interesting developmental expression pattern of stromelysin-3, and the strong evidence for its role in cancer, it will be important to understand the unique aspects of this expression.

IV. Gelatinase B

The single gene for gelatinase B has been cloned from humans (Huhtala *et al.*, 1991), mice (Masure *et al.*, 1993), and rabbits (Fini *et al.*, 1994b). The gene has been mapped to cytogenetic locus 20q11.2-13.1 in humans (Collier *et al.*, 1991; St Jean *et al.*, 1995; Linn *et al.*, 1996) and to locus 2h1-2 in mice (DuPont *et al.*, 1996).

A. Expression of Gelatinase B in Vivo

Two papers from the Tryggvason lab were the first to describe gelatinase B expression during mouse embryonic development (Reponen *et al.*, 1994, 1995). Expression was seen mostly in two types of cells: embryonic trophoblasts and cells of the osteoclast lineage. The trophoblast localizations have been confirmed in subsequent studies in mice (Alexander *et al.* 1996), and this expression also occurs in humans (Polette *et al.*, 1994). The osteoclast localization has also been found in humans (e.g., Hill *et al.*, 1994). A second set of *in situ* hybridization studies from the Muschel lab demonstrated many other important sites of gelatinase B expression during embryonic development (Canete-Soler *et al.*, 1995a, 1995b). Major sites within the embryo proper included the developing brain, eye, and skeleton. Expression was also seen at other sites including the liver, the primordial alveoli of the lungs, the epithelium of the thyroid gland, and the thymus. Extracts from the brain and liver contained a gelatinase activity of 105 kDa consistent with the size of mouse gelatinase B, providing further support for the validity of the localizations demonstrated with nucleic acid probes. It can be concluded that gelatinase B, like the other MMPs discussed thus far, plays a role in development.

Postnatally, gelatinase B is expressed at many sites of remodeling, as are collagenase and the stromelysins. However, there are some expression sites that are particularly representative of gelatinase B. The nervous system is one of these sites; for example, gelatinase B expression has been described in pyramidal neurons of the human hippocampus (Backstrom *et al.*, 1996) and in the ganglion cells of the mouse retina (Canete-Soler *et al.*, 1995a). Epithelial tissues and appendages constitute a second group of characteristic sites. Thus, in human lung, gelatinase B is expressed in ciliated epithelial cells, as well as endothelial cells, smooth muscle cells, macrophages, and pneumocytes (Hayashi *et al.*, 1996). In the eye, gelatinase B is expressed constitutively in epithelial cells of the lens capsule (Canete-Soler *et al.*, 1995b). As a third example, gelatinase B is expressed around the tooth in the gingival epithelium, glandular tissue, and acinar epithelial cells (Makela *et al.*, 1994). Expression of gelatinase B is also typically seen in endothelial cells of neovascularizing vessels (Cornelius *et al.*, 1995).

Expression of gelatinase B is induced at other sites under conditions that require tissue remodeling. Many of these conditions also result in expression of other MMPs such as collagenase, but sometimes gelatinase B is specifically expressed. For example, exposure to UV light is a strong inducer of gelatinase B and other MMPs such as collagenase in skin (Koivukangas *et al.*, 1994; Fisher *et al.*, 1996). Gelatinase B expression is induced in the epithelium migrating to heal a wound in cornea independently of collagenase (Matsubara *et al.*, 1991a, 1991b), while collagenase, stromelysin-1, and stromelysin-2 are expressed along with gelatinase B in the healing epidermis of skin (Oikarinen *et al.*, 1993; Salo *et al.*, 1994). In human females, gelatinase B is expressed in the amniochorion at the onset of labor in epithelial macrophages, trophoblasts, and decidual cells and neutralizing antibodies block the increase in gelatinase activity with labor, suggesting a role for this specific enzyme in rupture of the amniotic membranes (Vadillo-Ortega *et al.*, 1995).

The literature contains many examples of gelatinase B expression in pathological conditions of remodeling, and the specificity of these localizations in some cases suggests a role of gelatinase B in the disease process. Gelatinase B is expressed at high levels in the epithelium of nonhealing corneal ulcers (Matsubara *et al.*, 1991b; Fini *et al.*, 1996a).

Induced expression of gelatinase B in chondrocytes has been connected with damage to the knee joint (Mohtai *et al.*, 1993). Immunoreactive gelatinase B protein was detected in proximity to extracellular amyloid plaques in Alzheimer's disease patients and the enzyme was found to cleave β-amyloid peptide, the major plaque constituent (Backstrom *et al.*, 1996). Increased expression occurs in the lung undergoing pulmonary fibrosis (Hayashi *et al.*, 1996). Expression is also found in human abdominal aortic aneurysms (Thompson and Parks, 1996).

As is true for other MMPs, gelatinase B can be expressed by tumor cells or stromal cells surrounding the tumor. For example, one study found gelatinase B primarily localized to endothelial cells in human breast tumors (Heppner *et al.*, 1996). In a second study, gelatinase B expression was localized to carcinoma cells and expression was remarkably higher in the carcinoma cases with lymph node metastasis than in the nonmetastatic cases (Iwata *et al.*, 1996). Tumor cells in human giant cell tumor of bone express gelatinase B as do focal clusters of stromal cells; expression is associated with vascular invasion (Ueda *et al.*, 1996; Schoedel *et al.*, 1996). Gelatinase B is expressed in some brain gliomas where it is associated with higher malignancy (Nakano *et al.*, 1995). Expression of gelatinase B also appears to be an important characteristic of the malignant transformation of hepatocytes (Rao *et al.*, 1993; Ashida *et al.*, 1996). Many experimental studies using embryonic fibroblasts have connected gelatinase B expression with malignancy and suggest that this enzyme plays a key role (as compared to other MMPs) in metastatic tumor cell behavior. In a particularly direct demonstration, Muschel and colleagues (Bernhard *et al.*, 1994) found that cells transformed with Ha-ras were metastatic in nude mice and expressed gelatinase B. In contrast, those transformed with ha-ras plus the adenovirus E1A oncogene were not metastatic and did not release gelatinase B; however, transfection of a gelatinase B expression vector into these cells made them metastatic.

B. Signaling Pathways for Expression of Gelatinase B

Gelatinase B was first identified as a product of neutrophils and macrophages (Hibbs *et al.*, 1987); however, it was soon learned that the gene is expressed in many other cell types including both fibroblasts and epithelial cells (Wilhelm *et al.*, 1989; Fini *et al.*, 1990a, 1990b). Expression of gelatinase B is stimulated by many of the same agents and conditions that stimulate collagenase expression; however, collagenase and gelatinase B are not always coexpressed in response to these stimulators. Another observation is that expression of gelatinase B seems to be particularly sensitive to cell type and cross-species factors, as compared to collagenase. Some examples follow. TGF-β stimulates gelatinase B expression in cultured human mucosal and dermal keratinocytes, but not in a rat mucosal keratinocyte line, and expression is stimulated only slightly in human fibroblasts (Salo *et al.*, 1991). In corneal fibroblasts, TGF-β alone does not induce gelatinase B expression, but TGF-β enhances the inducing effect of PMA (Fini *et al.*, 1995). This enhancing effect occurs at the same time that expression of collagenase is undergoing repression in response to TGF-β. Another collagenase expression inhibitor, retinoic acid, stimulates expression of gelatinase B in mouse and human macrophage cell lines (Houde *et al.*, 1996). Expression of gelatinase B is induced by IL-1 in normal human articular cartilage (Mohtai *et al.*, 1993), in rat glomerular mesangial cells (Yokoo and Kitamura, 1996b), and in a rabbit epithelial cell line derived from the lens of the eye (J. A. West-Mays and M. E. Fini, unpublished). In contrast, expression of gelatinase B fails to be induced by IL-1 in other cell types including corneal fibroblasts, even though collagenase expression is induced (P. Bargagna-Mohan and M. E. Fini, in preparation). Agents that alter the actin cytoskeleton also fail to induce expression of gelatinase B in rabbit corneal fibroblast cultures (P. Bargagna-Mohan and M. E. Fini, in preparation), presumably because of the lack of gene responsiveness to the IL-1alpha intermediate. Gelatinase B is induced by PMA in human skin keratinocytes and HT1080 cells (Wilhelm *et al.*, 1989; Huhtala *et al.*, 1991), in rabbit synovial (Werb *et al.*, 1989) and corneal fibroblasts (Fini and Girard, 1990a) and in the human melanoma cell line A2058 (Lauricella-Lefebvre, 1993), indicating the participation of PKC in signaling. In rabbit corneal epithelial cells, PMA induces expression of gelatinase B but not collagenase (Fini and Girard, 1990a, 1990b). On the other hand, PMA treatment fails to induce gelatinase B expression in rat glomerular mesangial cells (Yokoo and Kitamura, 1996b) or in mouse skin keratinocytes (R. Mohan and M. E. Fini, unpublished).

Like collagenase, gelatinase B expression is stimulated by transformation with a variety of oncogenic viruses and oncogenes. For example, tumorigenic transformation of rat embryo fibroblasts with an activated Ha-ras oncogene induces expression of gelatinase B (Gum *et al.*, 1996), as does transformation of human lung fibroblasts with SV40 (Wilhelm *et al.*, 1989). Control by SV40 T antigen has also been reported to inhibit expression of gelatinase B (as well as stromelysin-1 and gelatinase A) when human placental trophoblast-like cells are used (Logan *et al.*, 1996). A newly recognized viral oncogene stimulator, the Moloney murine leukemia virus long terminal repeat, also induces expression of gelatinase B. Stimulation in this case occurs via an RNA polymerase III generated transcript, which does not encode a protein (Faller *et al.*, 1997).

Studies on fibroblast cultures have revealed that, although expression of both gelatinase B and collagenase is induced by PMA, each gene is induced with quite different kinetics (Fini et al., 1995). Thus, while collagenase expression is induced very slowly, gelatinase appears to be an early gene. Interestingly, in rat glomerular mesangial cells (Yokoo and Kitamura, 1996a), stimulation of gelatinase B expression by IL-1 is inhibited by antioxidants, suggesting the requirement for reactive oxygen species in the signaling pathway. As discussed in a preceeding section, reactive oxygen species are also required for expression of the IL-1α: an early gene whose product is required for collagenase expression. These data suggest the hypothesis that gelatinase B and IL-1α may be controlled by similar pathways in fibroblasts. They also suggest that gelatinase B expression is independent of intermediate cytokines, in contrast to collagenase.

C. Activation of the Gelatinase B Transcriptional Promoter in Cultured Cells

The examples cited of differences in regulation of gelatinase B and collagenase that occur both in vitro and in vivo indicate that the gelatinase B promoter must contain sequences that allow it to respond uniquely to extracellular signals. Sequence analysis of the 5' flank of the human gelatinase B gene (containing the putative transcriptional promoter) was first reported by Huhtala et al. (1991). The presence of four elements previously described in MMP promoters was noted: a TATA box located in the correct location for proper transcriptional initiation, a proximal TRE-like element located about 50 bases upstream in the same position as in other MMP genes, and an upstream TRE-like element linked directly to an ets-like response element. Functional studies of this promoter revealed that the proximal TRE, as well as sequence motifs homologous to the binding sites for transcription factors NF-κB and SP1, is necessary for efficient promoter activation by PMA and TNF-α in HT1080 fibroblasts. The NF-κB and SP1-like elements are not present in the collagenase or stromelysin promoters, suggesting that these sites help determine the unique aspects of gelatinase B regulation. The human promoter also contains a second SP1-like element, which is conserved in the mouse promoter and is located just downstream from the proximal TRE (Sato et al., 1993). This element is not required for response to TNF-α or PMA, but is essential for activation of the promoter by v-src. Gel mobility shifts demonstrated that SP1 proteins bind specifically to this element. These studies define independent pathways for activation of the promoter by v-src and by inflammatory stimuli. Expression studies further demonstrate the importance of transcription factors ets and c-jun in this activation.

All of the promoter elements described earlier are required for efficient activation in response to transfection by Ha-ras (Gum *et al.*, 1996).
The expression of a phosphatase (CL100) that inactivates multiple
MAP kinases abrogated the stimulation of the promoter by ras. In
contrast, expression of a dominant-negative form of MEK1 did not
prevent activation, indicating that MEK1 is not involved in promoter
activation. An activated c-raf expression vector also did not stimulate
promoter activity even though the raf kinase has been shown to be
involved in activation of AP-1 binding to the TRE in the collagenase
promoter in response to UV light or PMA (Radler-Pohl *et al.*, 1993).

Studies on the gelatinase B gene from rabbit have focused on identifying elements that allow this gene to be expressed preferentially over
collagenase in certain cell types (Fini *et al.*, 1994b). In corneal fibroblasts, basal activity of a portion of the gelatinase B promotor spanning
bases -520 to $+19$ is lower than activity of a collagenase promotor of
about 1800 bp. Nevertheless, activity of both promoters is similarly
stimulated by treatment of transfected cells with PMA, and stimulation
is enhanced in both cases by cotreatment with TGF-β. In contrast,
while basal activity of the gelatinase B promotor in epithelial cells is
somewhat lower than in fibroblasts, it is still higher than activity of the
collagenase promotor. Deletion analysis demonstrated that sequences
upstream of base -330 confer cell type-specific activity to the gelatinase
B promotor. Site-directed mutagenesis revealed that the TRE-like element within this region is utilized much more efficiently by fibroblasts
than by epithelial cells. This region also contains elements that confer
the capacity for activation by AP-2, a transcription factor expressed by
corneal epithelial cells but not by corneal fibroblasts. In contrast, AP-
2 does not activate the collagenase promotor. These results provide a
molecular basis for the unique, cell type-specific expression pattern of
gelatinase B.

D. Activity of the Gelatinase B Transcriptional Promoter in Transgenic Mice

Studies on MMP promoters in cultured cells are useful for defining
functional elements; however, *in vivo* studies are essential to learn
which of these elements are required for specific instances of biological
remodeling. Considering its importance, it is surprising that there have
been no confirmed reports of the activity of MMP promoters *in vivo*.
Currently unpublished studies from the one of our labs (Mohan *et al.*,
in preparation) provide the first demonstration of the activity of a MMP
promoter *in vivo*, in transgenic mice. In these studies, the rabbit DNA
between bases -520 and $+19$ was linked to the β-galactosidase reporter

gene and injected into mouse embryos. Two founder lines were obtained; both showed a similar expression pattern during embryonic development. Embryos examined before day 11 revealed little or no transgene expression in the embryo proper. At embryonic day 11 (Fig. 2, top) embryos stained as whole mounts showed prominent β-galactosidase expression in the developing brain and spinal cord. This expression pattern is similar to that described for the endogenous gene (Canete-Soler *et al.*, 1995a). A second prominant region of transgene expression

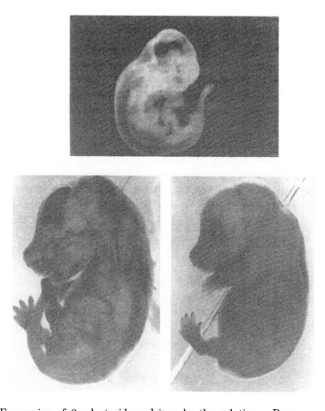

FIG. 2. Expression of β-galactosidase driven by the gelatinase B gene promoter in transgenic mouse embryos. Bases -520 to $+19$ were linked to a gene for β-galactosidase and this construct was injected into mouse blastocysts. Two independent transgenic mouse lines were derived, each of which showed a similar β-galactosidase expression pattern. Shown are embryos at two different stages of development, treated as whole mounts with the x-gal substrate for β-galactosidase. The blue stain indicates the location of β-galactosidase enzyme expression. Nontransgenic animals showed no staining (not shown). *Top:* Whole mount at embryonic day 11. *Bottom, left:* Whole mount at embryonic day 15, cut in half after staining. *Bottom, right:* The embryo shown in the middle panel before cutting.

is the margins of the somites. A third region is in the limb buds, where β-galactosidase expression is found in two lateral spots, which suggests the location of the anterior and posterior necrotic zones where programmed cell death occurs. These latter two sites of expression have not been previously observed for the endogenous gene, but may still be appropriate since the previous studies may not have covered these areas.

The β-galactosidase transgene is expressed at a large number of sites at embryonic day 15. At this stage, the skin is a barrier to penetration of the x-gal substrate. However, a few superficial sites show expression in stained whole mounts (Fig. 2, bottom right), including the eye, the nasal cavities, the ear, and the toes. Embryos cut in half (Fig. 2, bottom left) reveal strong expression in the brain, and the developing skeleton. Also there are a number of prominent sites in the mouth cavity including the developing incisors and the area around the tongue. Many of these sites have been described for expression of the endogenous gelatinase B gene in embryonic mouse and human (Canete-Soler *et al.*, 1995a, 1995b; Heikinheimo and Salo, 1995; Chin and Werb, 1997).

The transgene is also expressed appropriately in postnatal mice. An excellent example of this is the skin. No transgene expression occurs in newborn or adult skin, except in the underlying muscle layer. However, induction occurs following the initiation of wound healing, and expression is localized to basal cells in the migrating epidermis at the wound edge (Fig. 3). Expression also occurs in hair follicles after wounding or after the hairs are shaved or plucked. Expression is also induced by exposure of the skin to UV light. These mice should be useful for studies on MMP gene expression during biological remodeling since the β-galactosidase transgene provides such a convenient readout. Together, these studies reveal that bases −520 to +19 of the gelatinase B promoter are sufficient for appropriate expression of the gelatinase B gene *in vivo*. Studies are now under way to define the elements between bases −520 and +19 that are required for appropriate expression.

V. Gelatinase A

The gene for gelatinase A is present in a single copy in the human genome at cytogenetic locus 16q21 (Huhtala *et al.*, 1990a). The mouse gene for gelatinase A (Reponen *et al.*, 1992) has a similar structure as the human gene (Huhtala *et al.*, 1990b), and the coding sequences are 96% identical at the amino acid coding level. The gene has also been isolated from rat (Harendza *et al.*, 1995).

A. *Expression of Gelatinase A*

Reponen *et al.* (1992, 1995) reported on localization of gelatinase A during development as detected by *in situ* hybridization. This study

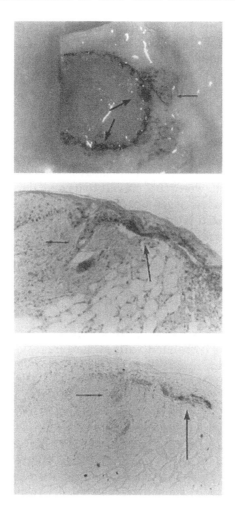

FIG. 3. Expression of β-galactosidase driven by the gelatinase B gene promoter in healing skin wounds created in transgenic mice. A full thickness portion of skin was removed from a mouse of the transgenic line described and allowed to heal for 3 days. The wound was then excised with a full thickness portion of surrounding tissue. This tissue was stained for β-galactosidase expression as a whole mount. The tissue was then embedded in paraffin and thin sections were prepared for histology. Blue color indicates the location of β-galactosidase expression. Staining for β-galactosidase is seen at the edge of the migrating epidermis within basal keratinocytes (heavy arrows) and within hair follicles (light arrows). *Top:* Tissue whole mount. *Middle:* Thin section from tissue whole mount. *Bottom:* Thin section without counterstain.

demonstrated expression of gelatinase A throughout the loose mesenchyme of the first, second, and third branchial arches of an embryonic day 10 mouse. The surface ectoderm was completely negative. Expression was not seen in implanting embryos or extraembryonic tissues. Also, decidual uterine stroma was negative. Some positive reaction was seen, however, in uterine mucosal stromal cells outside the decidual area. Furthermore, there was some positivity in the nonpregnant uterine stroma. The uterine surface epithelium was negative. At embryonic day 13, intense expression was evident throughout the head mesenchyme. In the eye, expression occurred in mesenchymal cells of the vitreous body as well as the corneal stroma. The lens epithelium and retina were negative. No expression occurred in the brain. Tongue, oral, and tooth bud epithelia were negative. The epithelial cells of the submaxillary gland were the only ectodermal derivatives that expressed the gelatinase A gene. A primarily mesenchymal pattern of expression occurred at later embryonic stages as well as in adults. For example, the skin epithelium was negative except for inner cells of hair follicles, which occasionally showed positivity. All stromal fibroblasts expressed the enzyme intensely.

Most studies on gelatinase A expression in postnatal animals have focused on tumors (Stetler-Stevenson *et al.*, 1996). However, the role of gelatinase A in angiogenesis has recently become an intense topic of interest (Stetler-Stevenson, 1996). As with the other MMPs, gelatinase A expression is correlated with a more metastatic state and can be induced experimentally by transfection of cells with oncogenes (Bernhard *et al.*, 1994). Induction by oncogenes aside, the expression characteristics of the gelatinase A gene in cultured cells are probably the most unique of the MMPs. Thus gelatinase A is usually expressed at high constitutive levels by normal diploid fibroblasts. Unlike the other MMPs, gelatinase A does not appear to be particularly responsive to stress stimuli (with the possible exception of stromelysin-3, which has not been studied). TGF-β typically stimulates expression of gelatinase A in fibroblasts at the same time that it is causing inhibition of collagenase expression (Overall *et al.*, 1989; Girard *et al.*, 1991). PKC does not appear to be involved in gelatinase A expression. Thus, the gene is typically not induced, and can even be repressed, by PMA (Fini and Girard, 1990b). In addition, inhibitors of PKC do not inhibit gelatinase A expression (Fabunmi *et al.*, 1996). There are a number of reports that expression is affected by cell–cell or cell–matrix adhesion; this type of regulation also occurs with other MMPs, however, the literature reflects a greater interest in this mode of regulation with respect to gelatinase A. Thus, transfection of poorly differentiated, nonadhesive colon carcinoma cells with a cDNA for the cell–cell adhesion molecule, e-cadherin, increases their adhesive interaction and also suppresses constitutive expression of gelatinase A (Miyaki *et al.*, 1995). Gelatinase A expression is induced in T cells upon adhesion to endothelial cells, and this is VCAM-1 dependent (Romanic and Madri, 1994). The basement membrane component, laminin-1, induces gelatinase A expression through phospholipase C (Reich *et al.*, 1995), a signaling component implicated in regulating expression of other MMPs. Gelatinase A expression is modulated via differential expression of $\alpha_5\beta_3$ and $\alpha_5\beta_1$ integrins during human melanoma cell invasion (Seftor *et al.*, 1993). Intracellular calcium influx also stimulates expression of gelatinase A, as it does other MMPs (Kohn *et al.*, 1994).

B. Activation of the Gelatinase A Transcriptional Promoter

As might be expected from the expression pattern of gelatinase B, the gelatinase A promoter is quite different from the promoters of the other MMP genes studied. In one report (Templeton and Stetler-Stevenson, 1991), the human gelatinase A gene was found to contain an element located at bases -26, which is a noncanonical TATA box. A transcriptional activation element was found between nucleotides -223 to -422 that produces a sevenfold increase in transcriptional activity in the highly metastatic A2058 melanoma cell line. The region immediately 5' of the basal promoter, upstream to position -422, contains a silencer and represses transcriptional activity in the nonmetastatic HT144 melanoma cell line. In a second report (Frisch and Morisaki, 1990), evidence for a potent silencer within the DNA upstream of the transcription start site of the human gene was also presented and a strong enhancer element was located about 1650 bp upstream of the start site for transcription. A third group (Harendza *et al.*, 1995) identified a strong activating element within a fragment of DNA up to base -1686 in the rat gene. The activity of the element was evaluated in several cell types with varying capabilities to synthesize gelatinase A including mesangial cells, glomerular epithelial cells, and the monocytic U937 cell. Although element binding activity (as assayed by band shifts) was present in all cell types studied, transciptional activation through this element was demonstrated only in mesangial and glomerular epithelial cells.

As is the case for the other MMP genes discussed, transformation of cells with the E1A oncogene inhibits expression of gelatinase A (Frisch *et al.*, 1990). Interestingly, E1A deletion mutants that could not bind to the AP-1 transcriptional coactivator, p300, could repress the gelatinase A promoter as efficiently as wild type protein (Somasundaram *et al.*, 1996). This indicates that the mechanism by which E1A represses gelatinase A gene expression is different from the mechanism for repression of the other MMP genes that have been studied. The enhancing activity of the AP-2-like element could be repressed by adenovirus oncoprotein E1A. E1A was found to interact with the DNA binding/ dimerization domain of transcription factor AP-2, suggesting a possible target for repression of promoter activity.

VI. SUMMARY AND FUTURE PERSPECTIVES

Many years worth of correlative data have provided overwhelming evidence for the essential role of MMPs in basic biological remodeling processes and in diverse pathologies. More recent technological advances have begun to allow experiments to test directly these hypotheses, as well as to elucidate the individual roles for specific MMPs. Regulation at the level of transcription is the most important of the many controls on activity of the matrix metalloproteinases, and these mechanisms need to be elucidated if we are to understand how remodeling is controlled. The past 10 years have seen rapid advances in our understanding of mechanisms controlling MMP gene expression. We now know that each of the MMP genes is regulated in unique ways. Determinants required in common for MMP transcription, as well as for unique expression patterns, have been identified by studies in cultured cells; more recently, these studies have advanced to transgenic mice. Much of this progress has followed closely on the heels of advances in our understanding of fundamental mechanisms of cell growth and differentiation, since MMP expression is so intimately entwined with these processes.

Synthetic MMP inhibitors currently under development are already offering new approaches for therapeutic intervention in such devastating diseases as cancer, arthritis, and corneal ulceration. The prolonged treatment periods necessary for efficacy, however, result in side effects due to inhibition of normal remodeling processes. Therefore, there is much interest in developing drugs that can selectively interfere with activity of individual MMPs. Since expression of individual MMPs involves unique mechanisms, it seems likely that novel therapeutic agents might be targeted to act at this level as well. In fact, such agents have already been identified. For example, a class of retinoids has been developed that dissociates the transactivation and AP-1-mediated transrepression functions of the retinoid receptors; these agents allow repression of MMPs such as collagenase, without affecting expression of stromelysin-3. We predict that the future holds exciting promise for further developments of this type which exploit the unique expression mechanisms of individual MMPs in the treatment of disease.

ACKNOWLEDGMENTS

The expert editing and graphics contributions made by Mr. Michael Goulston and Dr. Maureen Onuigbo are gratefully acknowledged. This work was supported by National

Institutes of Health grants EY09828 and AR42981 to MEF, and AR26599 to CEB; and by grants to CEB from the RGK Foundation of Austin, Texas, and the Council for Tobacco Research. MEF is a Jules and Doris Stein Research to Prevent Blindness Professor. JRC is an NIH postdoctoral fellow (EY06722).

REFERENCES

Aggeler, J., Frisch, S. M., and Werb, Z. (1984). Collagenase is a major gene product of induced rabbit synovial fibroblasts. *J. Cell. Biol.* **98,** 1662–1671.

Alexander, C. M., Hansell, E. J., Behrendtsen, O., Flannery, M. L., Kishnani, N. S., Hawkes, S. P., and Werb, Z., (1996). Expression and function of matrix metalloproteinases and their inhibitors at the maternal–embryonic boundary during mouse embryo implantation. *Development* **122,** 1723–1736.

Anderson, I. C., Sugarbaker, D. J., Ganju, R. K., Tsarwhas, D. G., Richards, W. G., Sunday, M., Kobzik, L., and Shipp, M. A. (1995). Stromelysin-3 is over-expressed by stromal elements in primary non-small cell lung cancers and is regulated by retinoic acid in pulmonary fibroblasts. *Cancer Res.* **55,** 4120–4126.

Angel P., and Karin, M. (1991). The role of Jun, Fos and the AP-1 complex in cell-proliferation and transformation. *Biochim. Biophys. Acta* **1072,** 129–157.

Angel, P., and Karin, M. (1992). Specific members of the Jun protein family regulate collagenase expression in response to various extracellular stimuli. *Matrix* Suppl. **1,** 156–164.

Angel, P., Poting, A., Mallick, U., Rahmsdorf, H. J., Schorpp, M., and Herrlich, P. (1986). Induction of metallothionein and other mRNA species by carcinogens and tumor promoters in primary human skin fibroblasts. *Mol. Cell Biol.* **6,** 1760–1766.

Angel, P., Imagawa, M., Chiu, R., Stein, B., Imbra, R. J., Rahmsdorf, H. J., Jonat, C., Herrlich, P., and Karin, M. (1987a). Phorbol ester-inducible genes contain a common *cis* element recognized by a TPA-modulated trans-acting factor. *Cell* **49,** 729–739.

Angel, P., Baumann, I., Stein, B., Delius, H., Rahmsdorf, H. J., Herrlich, P. (1987b). 12-o-tetradecanoyl-phorbol-13-acetate induction of the human collagenase gene is mediated by an inducible enhancer element located in the 5'-flanking region. *Mol. Cell Biol.* **7,** 2256–2266.

Angel, P., Allegretto, E. A., Okino, S. T., Hattori, K., Boyle, W. J., Hunter, T., and Karin, M. (1988). Oncogene jun encodes a sequence-specific trans-activation similar to AP-1. *Nature* **332,** 166–171.

Anglard, P., Melot, T., Guerin, E., Thomas, E., Thomas, G., and Basset, P. (1995). Structure and promoter characterization of the human stromelysin-3 gene. *J. Biol. Chem.* **270,** 20337–20344.

Arner, E. C., and Tortorella, M. D. (1995). Signal transduction through chondrocyte integrin receptors induces matrix metalloproteinase synthesis and synergizes with interleukin-1. *Arthritis Rheum.* **38,** 1304–1314.

Ashida, K., Nakatsukasa, H., Higashi, T., Ohguchi, S., Hino, N., Nouso, K., Urabe, Y., Yoshida, K., Kinugasa, N., and Tsuji, T. (1996). Cellular distribution of 92 kDa type IV collagenase/gelatinase B in human hepatocellular carcinoma. *Am. J. Pathol.* **149,** 1803–1811.

Auble, D. T., and Brinckerhoff, C. E. (1991). The AP-1 sequence is necessary but not sufficient for phorbol induction of collagenase in fibroblasts. *Biochemistry* **30,** 4629–4635.

Backstrom, J. R., Lim, G. P., Cullen, M. J., and Tokes, Z. A. (1996). Matrix metalloproteinase-9 (MMP-9) is synthesized in neurons of the human hippocampus and is capable of degrading the amyloid-beta peptide (1-40). *J. Neurosci.* **16,** 7910–7919.

Basset, P., Bellocq, J. P., Wolf, C., Stoll, I., Hutin, P., Limacher, J. M., Podhajcer, O. L., Chenard, M. P., Rio, M. C., and Chambon, P. (1990). A novel metalloproteinase gene specifically expressed in stromal cells of breast carcinomas. *Nature* **348,** 699–704.

340   M. ELIZABETH FINI *ET AL.*

Bauer, E. A., Cooper, T. W., and Huang, J. S. (1985). Stimulation of *in vitro* human skin collagenase expression by platelet-derived growth factor. *Proc. Natl. Acad. Sci. USA* **82,** 4132–4236.

Bazan, H. E. P., Tao, Y., and Bazan, N. G. (1993). Platelet-activating factor induces collagenase expression in corneal epithelial cells. *Proc. Natl. Acad. Sci.* **90,** 8678–8682.

Bernhard, E. J., Gruber, S. B., and Muschel, R. J. (1994). Direct evidence linking expression of matrix metalloproteinase 9 to the metastatic phenotype in transformed rat embryo cells. *Proc. Natl. Acad. Sci.* **91,** 4293–4297.

Biswas, C., Zhang, Y., DeCastro, R., Guo, H., Nakamura, T., Kataoka, H., and Nabeshima, K. (1995). The human tumor cell-derived collagenase stimulatory factor (renamed EMMPRIN) is a member of the immunoglobulin superfamily. *Cancer Res.* **55,** 434–439.

Breathnach, R., Matrisian, L. M., Gesnel, M-C., Staub, A., and Leroy, P. (1987). Sequences coding for part of oncogene-induced transin are highly conserved in a related rat gene. *Nucleic Acids Res.* **15,** 1149–1151.

Brenner, D. A., O'Hara, M., Angel, P., Chojkier, M., Karin, M. (1989). Prolonged activation of jun and collagenase genes by tumour necrosis factor-alpha. *Nature* **337,** 661–663.

Brinckerhoff, C. E., and Harris, E. D., Jr. (1981). Modulation by retinoic acid and corticosteroids of collagenase production by rabbit synovial fibroblasts treated with phorbol myristate acetate or polyethylene glycol. *Biochim. Biophys. Acta* **677,** 424–432.

Brinckerhoff, C. E., McMillan, R. M., Fahey, J. V., and Harris, E. D. Jr. (1979). Collagenase production by synovial fibroblasts treated with phorbol myristate acetate. *Arthritis Rheum.* **22,** 1109–1115.

Brinckerhoff, C. E., Gross, R. H., Nagase, H., Sheldon, L. A., Jackson, R. C., and Harris, E. D., Jr. (1982). Increased level of translatable collagenase messenger ribonucleic acid in rabbit synovial fibroblasts treated with phorbol myristate acetate or crystals of monosodium urate monohydrate. *Biochemistry* **21,** 2674–2679.

Brinckerhoff, C. E., Plucinska, I. M., Sheldon, L. A., and O'Connor G. T. (1986). Half-life of synovial cell collagenase mRNA is modulated by phorbol myristate acetate but not by all-*trans*-retinoic acid or dexamethasone. *Biochemistry* **25,** 6378–6384.

Brinckerhoff, C. E., Ruby, P. L., Austin, S. D., Fini, M. E., White, H. D. (1987). Molecular cloning of human synovial cell collagenase and selection of a single gene from the genomic DNA. *J. Clin. Invest.* **79,** 542–546.

Brinckerhoff, C. E., Mitchell, T. I., Karmilowicz, M. J., Kluve-Beckerman, B., and Benson, M. D. (1989). Autocrine induction of collagenase by serum amyloid A-like and beta2-microglobulin-like proteins. *Science* **243,** 655–657.

Brown, D. D., Wang, Z., Furlow, J. D., Kanamori, A., Schwartzman, R. A., Remo, B. F., and Pinder, A. (1996). The thyroid hormone-induced tail resorption program during *Xenopus laevis* metamorphosis. *Proc. Natl. Acad. Sci.* **93,** 1924–1929.

Buisson, A. C., Gilles, C., Polette, M., Zahm, J. M., Birembaut, P., and Tournier, J. M. (1996). Wound repair induced expression of a stromelysin is associated with the acquisition of a mesenchymal phenotype in human respiratory epithelial cells. *Lab. Invest.* **74,** 658–669.

Buttice, G., Quinones, S., Kurkinen, M. (1991). The AP-1 site is required for basal expression but is not necessary for TPA-response of the human stromelysin promoter. *Nucleic Acids Res.* **1,** 3723–3731.

Callaghan, M. M., Lovis, R. M., Rammohan, C., Lu, Y., and Pope, R. M. (1996). *J. Leukocyte Biol.* **59,** 125–132.

Canalis, E., Rydziel, S., Delany, A. M., Varghese, S., and Jeffrey, J. J. (1995). Insulin-like growth factors inhibit interstitial collagenase synthesis in bone cell cultures. *Endocrinology* **136,** 1348–1354.

Canete-Soler, R., Gui, Y-H., Linask, K. K., and Muschel, R. J. (1995a). MMP-9 (gelatinase B) mRNA is expressed during mouse neurogenesis and may be associated with vascularization. *Dev. Brain Res.* **88,** 37–52.

Canete-Soler, R., Gui, Y-H., Linask, K. K., and Muschel, R. J. (1995b). Developmental expression of MMP-9 (gelatinase B) mRNA in mouse embryos. *Develop. Dynamics* **204**, 30–40.

Case, J. P., Lafyatis, R., Kumkumian, G. K., Remmers, E. F., and Wilder, R. L. (1990). IL-1 regulation of transin/stromelysin transcription in rheumatoid synovial fibroblasts appears to involve two antagonistic transduction pathways. *J. Immunol.* **145**, 3755–3761.

Chamberlain, S. H., Hemmer, R. M., and Brinckerhoff, C. E. (1993). Novel phorbol ester response region in the collagenase promoter binds Fos and Jun. *J. Cell Biochem.* **52**, 337–351.

Chin, J. R., and Werb, Z. (1997). Matrix metalloproteinases regulate morphogenesis, migration, and remodelling of epithelium, tongue skeletal muscle, and cartilage in the mandibular arch. *Development* **124**, 1519–1530.

Chiu, R., Boyle, J. W., Meek, J., Smeal, T., Hunter, T., and Karin, M. (1988). The c-fos protein interacts with the c-jun/AP-1 to stimulate transcription of AP-1 responsive genes. *Cell* **54**, 541–552.

Chua, C. C., Geiman, D. E., Keller, G. H., and Ladda, R. L. (1985). Induction of collagenase secretion in human fibroblast cultures by growth promoting factors. *J. Biol. Chem.* **260**, 5213–5216.

Civitelli, R., Hruska, K. A., Jeffrey, J. J., Kahn, A. J., Avioli, L. V., and Partridge, N. C. (1989). Second messenger signalling in the regulation of collagenase production by osteogenic sarcoma cells. *Endocrinology* **124**, 2928–2934.

Collier, I. E., Smith, J., Kronberger, A., Bauer, E. A., Wilhelm, S. M., Eisen, A. Z., and Goldberg, G. I. (1988a). The structure of the human skin fibroblast collagenase gene. *J. Biol. Chem.* **263**, 10711–10713.

Collier, I. E., Wihelm, S. M., Eisen, A. Z., Marmer, B. L., Grant, G. A., Seltzer, J. L., Kronenberger, A., He, C., Bauer, E. A., and Goldberg, G. I. (1988b). H-ras oncogene-transforming human bronchial epithelial cells (TBE-1) secrete a single metalloprotease capable of degrading basement membrane collagen. *J. Biol Chem.* **263**, 6579–6587.

Collier, I. E., Bruns, G. A., Goldberg, G. I., Gerhard, D. S. (1991). On the structure and chromosome location of the 72- and 92-kDa human type IV collagenase genes. *Genomics* **9**, 429–434.

Cook, J. R., and Fini, M. E. Stromal cells from cornea are incompetent for NF-kappaB activation: A mechanism for maintaining tissue clarity. In preparation.

Corcoran, M. L., Stetler-Stevenson, W. G., Brown, P. D., and Wahl, L. M. (1992). Interleukin 4 inhibition of prostaglandin E2 synthesis blocks interstitial collagenase and 92-kDa type IV collagenase/gelatinase production by human monocytes. *J. Biol. Chem.* **267**, 515–519.

Cornelius, L. A., Nehring, L. C., Roby, J. D., Parks, W. C., and Welgus, H. G. (1995). Human dermal microvascular endothelial cells produce matrix metalloproteinases in response to angiogenic factors and migration. *J. Invest. Dermatol.* **105**, 170–176.

Dayer, J.-M., Russell, R. G. G., and Krane, S. M. (1977). Collagenase production by rheumatoid synovial cells: Stimulation by a human lymphocyte factor. *Science* **195**, 181–183.

De Cesare, D., Vallone, D., Caracciolo, A., Sassone-Corsi, P., Nerlov, C., and Verde, P. (1995). Heterodimerization of c-Jun with ATF-2 and c-Fos is required for positive and negative regulation of the human urokinase enhancer. *Oncogene* **11**, 365–376.

Delany, A. M., and Brinckerhoff, C. E. (1992). Post-transcriptional regulation of collagenase and stromelysin gene expression by epidermal growth factor and dexamethasone in cultured human fibroblasts. *J. Cell Biochem.* **50**, 400–410.

Delany, A. M., Jeffrey, J. J., Rydziel, S., Canalis, E. (1995). Cortisol increases interstitial collagenase expression in osteoblasts by post-transcriptional mechanisms. *J. Biol. Chem.* **270**, 26607–26612.

Desbois, C., Aubert, D., Legrand, C., Pain, B., and Samarut, J. (1991). A novel mechanism of action of v-ErbA: Abrogation of the inactivation of transcription factor AP-1 by retinoic acid and thyroid hormone receptors. *Cell* **67**, 731–740.

deSouza, S., Lochner, J., Machida, C. M., Matrisian, L. M., and Ciment, G. (1995). A novel growth factor-responsive element in the stromelysin-1 gene that is necessary and sufficient for gene expression in PC12 cells. *J. Biol. Chem.* **270**, 9106–9114.

Dorsman, J. C., Hagmeyer, B. M., Veenstra, J., Elfferich, P., Nabben, N., Zantema, A., van der Eb, A. J. (1995). The N-terminal region of the adenovirus type 5 E1A proteins can repress expression of cellular genes via two distinct but overlapping domains. *J. Virol.* **69**, 2962–2967.

Dupont, B. R., Linn, R., Knight, C. B., Roodman, G. D., Sakaguchi, A. Y., Lalley, P. A., Fourier, R. E., and Leach, R. J. (1996). Assignment of matrix metalloproteinase-9 (MMP9) to mouse chromosome 2 bands H1-H2. *Cytogenet. Cell Genet.* **74**, 118–119.

Edvardsen, K., Chen, W., Rucklidge, G., Walsh, F. S., Obrink, B., and Bock, E. (1993). Tansmembrane neural cell adhesion molecule (NCAM), but not glycosyl-phosphatidyl-inositol-anchored NCAM, down-regulates secretion of matrix metalloproteinases. *Proc. Natl. Acad. Sci.* **90**, 11463–11467.

Edwards, D. R., Murphy, G., Reynolds, J. J., Whitham, S. E., Docherty, A. J., Angel, P., and Heath, J. K. (1987). Transforming growth factor beta modulates the expression of collagenase and metalloproteinase inhibitor. *EMBO J.* **6**, 1899–1904.

Fabunmi, R. P., Baker, A. H., Murray, E. J., Booth, R. F., and Newby, A. C. (1996). Divergent regulation by growth factors and cytokines of 95 kDa and 72 kDa gelatinases and TIMPs in rabbit aortic smooth muscle cells. *Biochem. J.* **315**, 335–342.

Faller, D. V., Weng, H., and Choi, S. Y. (1997). Activation of collagenase IV gene expression and enzymatic activity by the Moloney murine leukemia virus long terminal repeat. *Virology* **227**, 331–342.

Fini, M. E. and Girard, M. T. (1990a). Expression of collagenolytic/gelatinolytic metalloproteinases by normal cornea. *Invest. Ophthalmol. Vis. Sci.* **31**, 1779–1788.

Fini, M. E., and Girard, M. T. (1990b). The pattern of metalloproteinase expression by corneal fibroblasts is altered with passage in cell culture. *J. Cell Sci.* **97**, 373–383.

Fini, M. E., Austin, S. D., Holt, P. T., Gross, R. H., White, D. H., and Brinckerhoff, C. E. (1986a). Homology between exon-containing portions of rabbit genomic clones for collagenase and human synovial cell mRNA. *Collagen Rel. Res.* **6**, 239–248.

Fini, M. E., Gross, R. H., and Brinckerhoff, C. E. (1986b). Characterization of rabbit genes for synovial cell collagenase. *Arthritis Rheum.* **29**, 1301–1315.

Fini, M. E., Plucinska, I. M., Mayer, A. S., Gross, R. H., and Brinckerhoff, C. E. (1987a). A gene for rabbit synovial cell collagenase: member of a family of metalloproteinases which degrades the connective tissue matrix. *Biochemistry* **26**, 6156–6164.

Fini, M. E., Karmilowicz, M. J., Ruby, P. L., Beeman, A. M., Borges, B. A., and Brinckerhoff, C. E. (1987b). Cloning of a cDNA for rabbit proactivator: A metalloproteinase that activates synovial cell collagenase, that shares homology with stromelysin and transin, and that is coordinately regulated with collagenase. *Arthritis Rheum.* **30**, 1255–1264.

Fini, M. E., Yue, B. Y. J. T., and Sugar, J. (1992). Collagenolytic/gelatinolytic enzymes in normal and keratoconus corneas. *Current Eye Res.* **11**, 849–862.

Fini, M. E., Strissel, K. J., Girard, M. T., West-Mays, J. A., and Rinehart, W. B. (1994a). Interleukin 1α mediates collagenase synthesis stimulated by phorbol myristate acetate. *J. Biol. Chem.* **269**, 11291–11298.

Fini, M. E., Bartlett, J. D., Matsubara, M., Rinehart, W. B., Mody, M., Girard, M. T., and Rainville, M. (1994b). The rabbit gene for 92 kDa matrix metalloproteinase: Role of AP1 and AP2 in cell type-specific transcription. *J. Biol. Chem.* **269**, 28620–28628.

Fini, M. E., Girard, M. T., Matsubara, M., and Bartlett, J. D. (1995). Unique regulation of the matrix metalloproteinase, gelatinase B. *Invest. Ophthalmol. Vis. Sci.* **36,** 622–636.

Fini, M. E., Parks, W., Rinehart, W. B., Matsubara, M., Girard, M. T., Cook, J. R., West-Mays, J. A., Sadow, P. M., Jeffrey, J. J., Burgeson, R. E., Raizman, M., Kreuger, R., and Zieske, J. (1996a). Role of matrix metalloproteinases in failure to re-epithelialize following corneal injury. *Am. J. Pathol.* **149,** 1287–1302.

Fini, M. E., Scott, S., Byrne, M., Krane, S., Wang, Z., and Brown, D. D. (1996b). Genbank accession number, L49412.

Fisher, G. J., Datta, S. C., Talwar, H. S., Wang, Z. Q., Varani, J., Kang, S., Voorhees, J. J. (1996). Molecular basis of sun-induced premature skin aging and retinoid antagonism. *Nature* **379,** 335–339.

Formstone, C. J., Byrd, P. J., Ambrose, H. J., Riley, J. H., Hernandez, D., McConville, C. M., and Taylor, A. M. (1993). The order and orientation of a cluster of MMP genes stromelysin-2, collagenase, and stromelysin together with D11S385 on chromosome 11q22-q23. *Genomics* **16,** 289–291.

Freije, J. M., Diaz-Itza, I., Balbin, M., Sanchez, L. M., Blasco, R., Tolivia, J., and Lopez-Otin, C. (1994). Molecular cloning and expression of collagenase-3, a novel human matrix metalloproteinase produced by breast carcinomas. *J. Biol. Chem.* **269,** 16766–16773.

Frisch, S. M., and Morisaki, J. H. (1990). Positive and negative transcriptional elements of the human type IV collagenase gene. *Mol. Cell Biol.* **10,** 6524–6532.

Frisch, S. M., and Ruley, H. E. (1987). Transcription from the stromelysin promoter is induced by interleukin-1 and repressed by dexamethason. *J Biol. Chem.* **262,** 16300–16304.

Frisch, S., Clark, E. J., and Werb, Z. (1987). Coordinate regulation of stromelysin and collagenase genes determined with cDNA probes. *Proc. Natl. Acad. Sci. USA* **84,** 2600–2604.

Frisch, S. M., Reich, R., Collier, I. E., Genrich, T., Martin, G., and Goldberg, G. I. (1990). Adenovirus E1A represses protease gene expression and inhibits metastasis of human tumor cells. *Oncogene* **5,** 75–83.

Furutani, Y., Notake, M., Fukui, T., Ohue, M., Nomura, H., Yamada, M., and Nakamura, S. (1986). Complete nucleotide sequence of the gene for human interleukin-1alpha. *Nucleic Acids Res.* **14,** 3167–3179.

Gack, S., Vallon, R., Schaper, J., Ruther, U., and Angel, P. (1994). Phenotypic alterations in fos-transgenic mice correlate with changes in Fos/Jun-dependet collagenase type I expression. regulation of mouse MMPs by carcinogens, cAMP, and Fos protein. *J. Biol. Chem.* **269,** 10363–10369.

Gack, S., Vallon, R., Schmidt, J., Grigoriadis, A., Tuckerman, J., Schenkel, J., Weiher, H., Wagner, E. F., and Angel, P. (1995). Expression of interstitial collagenase during skeletal development of the mouse is restricted to osteoblast-like cells and hypertrophic chondrocytes. *Cell Growth Differentiation* **6,** 759–767.

Ganser, G. L., Stricklin, G. P., and Matrisian, L. M. (1991). EGF and TGF-alpha influence in vitro lung development by the induction of matrix-degrading metalloproteinases. *Int. J. Dev. Biol.* **35,** 453–461.

Girard, M. T., Matsubara, M., and Fini, M. E. (1991). Transforming growth factor-β and Il-1 modulate expression of metalloproteinases by corneal stromal cells. *Invest. Ophthal. Vis. Sci.* **32,** 2441–2454.

Girard, M. T., Matsubara, M., Kublin, C., Tessier, M., Cintron, C., and Fini, M. E. (1993). Stromal fibroblasts synthesize collagenase and stromelysin during long-term remodeling of repair tissue. *J. Cell Sci.* **104,** 1001–1011.

Goldberg, G. I., Wilhelm, S. M., Kronberger, A., Bauer, E. A., Grant, G. A., and Eisen, A. Z. (1986). Human fibroblast collagenase: Complete primary structure, and homology to an oncogene transformation-induced protein. *J. Biol. Chem.* **261,** 6600–6605.

Gross, J., and Lapiere, C. M. (1962). Collagenolytic activity in amphibian tissues: A tissue culture assay. *Proc. Natl. Acad. Sci. USA* **48,** 1014–1022.

Gross, R. H., Sheldon, L. A., Fletcher, C. F., and Brinckerhoff, C. E. (1984). Isolation of a collagenase cDNA clone and measurement of changing collagenase mRNA levels during induction in rabbit synovial fibroblasts. *Proc. Natl. Acad. Sci. USA* **81,** 1981–1985.

Grove, J. R., Deutsch, P. J., Price, D. J., Habener, J. F., and Avruch, J. (1989). Plasmids encoding PKI(1-31), a specific inhibitor of cAMP-stimulated gene expression, inhibit the basal transcriptional activity of some but not all cAMP-regulated DNA response elements in JEG-3 cells. *J. Biol. Chem.* **264,** 19506–19513.

Guerin, E., Ludwig, M. G., Basset, P., and Anglard, P. (1997). Stromelysin-3 induction and interstitial collagenase repression by retinoic acid: Therapeutic implication of receptor-selective retinoids dissociating transactivation and AP-1-mediated transrepression. *J. Biol. Chem.* **272,** 11088–11095.

Gum, R., Lengyel, E., Jaurez, J., Chen, J. H., Sato, H., Seiki, M., and Boyd, D. (1996). Stimulation of 92-kDa gelatinase B promoter activity by ras is mitogen-activated protein kinase kinase-1 independent and requires multiple transcription factor binding sites including some closely spaced PEA3/ets and AP-1 sequences. *J. Biol. Chem.* **271,** 10672–10680.

Gutman, A., and Waslyk, B. (1990). The collagenase gene promoter contains a TPA and oncogene-responsive unit encomassing the PEA3 and AP-1 binding sites. *EMBO J.* **9,** 2241–2246.

Hagmeyer, B. M., Konig, H., Herr, I., Offringa, R., Zantema, A., van der Eb, A., Herrlich, P., and Angel, P. (1993). Adenovirus E1A negatively and positively modulates transcription of AP-1 dependent genes by dimer-specific regulation of the DNA binding and transactivation activities of Jun. *EMBO J.* **12,** 3559–3572.

Harendza, S., Pollock, A. S., Mertens, P. R., and Lovett, D. H. (1995). Tissue-specific enhancer-promoter interactions regulate high level constitutive expression of matrix metalloproteinase 2 by glomerular mesangial cells. *J. Biol. Chem.* **270,** 18786–18796.

Havarstein, L. S., Morgan, I. M., Wong, W. Y., and Vogt, P. K. (1992). Mutations in the Jun delta region suggest an inverse correlation between transformation and transcriptional activation. *Proc. Natl. Acad. Sci. USA* **89,** 618–622.

Hayashi T, Stetler-Stevenson W. G, Fleming, M. V., Fishback, N., Koss, M. N., Liotta, L. A., Ferrans, V. J., and Travis, W. D. (1996). Immunohistochemical study of metalloproteinases and their tissue inhibitors in the lungs of patients with diffuse alveolar damage and ideopathic pulmonary fibrosis. *Am. J. Pathol.* **149,** 1241–1256.

He, C., Wilhelm, S. M., Pentland, A. P., and Goldberg, G. I. (1989). Tissue cooperation in a proteolytic cascade activating human interstitial collagenase. *Proc. Natl. Acad. Sci. USA* **86,** 2632–2636.

Heck, S., Kullman, M., Gast, A., Ponta, H., Rahmsdorf, H. J., Herrlich, P., and Cato, A. C. (1994). A distinct modulating domain in glucocorticoid receptor monomers in the repression of activity of the transcription factor AP-1. *EMBO J.* **13,** 4087–4095.

Heikinheimo, K., and Salo, T. (1995). Expression of basement membrane type IV collagen and type IV collagenases (MMP-2 and MMP-9) in human fetal teeth. *J. Dental Res.* **74,** 1226–1234.

Henriet, P., Rousseau, G. G., and Eeckhout, Y.(1992). Cloning and sequencing of mouse collagenase cDNA: Divergence of mouse and rat collagenases from the other mammalian collagenases. *FEBS Letts.* **310,** 175–178.

Heppner, K. J., Matrisian, L. M., Jensen, R. A., and Rodgers, W. H. (1996). Expression of most matrix metalloproteinase family members in breast cancer represents a tumor-induced host response. *Am. J. Pathol.* **149,** 273–282.

Hibbs, M. S., Hoidal, J. R., and Kang, A. H. (1987). Expression of a metalloproteinase that degrades native type V collagen and denatured collagen by cultured human alveolar macrophages. *J. Clin. Invest.* **80**, 1644–1650.

Hill, P. A., Murphy, G., Docherty, A. J., Hembry, R. M., Millican, T. A., Reynolds, J. J., and Meikle, M. C. (1994). The effects of selective inhibitors of matrix metalloproteinases (MMPs) on bone resorption and the identification of MMPs and TIMP-1 in isolated osteoclasts. *J. Cell Sci.* **107**, 3055–3064.

Hitraya, E. G., Varga, J., and Jiminez, S. A. (1995). Heat shock of human synovial and dermal fibroblasts induces delayed up-regulation of collagenase gene expression. *Biochem. J.* **308**, 743–747.

Houde, M., Tremblay, P., Masure, S., Opdenakker, G., Oth, D., and Mandeville, R. (1996). Synergistic and selective stimulation of gelatinase B production in macrophages by LPS, *trans*-retinoic acid, and a protein kinase C regulator. *Biochim. Biophys. Acta* **1310**, 193–200.

Huhtala, P., Eddy, R. L., Fan, Y. S., Byers, M. G., Shows, T. B., and Tryggvason, K. (1990a). Completion of the primary structure of the human type IV collagenase preproenzyme and assignment of the gene to the q21 region of chromosome 16. *Genomics* **6**, 554–559.

Huhtala, P., Chow, L. T., and Tryggvason, K. (1990b). Structure of the human type IV collagenase gene. *J. Biol. Chem.* **265**, 11077–11082.

Huhtala, P., Tuuttila, A., Chow, L. Y., Lohi, J., Keski-Oja, J., and Tryggvason, K. (1991). Complete structure of the human gene for 92-kDa type IV collagenase. *J. Biol. Chem.* **266**, 16485–16490.

Huhtala, P., Humphries, M. J., McCarthy, J. B., Tremble, P. M., Werb, Z., and Damsky, C. H. (1995). Cooperative signalling by alpha5beta1 and alpha4beta1 integrins regulates metalloproteinase gene expression in fibroblasts adhering to fibronectin. *J. Cell Biol.* **129**, 867–879.

Huttenlocher, A., Werb, Z., Tremble, P., Huhtala, P., Rosenberg, L., and Damsky, C. H. (1996). Decorin regulates collagenase gene expression in fibroblasts adhering to vitronectin. *Matrix Biol.* **15**, 239–250.

Iwata, H., Kobayashi, S., Iwase, H., Masaoka, A., Fujimoto, N., and Okada, Y. (1996). Production of matrix metalloproteinases and tissue inhibitors of metalloproteinases in human breast carcinomas. *Jap. J. Cancer Res.* **87**, 602–611.

Johnson-Muller, B., and Gross, J. (1978). Regulation of corneal collagenase production: Epithelial-stromal cell interactions. *Proc. Natl. Acad. Sci. USA* **75**, 4417–4421.

Jonat, C., Rahmsdorf, H. J., Park, K. K., Cato, A. C., Gebel, S., Ponta, H., and Herrlich, P. (1990). Antitumor promotion and anti-inflammation: Down modulation of AP-1 activity by glucocorticoid hormone. *Cell* **62**, 1189–1204.

Jung, J. Y., Warter, S., and Rumpler, Y. (1990). Localization of stromelysin-2 gene to the q22.3-23 region of chromosome 11 by in situ hybridization. *Ann. Genet.* **33**, 21–23.

Kanamori, A., and Brown, D. D. (1993). Cultured cells as a model for amphibian metamorphosis. *Proc. Natl. Acad. Sci. USA* **90**, 6013–6017.

Kerr, L. D., Olashaw, N. E., and Matrisian, L. M. (1988). Transforming growth factor-beta1 and cAMP inhibit transcription of epidermal growth factor- and oncogene-induced transin RNA. *J. Biol. Chem.* **263**, 16999–17005.

Kerr, L. D., Miller, D. B., and Matrisian, L. M. (1990). TGF-beta1 inhibition of transin/stromelysin gene expression is mediated through a Fos binidng sequence. *Cell* **61**, 267–278.

Kim, S. J., Lafyatis, R., Kim, K. Y., Angel, P., Fujiki, H., Karin, M., Sporn, M. B., and Roberts, A. B. (1990). Regulation of collagenase gene expression by okadaic acid, an inhibitor of protein phosphatases. *Cell Regulation* **1**, 269–278.

Kirstein, M., Sanz, L., Quinones, S., Moscat, J., Diaz-Meco, M. T., and Saus, J. (1996). Cross-talk between different enhancer elements during mitogenic induction of the human stromelysin-1 gene. *J. Biol. Chem.* **271,** 18231–18236.

Kohn, E. C., Jacobs, W., Kim, Y. S., Alessandro, R., Stetler-Stevenson, W. G., and Liotta, L. A. (1994). Calcium influx modulates expression of matrix metalloproteinase-2. *J. Biol. Chem.* **269,** 21505–21511.

Koivukangas, V., Kallioinen, M., Autio-Harmainen, H., and Oikarinen, A. (1994). UV irradiation induces the expression of gelatinases in human skin *in vivo. Acta Dermato-Venereologica* **74,** 279–282.

Konig, H., Ponta, H., Rahmsdorf, H. J., and Herrlich, P. (1992). Interference between pathway-specific transcription factors: Glucocorticoids antagonize phorbol ester-induced AP-1 activity without altering AP-1 site occupation *in vivo. EMBO J.* **11,** 2241–2246.

Korzus, E., Nagase, H., Rydell, R., and Travis, J. (1997). The mitogen-activated protein kinase and JAK-STAT signalling pathways are required for an oncostatin M-responsive element-mediated activation of matrix metalloproteinase 1 gene expression. *J. Biol. Chem.* **272,** 1188–1196.

Kramer, M., Sachsenmaier, C., Herrlich, P., and Rahmsdorf, H. J. (1993). UV irradiation-induced interleukin-1 and basic fibroblast growth factor synthesis and release mediate part of the UV response. *J. Biol. Chem.* **268,** 6734–6741.

Krane, S. M., Byrne, M. H., Lemaitre, V., Henriet, P., Jeffrey, J. J., Witter, J. P., Liu, X., Wu, H., Jaenisch, R., and Eeckhout, Y. (1996). Different collagenase gene products have different roles in degradation of type I collagen. *J. Biol. Chem.* **271,** 28509–28515.

Kumar, S., Millis, A. J. T., and Baglioni, C. (1992). Expression of interleukin 1-inducible genes and production of interleukin 1 by aging human fibroblasts. *Proc. Natl. Acad. Sci. USA* **89,** 4683–4687.

Kuter, I., Johnson-Wint, B., Beaupre, N., and Gross, J. (1989). Collagenase secretion accompanying changes in cell shape occurs only in the presence of a biologically active cytokine. *J. Cell Sci.* **92,** 473–485.

Lauricella-Lefebvre, M. A., Castronovo, V., Sato, H., Seiki, M., French, D. L., and Merville, M. P. (1993). Stimulation of the 92-kD type IV collagenase promoter and enzyme expression in human melanoma cells. *Invas. Metastasis* **13,** 289–300.

Lee, J. S., See, R. H., Deng, T., and Shi, Y. (1996). Adenovirus E1A downregulates cJun- and JunB-mediated transcription by targeting their coactivator p300. *Mol. Cell Biol.* **16,** 4312–4326.

Lee, W., Mitchell, P., and Tjian, R. (1987). Purified transcription factor AP-1 interacts with TPA-inducible enhancer elements. *Cell* **49,** 741–752.

Lefebvre, O., Regnier, C., Chenard, M. P., Wendling, C., Chambon, P., Basset, P., and Rio, M. C. (1995). Developmental expression of mouse stromelysin-3 mRNA. *Development* **121,** 947–955.

Levy, A., Zucman, J., Delattre, O., Mattei, M. G., Rio, M. C., and Basset, P. (1992). Assignment of the human stromelysin-3 gene to the q11.2 region of chromosome 22. *Genomics* **13,** 881–883.

Linn, R., DuPont, B. R., Knight, C. B., Plaetke, R., and Leach, R. J. (1996). Reassignment of the 92-kDa type IV collagenase gene (CLG4B) to human chromosome 20. *Cytogenet. Cell Genet.* **72,** 159–61.

Liu, W., Hillman, A. G., and Harmon, J. M. (1995). Hormone-independent repression of AP-1-inducible collagenase promoter activity by glucocorticoid receptors. *Mol. Cell Biol.* **15,** 1005–1013.

Liu, X., Wu, H., Jeffrey, J., Krane, S., and Jaenische, R. (1995). A targeted mutation at the known collagenase cleavage site in mouse type I collagen impairs tissue remodelling. *J. Cell Biol.* **130,** 227–237.

Logan, S. K., Hansell E. J., Damsky, C. H., and Werb, Z. (1996). T-antigen inhibits metalloproteinase expression and invasion in human placental cells transformed with temperature-sensitive simian virus 40. *Matrix Biol.* **15**, 81–89.

Lopez, G., Schaufele, F., Webb, P., Holloway, J. M., Baxter, J. D., and Kushner, P. J. (1993). Positive and negative modulation of Jun action by thyroid hormone receptor at a unique AP-1 site. *Mol. Cell Biol.* **13**, 3042–3049.

Lyons, J. G., Birkedal-Hansen, B., Pierson, M. C., Whitelock, J. M., and Birkedal-Hansen, H. (1993). IL-1beta and TGF-alpha/epidermal growth factor induce expression of M(r) 95,000 type IV collagenase/gelatinase and interstitial fibroblast type collagenase by rat mucosal keratinocytes. *J. Biol. Chem.* **268**, 19143–19151.

Machida, C. M., Rodland, K. D., Matrisian, L., Magun, B. E., and Ciment, G. (1989). NGF induction of the gene encoding the protease transin accompanies neuronal differentiation in PC12 cells. *Neuron* **2**, 1587–1596.

MacNaul, K. L., Chartrain, N., Lark, M., Tocci, M. J., and Hutchinson, N. I. (1990). Discoordinate expression of stromelysin, collagenase, and TIMP in rheumatoid human synovial fibroblasts. *J. Biol. Chem.* **265**, 17238–17245.

Makela, M., Salo, T., Uitto, V. J., and Larjava, H. (1994). Matrix metalloproteinases (MMP-2 and MMP-9) of the oral cavity: Cellular origin and relationship to periodontal status. *J. Dental Res.* **73**, 1397–1406.

Malik, N., Greenfield, B. W., Wahl, A. F., and Kiener, P. A. (1996). Activation of human monocytes through CD40 induces matrix metalloproteinases. *J. Immunol.* **156**, 3952–3960.

Masure, S., Nys, G., Fiten, P., Van Damme, J., and Opdenakker, G. (1993). Mouse gelatinase B. cDNA cloning, regulation of expression and glycosylation in WEHI-3 macrophages and gene organization. *Europ. J. Biochem.* **218**, 129–141.

Matrisian, L. M., Glaichenhaus, N., Gesnel, M-C., and Breathnach, R. (1985). Epidermal growth factor and oncogenes induce transcription of the same cellular mRNA in rat fibroblasts. *EMBO J.* **4**, 1435–1440.

Matrisian, L. M., Leroy, P., Ruhlmann, C., Gesnel, M.-C., and Breathnach, R. (1986). Isolation of the oncogene and epidermal growth factor-induced transin gene: Complex control in rat fibroblasts. *Mol. Cell. Biol.* **6**, 1679–1686.

Matsubara, M., Girard, M. T., Kublin, C. L., Cintron, C., and Fini, M. E. (1991a). Differential roles for two gelatinolytic enzymes of the matrix metalloproteinase family in the remodelling cornea. *Develop. Biol.* **147**, 425–439.

Matsubara, M., Zieske, J., and Fini, M. E. (1991b). Mechanism of basement membrane dissolution preceding corneal ulceration. *Invest. Ophthalmol. Vis. Sci.* **32**, 3221–3237.

Mauviel, A., Chung, K .Y., Agarwal, A., Tamai, K., and Uitto, J. (1996). Cell-specific induction of distinct oncogenes of the Jun family is responsible for differential regulation of collagenase gene expression by transforming growth factor-beta in fibroblasts and keratinocytes. *J. Biol. Chem.* **271**, 10917–10923.

Mawatari, M., Kohno, K., Mizoguchi, H., Matsuda, T., Asoh, K., Van Damme, J., Welgus, H. G., and Kuwano, M. (1989). Effects of TNF and EGF on cell morphology, cell surface receptors, and the production of tissue inhibitor of metalloproteinases and IL-6 in human microvascular endothelial cells. *J. Immunol.* **143**, 1619–1627.

McCachren, S. S., Greer, P. K., and Niedel, J. E. (1989). Regulation of human synovial fibroblast collagenase messenger RNA by interleukin-1. *Arthritis Rheum.* **32**, 1539–1545.

McCarthy, G. M., Mitchell, P. G., and Cheung, H. S. (1991). The mitogenic response to stimulation with basic calcium phosphate crystals is accompanied by induction and secretion of collagenase in human fibroblasts. *Arthritis Rheum.* **34**, 1021–1030.

McDonnell, S. E., Kerr, L. D., and Matrisian, L. M. (1990). Epidermal growth factor stimulation of stromelysin mRNA in rat fibroblasts requires induction of protooncogenes c-fos and c-jun and activation of protein kinase C. *Mol. Cell Biol.* **10**, 4284–4293.

McGowan, K. A., Bauer, E. A., and Smith, L. T. (1994). Localization of type I human skin collagenase in developing embryonic and fetal skin. *J. Invest. Dermatol.* **102,** 951–957.

Meikle, M. C., Bord, S., Hembry, R. M., Compston, J., Crocher, P. I., and Reynolds, J. J. (1992). Human osteoblasts in culture synthesize collagenase and other matrix metalloproteinases in response to osteotropic hormones and cytokines. *J. Cell Sci.* **103,** 1093–1099.

Miltenburg, A. M., Lacraz, S., Welgus, H. G., and Dayer, J. M. (1995). Immobilized anti-CD3 antibody activates T cell clones to induce the production of interstitial collagenase, but not tissue inhibitor of metalloproteinases, in monocytic THP-1 cells and dermal fibroblasts. *J. Immunol.* **154,** 2655–2667.

Mitchell, P. G, and Cheung, H. S. (1991). TNFalpha and EGF regulation of collagenase and stromelysin in adult porcine articular chondrocytes. *J. Cell Physiol.* **149,** 132–140.

Miyaki, M., Tanaka, K., Kikuchi-Yanoshita, R., Muraoka, M., Konishi, M., and Takeichi, M. (1995). Increased cell-substratum adhesion, and decreased gelatinase secretion and cell growth, induced by e-cadherin transfection of human colon carcinoma cells. *Oncogene* **11,** 2547–2552.

Mizel, S. B., Dayer, J. M., Krane, S. M., and Mergenhagen, S. E. (1981). Stimulation of rheumatoid synovial cell collagenase and prostaglandin production by partially-purified lymphocyte-activating factor (Interleukin-1). *Proc. Natl. Acad. Sci. USA* **78,** 2474–2477.

Mohan, R., Mohan, P., and Fini, M. E. Differential regulation of the expression of matrix metalloproteinases, gelatinase B and collagenase via an autocrine IL-1alpha loop. In preparation.

Mohtai, M., Smith, R. L., Schurman, D. J., Tsuji, Y., Torti, F. M., Hutchinson, N. I., Stetler-Stevenson, W. G., and Goldberg, G. I. (1993). Expression of 92-kD type IV collagenase/gelatinase (gelatinase B) in osteoarthritic cartilage and its induction in normal human articular cartilage by interleukin 1. *J. Clin. Invest.* **92,** 179–185.

Murphy, G., Nagase, H., and Brinckerhoff, C. E. (1988). Relationship of procollagenase activator, stromelysin, and matrix metalloproteinase 3. Collagen Rel. Res. **8,** 389–391.

Nagase, H., Jackson, R. C., Brinckerhoff, C. E., Vater, C. A., and Harris, E. D., Jr. (1981). A precursor form of latent collagenase produced in a cell-free system with mRNA from rabbit synovial cells. *J. Biol. Chem.* **256,** 11951–11954.

Nagase, H., Brinckerhoff, C. E., Vater, C. A., and Harris, E. D. Jr. (1983). Biosynthesis and secretion of procollagenase by rabbit synovial fibroblasts. *Biochem J.* **214,** 281–288.

Nakano, A., Tani, E., Miyazaki, K., Yamamoto, Y., and Furuyama, J. (1995). Matrix metalloproteinases and tissue inhibitors of metalloproteinases in human gliomas. *J. Neurosurg.* **83,** 298–307.

Newsome, D. A. and Gross, J. (1979). Regulation of corneal collagenase production: stimulation of serially passaged stromal cells by blood mononuclear cells. *Cell* **16,** 895–900.

Nicholson, R. C., Mader, S., Nagpal, S., Leid, M., Rochette-Egly, C., and Chambon, P. (1990). Negative regulation of the rat stromelysin promoter by retinoic acid is mediated by an AP1 binding site. *EMBO J.* **9,** 4443–4454.

Offringa, R., Smits, A. M., Houweling, A., Bos, J. L., and van der Eb, A. J. (1988). Similar effects of adenovirus E1A and glucocorticoid hormones on the expression of the metalloproteinase, stromelysin. *Nucleic Acids Res.* **16,** 10973–10984.

Okada, A., Tomasetto, C., Lutz, Y., Bellocq, J. P., Rio, M. C., and Basset, P. (1997). Expression of matrix metalloproteinases during rat skin wound healing: Evidence that membrane type-1 matrix metalloproteinase is a stromal activator of pro-gelatinase A. *J. Cell Biol.* **137,** 67–77.

Oikarinen, A., Kylmaniemi, M., Autio-Harmainen, H., Autio, P., and Salo, T. (1993). Demonstration of 72 kDa and 92 kDa forms of type IV collagenase in human skin. *J. Invest. Dermatol.* **101,** 205–210.

Oofusa, K., and Yomori, S. (1994). Regionally and hormonally regulated expression of genes of collagen and collagenase in the anuran larval skin. *Int. J. Devel. Biol.* **38**, 345–350.

Overall, C. M., Wrana, J. L., and Sodek, J. (1989). Independent regulation of collagenase, 72-kDa progelatinase, and metalloendoproteinase inhibitor expression in human fibroblasts by transforming growth factor-beta. *J. Biol. Chem.* **264**, 1860–1869.

Overall, C. M.., Wrana, J. L., and Sodek, J. (1991). Transcriptional and post-transcriptional regulation of 72-kDa gelatinase/type IV collagenase by transforming growth factor-beta 1 in human fibroblasts. *J. Biol. Chem.* **266**, 14064–14071.

Pages, G., Lenormand, P., L'Allemain, G., Chambard, J. C., Meloche, S., and Pouyssegur, J. (1993). Mitogen-activated protein kinases p42mapk and p44mapk are required for fibroblast proliferation. *Proc. Natl. Acad. Sci. USA* **90**, 8319–8323.

Pan, L., Chamberlain, S. H., Auble, D. T., and Brinckerhoff, C. E. (1992). Differential regulation of collagenase gene expression by retinoic acid receptors—alpha, beta, and gamma. *Nucleic Acids Res.* **20**, 3105–3111.

Pan, L., Eckhoff, C., and Brinckerhoff, C. E. (1995). Suppression of collagenase gene expression by all-*trans* and 9-*cis* retinoic acid is ligand dependet and requires both RARs and RXRs. *J. Cell Biochem.* **57**, 575–578.

Patterton, D., Hayes, W. P., and Shi, Y. B. (1995). Transcriptional activation of the matrix metalloproteinase gene stromelysin-3 coincides with the thyroid hormone-induced cell death during frog metamorphosis. *Dev. Biol.* **167**, 252–262.

Pendas, A. M., Matilla, T., Estivill, X., and Lopez-Otin, C. (1995). The human collagenase-3 gene is located on chromosome 11q22.3 clustered to other members of the matrix metalloproteinase gene family. *Genomics* **26**, 615–8.

Pendas, A. M., Balbin, M., Llano, E., Jiminez, M. G., and Lopez-Otin, C. (1997). Structural analysis and promoter characterization of the human collagenase-3 gene (MMP-13). *Genomics* **40**, 222–233.

Petersen, M. J., Hansen, C., and Craig, S. (1992). Ultraviolet A irradiation stimulates collagenase production in cultured human fibroblasts. *J. Invest. Dermatol.* **99**, 440–444.

Pierce, R. A., Sandefur, S., Doyle, G. A., and Welgus, H. G. (1996). Monocytic cell type-specific transcriptional induction of collagenase. *J. Clin. Invest.* **97**, 1890–1899.

Pilcher, B. K., Gaither-Ganim, J., Parks, W. C., and Welgus, H. G. (1997). Cell type-specific inhibition of keratinocyte collagenase-1 expression by FGF and FGF-7: A common receptor pathway. *J Biol. Chem.* **272**, 18147–18154.

Pilcher, B. K., Gaither-Ganim, J., Parks, W. C., and Welgus, H. G. Type I collagen-mediated induction of keratinocyte collagenase-1 requires an epidermal growth factor receptor autocrine loop. In preparation.

Polette, M., Nawrocki, B., Pintiaux, A., Massenat, C., Maquoi, E., Volders, L., Schaaps, J. P., Birembaut, P., and Foidart, J. M. (1994). Expression of gelatinases A and B and their tissue inhibitors by cells of early and term human placenta and gestational endometrium. *Lab. Invest.* **71**, 838–846.

Porras-Reyes, B. H., Blair, H. C., Jeffrey, J. J., and Mustoe, T. A. (1991). Collagenase production at the border of granulation tissue in a healing wound: Macrophage and mesenchymal collagenase production *in vivo*. *Connective Tissue Res.* **27**, 63–71.

Postlethwaite, A. E., Lachman, L. B., Mainardi, C. L., and Kang, A. H. (1983). Interleukin 1 stimulation of collagenase production by cultured fibroblasts. *J. Exp. Med.* **157**, 801–806.

Quinn, C. O., Scott, D. K., Brinckerhoff, C. E., Matrisian, L. M., Jeffrey, J. J., and Partridge, N. C. (1990). Rat collagenase: Cloning, amino acid sequence comparison, and parathyroid hormone regulation in osteoblastic cells. J. Biol. Chem. **265**, 22342–22347.

Quinones, S., Saus, J., Otani, Y., Harris, E. D. Jr., and Kurkinen, M. (1989). Transcriptional regulation of human stromelysin. *J. Biol. Chem.* **264,** 8339–8344.

Quinones, S., Buttice, G., and Kurkinen, M. (1994). Promoter elements in the transcriptional activation of the human stromelysin-1 gene by the inflammatory cytokine, interleukin-1. *Biochem. J.* **302,** 471–477.

Radler-Pohl, A., Sachsenmaier, C., Gebel, S., Auer, H. P., Bruder, J. T., Rapp, U., Angel, P., Rahmsdorf, H. J., and Herrlich, P. (1993). UV-induced activation of AP-1 involves obligatory extranuclear steps including Raf-1 kinase. *EMBO J.* **12,** 1005–1012.

Rajabi M., Solomon S., and Poole, A. R. (1991). Hormonal regulation of interstitial collagenase in the uterine cervix of the pregnant guinea pig. *Endocrinology* **128,** 863–871.

Rajakumar, R. A., and Quinn, C. O. (1996). Parathyroid hormone induction of rat interstitial collagenase mRNA in osteosarcoma cells is mediated through an AP-1-binding site. *Mol. Endocrinol.* **10,** 867–878.

Rao, J. S., Steck, P. A., Mohanam, M., Stetler-Stevenson, W. G., Liotta, L. A., and Sawaya, R. (1993). Elevated levels of M(r) 92,000 type IV collagenase in human brain tumors. *Cancer Res.* **53,** 2208–2211.

Reich R., Blumenthal, M., and Liscovitch, M. (1995). Role of phospholipase D in laminin-induced production of gelatinase A (MMP–2) in metastatic cells. *Clin. Exp. Metastasis* **13,** 134–140.

Reifel-Miller, A. E., Conarty, D. M., Valasek, K. M., Iversen, P. W., Burns, D. J., and Birch, K. A. (1996). Protein kinase C isozymes differentially regulate promoters containing PEA-3 TPA response element motifs. *J. Biol. Chem.* **271,** 21666–21671.

Reitamo, S., Remitz, A., Tamai, K., and Uitto, J. (1994). Interleukin-10 modulates type I collagen and matrix metalloproteinase gene expression in cultured human skin fibroblasts. *J. Clin. Invest.* **94,** 2489–2492.

Reponen, P., Sahlberg, C., Huhtala, P., Hurskainen, T., Thesleff, I., and Tryggvason, K. (1992). Molecular cloning of murine 72 kDa type IV collagenase and its expression during mouse development. *J. Biol. Chem.* **267,** 7856–7862.

Reponen, P., Sahlberg, C., Munaut, C., Thesleff, I., and Tryggvason, K. (1994). High expression of 92-kD type IV collagenase (gelatinase B) in the osteoclast lineage during mouse development. *J. Cell Biol.* **124,** 1091–1102.

Reponen, P., Leivo, I., Sahlberg, C., Apte, S., Olsen, B., Thesleff, I., and Tryggvason, K. (1995). Expression of 72-kDa and 92 kDa type IV collagenases (MMP-1 and MMP-9) and TIMPs 1, 2, and 3 during mouse embryo implantation. *Dev. Dynamics* **202,** 388–396.

Rodgers, W. H., Matrisian, L. M., Giudice, L. C., Dsupin, B., Cannon, P., Svitek, C., Gorstein, F., and Osteen, K. G. (1994). Patterns of matrix metalloproteinase expression in cycling endometrium imply differential functions and regulation by steroid hormones. *J. Clin. Invest.* **94,** 946–953.

Romanic, A. M., and Madri, J. A. (1994). The induction of 72 kDa gelatinase in T cells upon adhesion to endothelial cells is VCAM-1 dependent. *J. Cell Biol.* **125,** 1165–1178.

Rouyer, N., Wolf, C., Chenard, M. P., Rio, M.C., Chambon, P., Bellocq, J. P., and Basset, P. (1994). Stromelysin-3 gene expression in human cancer: An overview. *Invasion Metastasis* **14,** 269–275.

Saarialho-Kere, U. K., Kovacs, S. O., Pentland, A. P., Olerud, J. E., Welgus, H. G., and Parks, W. C. (1993a). Cell-matrix interactions modulate interstitial collagenase expression by human keratinocytes actively involved in wound healing. *J. Clin. Invest.* **92,** 2858–2866.

Saarialho-Kere, U. K., Chang, E. S., Welgus, H. G., and Parks, W. C. (1993b). Distinct localization of collagenase and tissue inhibitor of metalloproteinases expression in wound healing associated with ulcerative pyogenic granuloma. *J. Clin. Invest.* **90,** 1952–1957.

Saarialho-Kere, U. K., Pentland, A. P., Birkedal-Hansen, H., Parks, W. C., and Welgus, H. G. (1994). Distinct populations of basal keratinocytes express stromelysin-1 and stromelysin-2 in chronic wounds. *J. Clin. Invest.* **94,** 79–88.

Sachsenmaier, C., Radler-Pohl, A., Zinck, R., Nordheim, A., Herrlich, P., and Rahmsdorf, H. J. (1994). Involvement of growth factor receptors in the mammalian UVC response. *Cell* **78,** 963–972.

Saklatvala, J., Pilsworth, L. M., Sarsfield, S. J., Gavrilovic, J., and Heath, J. K. (1984). Pig catabolin is a form of interleukin-1. *Biochem. J.* **224,** 461–466.

Salo, T. Lyons, J. G., Rahemtulla, F., Birkedal-Hansen, H., and Larjava, H. (1991). Transforming growth factor-beta1 up-regulates type IV collagenase expression in cultured human keratinocytes. *J. Biol. Chem.* **266,** 11436–11441.

Salo, T., Makela, M., Kylmaniemi, M., Autio-Harmainen, H., and Larjava, H. (1994). Expression of matrix metalloprtoeinase-2 and -9 during early human wound healing. *Lab. Invest.* **70,** 176–182.

Salvatori, R., Guidon, P. T., Jr., Rapuano, B. E., and Bockman, R. S. (1992). Prostaglandin E1 inhibits collagenase gene expression in rabbit synoviocytes and human fibroblasts. *Endocrinology* **131,** 21–28.

Sanz, L., Moscat, J., and Diaz-Meco, M. T. (1995). Molecular characterization of a novel transcription factor that controls stromelysin expression. *Mol. Cell Biol.* **15,** 3164–3170.

Sato, H., Takeshita, H., Furukawa, M., and Seiki, M. (1992). Epstein-Barr virus BZLF1 transactivator is a negative regulator of Jun. *J. Virol.* **66,** 4732–4736.

Sato, H., Kita, M., and Seiki, M. (1993). v-Src activates the expression of 92-kDa type IV collagenase gene through the AP-1 site and the GT box homologous to retinoblastoma control elements. A mechanism regulating gene expression independent of that by inflammatory cytokines. *J. Biol. Chem.* **268,** 23460–23468.

Sato, H., Takino, T., Okada, Y., Cao, J., Shinagawa, A., Yamamoto, E., and Seiki, M. (1994). A matrix metalloproteinase expressed on the surface of invasive tumor cells. *Nature* **370,** 61–65.

Sato, M., Miyazaki, T., Nagaya, T., Murata, T., Ida, N., Maeda, K., and Seo, H. (1996). Antioxidants inhibit tumor necrosis alpha mediated stimulation of interleukin-8, monocyte chemoattractant protein-1, and collagenase expression in cultured human synovial cells. *J. Rheumatol.* **23,** 432–438.

Saus, J., Quinones, S., Otani, Y., Nagase, H., Harris, E. D., Jr., and Kurkinen, M. (1988). The complete primary structure of matrix metalloproteinase-3: Identity with stromelysin. *J. Biol. Chem.* **263,** 6742–6745.

Sciavolino, P. J., Lee, T. H., and Vilcek, J. (1994). Interferon-beta induces metalloproteinase mRNA expression in human fibroblasts. Role of activator protein-1. *J. Biol. Chem.* **269,** 21627–21634.

Schoedel, K. E., Greco, M. A., Stetler-Stevenson, W. G., Ohori, N. P., Goswami, S., Present, D., and Steiner, G. C. (1996). Expression of metalloproteinasesin giant cell tumor of bone: An immunohistochemical study with clinical correlation. *Hum. Pathol.* **27,** 1144–1148.

Schonthal, A., Herrlich, P., Rahmsdorf, H. J., and Ponta, H. (1988). Requirement for fos gene expression in the transcriptional activation of collagenase by other oncogenes and phorbol esters. *Cell* **54,** 325–340.

Schonthal, A., Alberts, A. S., Frost, J. A., and Ferimisco, J. R. (1991). Differential regulation of jun family gene expression by the tumor promoter okadaic acid. *New Biologist* **3,** 977–986.

Schorpp, M., Mattai, M. G., Herr, I., Gack, S., Schaper, J., and Angel, P. (1995). Structural organization and chromosomal localization of the mouse collagenase type I gene. *Biochem. J.* **308,** 211–217.

Schreiber, M., Baumann, B., Cotten, M., Angel, P., and Wagner, E. F. (1995). Fos is an essential component of the mammalian UV response. *EMBO J.* **14,** 5338–5349.

Schroen, D. J., and Brinckerhoff, C. E. (1996). Inhibition of rabbit collagenase (matrix metalloproteinase-1; MMP-1) transcription by retinoid receptors: Evidence for binding of RARs/RXRs to the -77 AP-1 site through interactions with c-Jun. *J. Cell Phys.* **169,** 320–332.

Schule, R., Rangarajan, P., Kliewer, S., Ransone, L. J., Bolado, J., Yang, N., Verma, I. M., and Evans, R. M. (1990). Functional antagonism between oncoprotein c-Jun and the glucocorticoid receptor. *Cell.* **62,** 1217–26.

Scott, D. K., Brakenhoff, K. D., Clohisy, J. C., and Quinn, C. O. (1992). Parathyroid hormone induces transcription of collagenase in rat osteoblastic cells by a mechanism using cAMP and requiring protein synthesis. *Mol. Endocrinol.* **6,** 2153–2159.

Seftor, R. E., Seftor, E. A., Stetler-Stephenson, W. G., and Hendrix, M. J. (1993). The 72 kDa type IV collagenase is modulated via differential expression of alphaVbeta3 and alpha5beta1 integrins during human melanoma cell invasion. *Cancer Res.* **53,** 3411–3415.

Sirum, K. L., and Brinckerhoff, C. E. (1989). Cloning of the gene for human stromelysin and stromelysin-2: Differential expression in rheumatoid synovial cells. *Biochem.* **28,** 8691–8698.

Sirum-Connolly, K. L., and Brinckerhoff, C. E. (1991). Interleukin-1 or phorbol induction of the stromelysin promoter requires an element that cooperates with AP-1. *Nucleic Acids Res.* **19,** 335–341.

Sistonen, L., Holtta, E., Makela, T. P., Keski-Oja, J., and Alitalo, K. (1989). The cellular response to induction of the p21 c-Ha-ras oncoprotein includes stimulation of Jun gene expression. *EMBO J.* **8,** 815–822.

Smith, S. E., Papavassiliou, A. G., and Bohmann, D. (1993). Different TRE-related elements are distinguished by sets of DNA binding proteins with overlapping sequence specificity. *Nucleic Acids Res.* **21,** 1581–1585.

Smits, P. H., de Wit, L., van der Eb, A. J., and Zantema, A. (1996). The adenovirus E1A-associated 300 kDa adaptor protein counteracts the inhibition of the collagenase promoter by E1A and represses transformation. *Oncogene* **12,** 1529–1535.

Somasundaram, K., Jayaraman, G., Williams, T., Moran, E., Frisch, S., and Thimmapaya, B. (1996). Repression of a matrix metalloproteinase gene by E1A correlates with its ability to bind to cell type-specific transcription factor AP-2. *Proc. Natl. Acad. Sci. USA* **93,** 3088–3093.

Song, C. Z., Tierney, C. J., Loewenstein, P. M., Pusztai, R., Symington, J. S., Tang, Q. Q., Toth, K., Nishikawa, A., Bayley, S. T., and Green, M. (1995). Transcriptional repression by human adenovirus E1A N terminus/conserved domain 1 polypeptides *in vivo* and *in vitro* in the absence of protein synthesis. *J. Biol. Chem.* **270,** 23263–23267.

St Jean, P. L., Zhang, X. C., Hart, B. K., Lamlum, H., Webster, M. W., Steed, D. L., Henney, A. M., and Ferrell, R. E. (1995). Characterization of a dinucleotide repeat in the 92 kDa type IV collagenase gene (CLG4B), localization of CLG4B to chromosome 20 and the role of CLG4B in aortic aneurysmal disease. *Ann. Hum. Genet.* **59,** 7–24.

Stein, B., Rahmsdorf, H. J., Steffen, A., Litfin, M., and Herrlich, P. (1989). UV-induced DNA damage is an intermediate step in UV-induced expression of human immunodeficiency virus type 1, collagenase, c-fos, and metallothionein. *Mol. Cell Biol.* **9,** 5169–5181.

Stetler-Stevenson, W.G. (1996). MMP-2: Expression, activation, and inhibition. *Enzyme Protein* **49,** 7–19.

Stetler-Stevenson, W. G., and Corcoran, M. L. (1997). Tumor angiogenesis: Functional similarities with tumor invasion. *EXS* **79**, 413–418.

Stetler-Stevenson, W. G., Hewwitt, R., and Corcoran, M. (1996). Matrix metalloproteinases and tumor invasion: From correlation and causality to the clinic. *Semin. Cancer Biol.* **7**, 147–154.

Stolow, M. A., Bauzon D. D., Li, J., Sedgewick, T., Liang, V.C-T., Sang Q. A., and Shi, Y.-B. (1996). Identification and characterization of a novel collagenase in *Xenopus laevis:* Possible roles during frog development. *Mol. Biol. Cell* **7**, 1471–1483.

Strissel, K. J., Rinehart, W. B., Girard, M. T., and Fini, M. E., (1997a). Regulation of paracrine cytokine balance controlling collagenase synthesis by corneal cells. *Invest. Ophthalmol. Vis. Sci.* **38**, 546–552.

Strissel, K. J., Shams, N. B., Grabbe, S., Gross, J., and Fini, M. E. (1997). Frog PNKT-4B cells express specific extracellular matrix-degrading enzymes and cytokines correlated with an invasive phenotype. *J. Exp. Zool.* **278**, 201–214.

Sudbeck, B. D., Parks, W. C., Welgus, H. G., and Pentland, A. P. (1994). Collagen-stimulated induction of keratinocyte collagenase is mediated via tyrosine kinase and protein kinase C activities. *J. Biol. Chem.* **269**, 30022–30009.

Sudbeck, B. D., Pilcher, B. K., Pentland, A. P., and Parks, W. C. (1997). Modulation of intracellular calcium levels inhibits secretion of collagenase 1 by migrating keratinocytes. *Mol. Biol. Cell.* **8**, 811–824.

Talhouk, R. S., Bissel, M. J., and Werb, Z. (1992). Coordinated expression of extracellular matrix-degrading proteinases and their inhibitors regulates mammary epithelial function during involution. *J. Cell Biol.* **118**, 1271–1282.

Tamai, K., Ishikawa, H., Mauviel, A., and Uitto, J. (1995). Interferon-gamma coordinately upregulates MMP-1 and MMP-3, but not tissue inhibitor of MMPs (TIMP), expression in cultured keratinocytes. *J. Invest. Dermatol.* **104**, 384–390.

Tan, E. M., Qin, H., Kennedy, S. H., Rouda, S., Fox, J. W., 4th, and Moore, J. H., Jr, (1995). Platelet-derived growth factor-AA and -BB regulate collagen and collagenase gene expression differentially in human fibroblasts. *Biochem. J.* **310**, 585–588.

Templeton, N. S., and Stetler-Stevenson, W. G. (1991). Identification of a basal promoter for the human Mr 72,000 type IV collagenase gene and enhanced expression in a highly metastatic cell line. *Cancer Res.* **51**, 6190–6193.

Thompson, R. W., and Parks, W. C. (1996). Role of matrix metalloproteinases in abdominal aortic aneurysms. *Ann. N.Y. Acad. Sci.* **800**, 157–174.

Tremble, P., Lane, T. F., Sage, E. H., and Werb, Z. (1993). SPARC, a secreted protein associated with morphogenesis and tissue remodelling, induces expression of metalloproteinases in fibroblasts through a novel extracellular matrix-dependent pathway. *J. Cell Biol.* **121**, 1433–1444.

Tremble, P., Chiquet-Ehrismann, R., and Werb, Z. (1994). The extracellular matrix ligands fibronectin and tenascin collaborate in regulating collagenase gene expression in fibroblasts. *Mol. Biol. Cell* **5**, 439–453.

Tremble, P., Damsky, C. H., and Werb, Z. (1995). Components of the nuclear signalling cascade that regulate collagenase gene expression in response to integrin-derived signals. *J. Cell Biol.* **129**, 1707–1720.

Ueda, Y., Tsuchiya, H., Fujimoto, N., Nakanishi, I., Katsuda, S., Seiki, M., and Okada, Y. (1996). Matrix metalloproteinase 9 (gelatinase B) is expressed in multinucleated giant cells of human giant cell tumor of bone and is associated with vascular invasion. *Am. J. Pathol.* **148**, 611–622.

Unemori, E. N., and Werb, Z. (1986). Reorganization of polymerized actin: A possible trigger for induction of procollagenase in fibroblasts cultured in and on collagen gels. *J. Cell Biol.* **103**, pp. 1021–1031.

Unemori, E. N., and Werb, Z. (1988). Collagenase expression and endogenous activation in rabbit synovial fibroblasts stimulated by the calcium ionophor A23187. *J. Biol. Chem.* **263**, 16252–16259.

Unemori, E. N., Ferrara, N., Bauer, E. A., and Ameato, E. P. (1992). Vascular endothelial growth factor induces interstitial collagenase expression in human endothelial cells. *J. Cell. Physiol.* **153**, 557–562.

Vaalamo, M., Mattila L., Johansson, N., Kariniemi, A.-L., Karjalainen-Lindsberg, M.-L., Kahari, V.-M., and Saarialho-Kere, U. (1997). Distinct populations of stromal cells express collagenase-3 (MMP-13) and collagenase-1 (MMP-1) in chronic ulcers but not in normally healing wounds. *J. Invest. Dermatol.* **109**, 97–101.

Vadillo-Ortega, F., Gonzalez-Avila, G., Furth, E. E., Lei, H., Muschel, R. J., Stetler-Stevenson, W. G., Strauss, J. F., 3rd. (1995). 92-kd type IV collagenase (matrix metalloproteinase-9) activity in human amniochorion increases with labor. *Am. J. Pathol.* **146**, 148–156.

van Dam, H., Offringa, R., Smits, A. M. M., Bos, J. L., Jones, N. C., and van der Eb, A. J. (1989). The repression of the growth factor-inducible genes JE, c-myc and stromelysin by adenovirus E1A is mediated by conserved region 1. *Oncogene* **4**, 1207–1212.

van Dam, H., Duyndam, M., Rottier, R., Bosch, A., de Vries-Smits, L., Herrlich, P., Zantema, A., Angel, P., and van der Eb, A. J. (1993). Heterodimer formation of cJun and ATF-2 is responsible for induction of c-Jun by the 243 amino acid adenovirus E1A protein. *EMBO J.* **12**, 479–487.

van Dam, H., Wilhelm, D., Herr, I., Steffen, A., Herrlich, P., and Angel, P. (1995). ATF-2 is preferentially activated by stress-induced protein kinases to mediate c-jun induction in response to genotoxic agents. *EMBO J.* **14**, 1798–1811.

Vance, B. A., Kowalski, C. G., and Brinckerhoff, C. E. (1989). Heat shock of rabbit synovial fibroblasts increases expression of mRNAs for two metalloproteinases, collagenase and stromelysin. *J. Cell Biol.* **108**, pp. 2037–2043.

Vincenti, M. P., Coon, C. I., and Brinckerhoff, C. E. (1994). Regulation of collagenase gene expression by IL-1 beta requires transcriptional and post-transcriptional mechanisms. *Nucleic Acids Res.* **22**, 4818–4827.

Vincenti, M. P., Coon, C. I., White, L. A., Barchowsky, A., and Brinckerhoff, C. E. (1996a). Src-related tyrosine kinases regulate transcriptional activation of the interstitial collagenase gene, MMP-1, in interleukin-1-stimulated synovial fibroblasts. *Arthritis Rheum.* **39**, 574–582.

Vincenti, M. P., White, L. A., Schroen, D. J., Benbow, U., and Brinckerhoff, C. E. (1996b). Regulating expression of the gene for matrix metalloproteinase-1 (collagenase): Mechanisms that control enzyme activity, transcription, and mRNA stability. *Crit. Rev. Eukaryotic Gene Expression* **6**, 391–411.

Wahl, L. M., Wahl, S. M., Mergenhagen, S. E., and Martin, G. R. (1975). Collagenase production by lymphokine-activated macrophages. *Science* **187**, 261–263.

Wang, Z., and Brown, D. D. (1993). Thyroid hormone-induced gene expression program for amphibian tail resorption. *J. Biol. Chem.* **268**, 16270–16278.

Wasylyk, C., Gutman, A., Nicholson, R., and Wasylyk, B. (1991). The c-ets oncoprotein activates the stromelysin promoter through the same elements as several on-nuclear oncoproteins. *EMBO J.* **10**, 1127–1134.

Ways, D. K., Qin, W., Cook, P., Parker, P. J., Menke, J. B., Hao, E., Smith, A. M., Jones, C., Hershman, J. M., and Geffner, M. E. (1993). Dominant and nondominant negative C-erbA beta 1 receptors associated with thyroid hormone resistance syndromes augment PMA induction of the collagenase promoter and exhibit defective thyroid hormone mediated repression. *Mol. Endocrinol.* **7**, 1112–1120.

Webb, P., Lopez, G. N., Uht, R. M., and Kushner, P. J. (1995). Tamoxifen activation of the estrogen receptor/AP-1 pathway: Potential origin for the cell-specific estrogen-like effects of antiestrogens. *Mol. Endocrinol.* **9**, 443–456.

Werb, Z., Tremble, P., Berendtsen, O., Crowley, E., and Damsky, C. H. (1989). Signal transduction through the fibronectin receptor induces collagenase and stromelysin gene expression. *J. Cell Biol.* **109**, 877–889.

Westermarck, J., Lohi, J., Keski-Oja, J., and Kahari, V. M. (1994). Okadaic acid-elicited transcriptional activation of collagenase gene expression in HT-1080 fibrosarcoma cells is mediated by JunB. *Cell Growth Differentiation* **5**, 1205–1213.

West-Mays, J. A., Strissel, K. J., Sadow, P. M., and Fini, M. E. (1995). Competence for collagenase gene expression by tissue fibroblasts requires activation of an IL-1alpha autocrine loop. *Proc. Natl. Acad. Sci. USA* **92**, 6768–6772.

West-Mays, J. A., Sadow, P. M., Tobin, T. W., Strissel, K. J., Cintron, C., and Fini, M. E. (1997). Regulation of repair phenotype in corneal fibroblasts by an IL-1alpha feedback loop. *Invest. Ophthalmol. Vis. Sci.* **38**, 1367–1379.

West-Mays, J. A., Cook, J. R., Sadow, P. M., Mohan, P., Mullady, D. K., and Fini, M.E. Suramin and other inhibitors of collagenase synthesis operate through IL-1alpha dependent and independent pathways. Submitted for publication.

White, L. A., and Brinckerhoff, C. E. (1995). Two activator protein-1 elements in the matrix metalloproteinase-1 promoter have different effects on transcription and bind Jun D, c-Fos, and Fra-2. *Matrix Biol.* **14**, 715–725.

Whitham, S. E., Murphy, G., Angel, P., Rahmsdorf, H. J., Smith, B. J., Lyons, A., Harris, T. J. R., Reynolds, J. J., Herrlich, P., and Docherty, A. P. (1986). Comparison of human stromelysin and collagenase by cloning and sequence analysis. *Biochem. J.* **240**, 913–916.

Wilcox, B. D., Dumin, J. A., and Jeffrey, J. J. (1994a) Serotonin regulation of IL-1 mRNA in rat uterine smooth muscles cells: Relationship to the production of collagenase. *J. Biol. Chem.* **269**, 29658–29664.

Wilcox, B. D., Rydelek-Fitzgerald L., and Jeffrey, J. J. (1994b). Regulation of collagenase gene expression by serotonin and progesterone in rat uterine smooth muscle cells. *J. Biol. Chem.* **267**, 20752–20757.

Wilhelm, S. M., Collier, I. E., Kronberger, A., Eisen, A. Z., Marmer, B .L., Grant, G. A., Bauer, E. A, and Goldberg, G. I. (1987). Human skin fibroblast stromelysin: Structure, glycosylation, substrate specificity, and differential expression in normal and tumorigenic cells. *Proc. Natl. Acad. Sci. USA* **84**, 6725–6729.

Wilhelm, S. M., Collier, I. E., Marmer, B. L., Eisen, A. Z., Grant, G. A., and Goldberg, G. I. (1989). SV-40 transformed human lung fibroblasts secrete a 92-kDa type IV collagenase which is identical to that secreted by normal human macrophages. *J. Biol. Chem.* **264**, 17213–17221.

Windsor, L. J., Grenett, H., Birkedahl, Hansen, B., Bodden, M. K., Engler, J. A., and Birkedal-Hansen, H. (1993). Cell type specific regulation of SL-1 and SL-2 genes. *J. Biol. Chem.* **268**, 17341–1734-7.

Witty, J. P., Wright, J. H., and Matrisian, L. M. (1995). Matrix metalloproteinases are expressed during ductal and alveolar mammary morphogenesis, and misregulation of stromelysin-1 in transgenic mice induces unscheduled alveolar development. *Mol. Biol. Cell* **6**, 1287–1303.

Wlaschek, M., Bolsen, K., Hermann, G., Schwarz, A., Wilmroth, F., Heinrich, P., Goerz, C., and Schreffetter-Kochanek, G. (1993). UVA-induced autocrine stimulation of fibroblast-derived collagenase by IL-6: A possible mechanism in dermal photodamage? *J. Invest. Dermatol.* **101**, 164–168.

Wolf, C., Chaenard, M. P., Durand, de Grossouvre, P., Bellocq, J. P., Chambon, P., and Basset, P. (1992). Breast cancer-associated stromelysin-3 gene is expressed in basal carcinoma and during cutaneous wound healing. *J. Invest. Dermatol.* **99**, 870–872.

Yang-Yen, H. F., Chambard, J. C., Sun, Y. L., Smeal, T., Schmidt, T. J., Drouin, J., and Karin, M. (1990). Transcriptional interference between c-Jun and the glucocorticoid receptor: Mutual inhibition of DNA-binding due to direct protein-protein interaction. *Cell* **62**, 1205–1215.

Yokoo, T., and Kitamura, M. (1996a). Antioxidant PDTC induces stromelysin expression in mesangial cells via a tyrosine kinase-AP-1 pathway. *Am. J. Physiol.* **270**, F806–F811.

Yokoo, T., and Kitamura, M. (1996b). Dual regulation of Il-1 beta-mediated matrix metalloproteinase-9 expression in mesangial cells by NF-kappaB and AP-1. *Am. J. Physiol.* **270**, F123–F130.

Yoshioka, K., Deng, T., Cavigelli, M., and Karin, M. (1995) Antitumor promotion by phenolic antioxidants: Inhibition of AP-1 activity through induction of Fra expression. *Proc. Natl. Acad. Sci. USA* **92**, 4972–4976.

Index

Printed and bound by CPI Group (UK) Ltd, Croydon, CR0 4YY

08/05/2025

01864990-0001